ADVANCES IN CHEMISTRY SERIES **252**

D0208267

Toughened Plastics II

Novel Approaches in Science and Engineering

C. Keith Riew, EDITOR
University of Akron

Anthony J. Kinloch, EDITOR
University of London

American Chemical Society, Washington, DC

Library of Congress Cataloging-in-Publication Data
Toughened plastics II : novel approaches in science and engineering / C. Keith Riew, editor, Anthony J. Kinloch, editor.

    p.   cm. — (Advances in chemistry series, ISSN 0065-2393 ; 252)

    "Developed from a symposium sponsored by the Division of Polymeric Materials: Science and Engineering, at the 207th National Meeting of the American Chemical Society, San Diego, California, March 13–18, 1994."

    Includes bibliographical references and index.

    ISBN 0–8412–3151–6

    1. Plastics—Additives—Congresses. 2. Elastomers—Congresses. I. Riew, C. Keith. II. Kinloch, A. J. III. American Chemical Society. Division of Polymeric Materials: Science and Engineering. IV. American Chemical Society. Meeting (207th : 1994 : San Diego, Calif.) V. Series.

QD1.A355 no. 252 [TP142]

540 s—dc20 [668.4]

         96–38478

         CIP

Copyright ©1996 American Chemical Society

FOREWORD

The ADVANCES IN CHEMISTRY SERIES was founded in 1949 by the American Chemical Society as an outlet for symposia and collections of data in special areas of topical interest that could not be accommodated in the Society's journals. It provides a medium for symposia that would otherwise be fragmented because their papers would be distributed among several journals or not published at all.

Papers are reviewed critically according to ACS editorial standards and receive the careful attention and processing characteristic of ACS publications. Volumes in the ADVANCES IN CHEMISTRY SERIES maintain the integrity of the symposia on which they are based; however, verbatim reproductions of previously published papers are not accepted. Papers may include reports of research as well as reviews, because symposia may embrace both types of presentation.

Toughened Plastics II

ABOUT THE EDITORS

CHANGKIU KEITH RIEW is a consultant in the field of polymer science and engineering. He is an adjunct professor and has served as a research professor in the Department of Chemical Engineering, College of Engineering, University of Akron. He was an R&D Fellow at, and has retired from, the Corporate Research Division, BF-Goodrich Company, Brecksville, Ohio. He has served as a research scientist or manager for new products research in the industry for more than 25 years. He received M.S. and Ph.D. degrees in physical organic chemistry from Wayne State University, Detroit, Michigan, and a B.S. degree in chemistry from Seoul National University, Seoul, Korea.

Dr. Riew has presented more than 50 technical papers and holds more than 25 patents on emulsion polymers, hydrophilic polymers, synthesis and application of telechelic polymers, and toughened plastics for adhesives and composites. His latest research is in the synthesis, characterization, and performance evaluation of impact modifiers for thermosets and engineering thermoplastics. His research interests include correlating polymer chemistry and physics, morphology, engineering, and static and dynamic thermomechanical properties to the failure mechanisms of toughened plastics.

He has organized or co-organized four international symposia at the National Meetings of the American Chemical Society, Division of Polymeric Materials: Science and Engineering, Inc. The first symposium was on "Rubber-Modified Thermoset Resins" in 1983, the second on "Rubber-Toughened Plastics" in 1987, and the third and fourth on "Toughened Plastics: Science and Engineering" in 1990 and 1994. He organized a workshop on "Toughened Plastics" in 1990. He also chaired the North American Research Conference on Toughening of Polymers, sponsored by the American Chemical Society, Division of Polymeric Materials: Science and Engineering, Inc., at Hilton Head Island, SC, in 1996.

He has edited or co-edited several volumes in the ACS Advances in Chemistry Series: *Rubber-Modified Thermoset Resins*, Volume 208 (1984); *Rubber-Toughened Plastics*, Volume 222 (1989); *Toughened Plastics I: Science and Engineering*, Volume 233 (1993); and the current volume.

TONY KINLOCH has a personal chair as Professor of Adhesion in the Department of Mechanical Engineering at Imperial College of Science, Technology and Medicine, University of London. He obtained a Ph.D. degree in materials science from Queen Mary College, University of London, in 1972. He leads the Engineering Adhesives and Composites Research Group at Imperial College and has published more than 150 patents and papers in the areas of adhesion and adhesives, toughened polymers, and fracture of polymers and fiber-composites. He has also authored the books *Fracture Behavior of Polymers* (with R. J. Young) and *Adhesion and Adhesives.* He was co-editor (with C. Keith Riew) of *Toughened Plastics I: Science and Engineering,* Advances in Chemistry Series, Volume 233. In 1992 he received the U.S. Adhesion Society Award for Excellence in Adhesion Science, and in 1995 he was made an R. L. Patrick Fellow. In 1994 he received the Adhesion Society of Japan Award for Distinguished Contributions to the Development of Adhesion Science and Technology. He was awarded the Griffith Model and Prize from the Institute of Materials, UK, in 1996 for his work in developing and applying fracture mechanics to the failure of adhesives and interfaces.

CONTENTS

PREFACE

TOUGHENED PLASTICS ARE USUALLY MULTIPHASE POLYMERS; the dispersed phase consists of rubbery or thermoplastic domains, and the continuous phase is a cross-linked thermosetting or thermoplastic polymer matrix. The basic reason for toughening a plastic via this route is to improve its toughness, or crack resistance, without significantly decreasing other important properties such as the modulus and creep resistance. Also, the general mechanical properties of the polymer at elevated temperatures are usually not greatly affected by adopting this approach to polymer toughening. Such toughened plastics are widely used in many diverse industries and form the basis for engineering plastics, structural adhesives, and matrices for fiber-composite materials.

The importance of the science and engineering of toughened plastics is reflected in the successful series of symposia held on the topic under the auspices of the American Chemical Society. The first, on "Rubber-Modified Thermoset Resins," was held in Washington, DC, in 1983; the papers from that conference were published in 1984 as Volume 208 of the Advances in Chemistry Series. The theme of the 1988 symposium, "Rubber-Toughened Plastics," was broadened to cover both thermosets and thermoplastics. The papers from that symposium, held in New Orleans, LA, were published in 1989 as Volume 222 of the Advances in Chemistry Series. In 1990 the symposium returned to Washington, DC, and was titled "Toughened Plastics: Science and Engineering." The papers were published in 1993 as Volume 233 of the Advances in Chemistry Series.

Most of the chapters in this book are based on papers presented at the 1994 symposium on "Toughened Plastics II: Science and Engineering," held at the 207th ACS National Meeting, Division of Polymeric Materials: Science and Engineering, Inc., San Diego, CA, March 13–18, 1994.

This book is divided into two main sections. The first consists of fourteen chapters and is concerned with thermosetting polymers. The first chapter deals with modeling the toughness of rubber-toughened epoxy polymers. The theme of modeling is continued in Chapter 2, where computer modeling of the microstructure of rubber-toughened plastics is discussed. These two chapters were chosen to introduce the book because the modeling of the microstructure and properties of rubber-toughened plastics is a relatively new area of research, but one that is growing rapidly and holds much promise. Obviously, if theoretical modeling studies could provide a guide to the best microstructure needed for the multiphase polymer to possess particular properties, such studies would greatly reduce the need for extensive laboratory trials.

The next two chapters consider the toughening of epoxy resins by the inclusion of rigid particles. In Chapter 3 the use of preformed, rigid, core–shell polymers is described. The use of core–shell polymers to toughen thermoplastics has been practiced in industry for the past few decades, but now it is becoming a more popular route for the toughening of thermosetting polymers. However, the use of core–shell particles does pose some dispersion and rheological problems during blending with the liquid thermoset. Chapter 4 describes the mechanisms of crack trapping. These mechanisms may be observed when an epoxy polymer is modified using rigid thermoplastic rods or spherical particles. The influence of adhesion between the epoxy matrix and inclusion is examined. The next three chapters consider the cure chemistry of the phase separation and the use of modified liquid polymers, or an epoxidized oil, as toughening agents for epoxy resins. This area of research is receiving renewed interest because previous work has shown that in order to attain the toughest thermosets, it is not only necessary to obtain phase-separated rubbery particles but also to relieve the cross-link density of the epoxy matrix. Obviously, the secret to successful materials of this type is to achieve the latter goal without a significant loss of modulus and glass-transition temperature.

Chapters 8 and 9 consider the mechanical properties of rubber- and ceramic-particle toughened-epoxy materials. The importance of rubber cavitation is highlighted in Chapter 8. It is well known that this mechanism can relieve the high degree of triaxiality at a crack tip in the material and enable subsequent plastic hole growth of the epoxy resin, which is a major toughening mechanism. We return to rigid particles in Chapter 9, which examines their use to increase the thermal shock resistance of epoxy resins.

The next two chapters are concerned with the toughening of unsaturated polyester resins. These thermoset polymers have attracted increasing industrial interest because of their use as the matrix for fiber-reinforced panels in vehicles. However, this application has raised the need for tougher resins. Chapters 10 and 11 consider the chemistry, mechanical properties, and toughening mechanisms of unsaturated polyester resins toughened using rubber particles. The next chapters describe the use of newer, and rather different types of, cross-linked polymers. Chapter 12 discusses a 1,2-dihydrobenzocyclobutene and maleimide resin system, and Chapter 13 considers poly(cyanurate)s. The poly(cyanurate)s have considerable potential, and the use of rubber and thermoplastic modifiers to toughen such polymers is an exciting new area of research. Finally, Chapter 14 describes the mechanical behavior of a solid composite propellant, which is based on hydroxy-terminated polybutadiene, during the ignition of a motor.

The second section of the book consists of nine chapters and addresses the topic of thermoplastic polymers. Chapters 15 and 16 are concerned with a rubber-toughened poly(methyl methacrylate). In both chapters, the relationships between the microstructure and mechanical properties are discussed and the toughening mechanisms identified. These two chapters highlight how much our understanding of these aspects has increased since the symposium was first held in 1983. Chapter 17 then examines the influence of the microstructure of rubber-toughened thermoplastics on the toughening mechanisms, and reviews the issues of whether a critical particle diameter and interparticle distance exist. There has been much debate on these issues, and a detailed consideration of these aspects is timely. Next come three chapters on the toughening of nylon. Chapters 18 and 19 consider the compatibilizing of nylon blends, while Chapter 20 emphasizes the toughening mechanisms. Our understanding of the technology and science of toughened nylon has increased considerably in recent years, and these chapters all draw upon this fact. Chapters 21 and 22 both consider failure mechanisms in blends of linear low-density polyethylene and polystyrene. Again, mechanical models for describing the deformation and failure mechanisms are highlighted. Finally, Chapter 23 also considers the chemistry and properties of polystyrene blends, but here polypropylene is the other polyolefin involved.

We are indebted to each contributor to the book and to the referees who provided valuable advice. We thank the Petroleum Research Fund and the Division of Polymeric Materials: Science and Engineering, Inc., of the American Chemical Society for sponsoring the symposium and this book. We also thank Mrs. H. Kim Riew for her secretarial contributions to this book. Finally, we thank our families for their support and patience.

C. Keith Riew
Riewkim Research and Associates
2174 Rickel Drive
Akron, OH 44333–2917

A.J. Kinloch
Department of Mechanical
  Engineering
Imperial College of Science,
  Technology and Medicine
Exhibition Road, London
SW7 2BX, United Kingdom

Predictive Modeling of the Properties and Toughness of Rubber-Toughened Epoxies

Anthony J. Kinloch[1] and Felicity J. Guild[2]

[1]Department of Mechanical Engineering, Imperial College of Science, Technology, and Medicine, Exhibition Road, London SW7 2BX, United Kingdom
[2]Department of Mechanical Engineering, University of Bristol, Queen's Building, Bristol, BS81TR, United Kingdom

A finite-element model for rubber particles in a polymeric matrix has been proposed that is based on a collection of spheres, each consisting of a sphere of rubber surrounded by an annulus of matrix. We have used this model to investigate stress distributions in and around the rubber particle, or around a void, in a matrix of epoxy polymer. This chapter describes the modeling of stress concentrations in rubber-toughened epoxy and gives a simple model for predicting the fracture energy, G_C, of such a material.

EPOXY RESINS ARE FREQUENTLY TOUGHENED by the addition of rubber particles, and such two-phase polymers are important materials, both as structural adhesives and as matrices for fiber- and particulate-composites (*1–8*). The rubber particles are typically about 0.5 to 5 μm in diameter and are present at a volume fraction of 5 to 30%. These particles greatly increase the toughness of the epoxy polymer but do not significantly degrade other important properties of the material. The mechanism of toughening in rubber-toughened epoxy polymers is therefore an important area for both experimental studies and predictive modeling. In particular, the use of predictive modeling as an investigative tool can lead to both the elucidation of experimental observations, and suggestions for routes to improved materials.

Two important toughening mechanisms have been identified for such two-phase materials. The first is localized shear yielding, or shear banding, which occurs between rubber particles (*1–8*). Because of the large number of particles involved, the volume of thermoset matrix material that can undergo plastic yielding is effectively increased compared with the single-phase polymer. Consequently, far more irreversible energy dissipation is involved, and the toughness of the material is improved. The second mechanism is internal cavitation, or interfacial debonding, of the rubber particles, which may then enable the growth of these voids by plastic deformation of the epoxy matrix (7–9). The importance of this mechanism is that the irreversible hole-growth process of the epoxy matrix also dissipates energy and so contributes to the enhanced fracture toughness.

The mechanisms just described are triggered by the different types of stress concentrations that act within the overall stress field in the two-phase material. For example, initiation and growth of the shear bands are largely governed by the concentration of von Mises stress in the matrix, whereas cavitation, or interfacial debonding, of the rubber particles is largely controlled by the hydrostatic (dilatational) tensile stresses that are acting. To accurately analyze the stress field, it is necessary to employ numerical methods such as the finite-element method. The first study of rubber-modified polymers that used such methods, an elastic analysis, was reported by Broutman and Panizza (10). They simplified the two-phase material into an assembly of axisymmetric cylindrical cells. Their study revealed that the maximum direct and shear stresses were located at the equator of the particle, indicating that yielding of the matrix would begin there. They also found that the stress concentration increased as the volume fraction of rubber particles was increased. Later, Agarwal and Broutman (11) developed a three-dimensional model assuming regular packing of the rubbery particles. The results obtained from that model were compared with previous results obtained from employing the axisymmetric analysis. The two sets of results were found to agree well when presented as a function of interparticle spacing. Because a three-dimensional analysis was more complicated, and more costly in terms of computer resources, the authors subsequently concluded that the axisymmetric model could be used without a significant loss in accuracy.

In a more recent study, a spatial statistical technique developed by Davy and Guild (12) was incorporated into the axisymmetric model by Guild and Young (13) to study the influence of particle distribution on stress states around rubber particles. Their study suggested that particle distribution does not significantly change the calculated stress states around rubber particles when the rubbery volume fraction is below 0.3, which is usually the upper limit in rubber-toughened epoxy polymers (1, 3). In a later paper, Guild et al. (14) again used the axisymmetric model but modeled the effect of particle morphology on stress distribution around the particle.

The aforementioned analyses were essentially elastic in nature. However, Huang and Kinloch (7, 8) developed a two-dimensional, plane-strain model to analyze the stress fields around the dispersed rubbery particles in multiphase, rubber-modified epoxy polymers. The epoxy matrix was modeled as either an elastic or elastic–plastic material. Their work revealed that the plane-strain model predicted higher stress concentrations within the glassy polymeric matrix than the axisymmetric model. Furthermore, they successfully applied their model to simulate the initiation and growth of the localized shear zones in the epoxy matrix that were initiated around rubber particles or voids. However, because this earlier model developed by Huang and Kinloch was essentially two-dimensional in nature, it could not accurately model the mechanism of cavitation–debonding of the rubber particle and the subsequent plastic-hole growth in the epoxy matrix. Their finite-element analysis studies were also limited by the fact that their code was unable to incorporate relatively high values of the Poisson ratio, v, of the rubber particles. Therefore, these authors were forced to combine the finite-element analysis with an analytical model in order to predict the fracture energy, G_C, of the rubber-modified epoxy. This work emphasized the need for better, and more realistic, finite-element analysis models.

In the present work, a more accurate material model has been developed that consists of a collection of spheres (*15*). The greater accuracy of this model arises from proper consideration of material modeling, as described in the next section. The method of application of load in this model is more complex than that required for the less accurate cylindrical-material model. The importance of this new material model is demonstrated for the quantitative modeling of the toughening processes. A preliminary model of the material's toughness shows good agreement with experimental results.

The Predictive Models

Development of the Material Models. The method of finite-element analysis is a powerful tool for predictive modeling of composite materials. However, such an analysis cannot be properly carried out without due attention to the deduction of the material model. The finite-element analysis method is based on the analysis of a representative cell of the material, which is deduced from an assumption about the distribution of the filler material. The deduction of the representative cell may be based either on the assumption of a fixed, regular distribution of the filler particles (e.g., *7*) or on the assumption of a random distribution (e.g., *12–14*). The present work assumes a random distribution. This material model is based on the concept that the interactions of neighboring particles on the given particle are not directional; the overall effect is an "average" arising from all the neighboring particles. Thus, the overall material can be divided into cells, each containing a single filler sphere with surrounding matrix. It should be noted that this division of the material into representative cells is a distinct step in the analysis procedure, separate from the subsequent numerical analysis. The overall composite material should not be visualized as a simple conglomeration of the representative cell.

The boundary of a given cell is the region of the matrix closer to that particle than any other. These cells are the Voronoi cells; assuming a random distribution of particles, the distribution functions describing the interparticle distances have been calculated for both two dimensions (continuous fibers) and three dimensions (spheres) (*12*). The calculated distributions have been verified by numerical simulations (*12*) and by experimental measurements in two dimensions (*16*). It is important to note that this method of describing composite materials includes the effect of finite particle size; there are no underlying gross assumptions such as those included in percolation theory (e.g., *17*) where, for example, the "ligament" lengths are treated as independent.

On "average," the correct shape of the Voronoi cell is spherical. Thus, the correct overall material model for the assumption of a random distribution is a collection of spherical cells of different sizes, each containing a single sphere. Strictly speaking, any overall property of the material should be obtained by summing the contributions from the different cell sizes. This summing may be carried out by application of a "dispersion factor" to the property value found for the cell describing the overall volume fraction (*12*). The results presented here were obtained for the cells that describe the overall volume fraction. The application of the spatial statistical model to take into account the effect of the variable cell size is the subject of current work.

Previous work based on the assumption of a random distribution assumed that the material consists of a collection of cylinders, each containing a sphere at its center (*12–14*). Such an analysis is relatively straightforward because the constraints required to model the interactions of surrounding particles may be simply applied by forcing the sides of the cylinder to remain straight. However, as described in the preceding section, this material model is inherently inaccurate; it does not reflect either the assumption of a regular distribution or the assumption of a random distribution. The work described here is based on the new spherical-material model; some results obtained using the cylindrical model are given for comparison.

The Cylindrical-Material Model. As shown in Figure 1, the model consists of a rubber sphere centered in a cylinder of matrix; the cylinder has equal height and diameter. The cylinder can be represented by the plane ABCD using axisymmetric elements; the axis of symmetry is the y axis. The interactions of neighboring particles are modeled by constraining the shape of the deformed grid as shown in Figure 1. These constraints are achieved by loading via prescribed displacements applied to the nodes along the top of the cylinder, CD. Constraint equations are used to force the edge, BC, to remain straight and parallel to its original direction. The sides AB and AD remain stationary because these are lines of symmetry in the cylinder. The grid was drawn using eight-noded axisymmetric elements, except at the apex where six-noded, triangular, axisymmetric elements were used. A typical grid is shown in Figure 1; it contains 1312 elements. The mesh density was refined until stress continuity across the interface, for both radial direct and shear stresses, was attained for a wide range of particle and matrix properties.

Young's modulus can be calculated from the sum of the reactions along CD. The value of the Poisson ratio is calculated from the ratio of the displace-

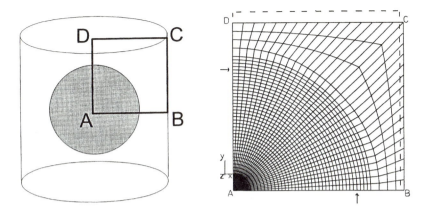

Figure 1. Development of the finite-element mesh for the cylindrical-material model, showing the shape of the deformed grid for application of unidirectional load (---). The interface on the grid is indicated by the arrows.

ments of BC and CD. Thus the full set of elastic properties can be immediately gained using this material model.

The Spherical-Material Model. The finite-element model used for the spherical-material model is a single rubber sphere surrounded by an annulus of epoxy resin. The complete cell can be represented by axisymmetric elements in an analogous manner to the cylindrical model, as shown in Figure 2. The same number and types of elements were used for this model. Analyses were also undertaken assuming a hole instead of the rubber particle; the grid then consisted of the epoxy annulus alone.

The constraints used for this model are analogous to those used for the cylindrical model. The sides AB and CA remain stationary because these are lines of symmetry in the spherical cell. The shape of the deformed cell is defined by the deformed shape of BC. Hydrostatic loading can be simply applied to this material model. The shape of the deformed cell is automatically a sphere, which is the required shape. This loading regime was used for many results presented here. However, hydrostatic loading generates values of bulk modulus only; unidirectional loading must be applied to derive the full set of elastic properties. The derivation of all the elastic properties is useful for comparison with experimental results and for the application of more complex triaxial loading, such as that required to model the conditions at a crack tip. However, the application of unidirectional load to the spherical-material model is not straightforward.

Under unidirectional loading, the shape of the deformed sphere of isotropic material is an ellipsoid. This deformed shape must also be attained by a cell of rubber-toughened epoxy because the overall material is isotropic. This shape would not be attained from application of the load alone; constraints must be applied to force this shape. The application of these constraints models the interactions between neighboring rubber spheres. These constraints are analogous to those applied to force the edges of the cylindrical model to remain straight. The deformed shape is shown in Figure 2. Unidirectional load is applied in the y direction via a prescribed displacement of each node along BC. For a sphere of radius R, centered at coordinates 0, 0 and subjected to y-directional strain, ϵ, the prescribed displacements, Δx and Δy, for the node at x, y are given by

$$\Delta x = -\nu\epsilon x \tag{1}$$

$$\Delta y = \epsilon y \tag{2}$$

where ν is the Poisson ratio. Young's modulus, E, can be calculated from the sum of the y-reactions-to-earth along AB and the y strain, ϵ (the ratio of the displacement at the pole and the radius). The ratio of the displacement at the equator to the displacement at the pole is the Poisson ratio.

The application of pure hydrostatic stress leads to the prediction of the bulk modulus, K, for the rubber-toughened epoxy. For an isotropic material, the value of the bulk modulus is related to the value of Young's modulus, E, and the Poisson ratio, ν, by

$$K = \frac{E}{3(1-2\nu)} \tag{3}$$

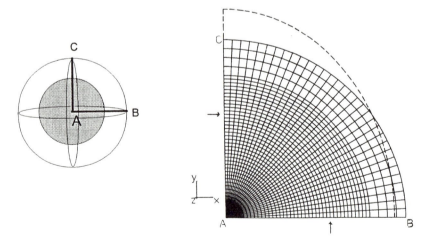

Figure 2. Development of the finite-element mesh for the spherical-material model, showing the shape of the deformed grid for application of unidirectional load (---). The interface on the grid is indicated by the arrows.

As just described, the deformed shape of the rubber sphere surrounded by an annulus of epoxy must be a perfect ellipsoid because the overall material is isotropic. As shown by equations 1 and 2, the deformed shape BC is defined from the Poisson ratio, and Young's modulus can be calculated from summing the reactions. However, for the toughened material, the value of the Poisson ratio is not known; only a relationship between E and ν is known, as defined by K. Thus, the analysis of unidirectional loading for the spherical cell must be carried out using an iterative procedure. The method has been described in detail elsewhere (*15*). The results of the iteration can be verified by checking the reduction of the x reactions-to-earth to zero; this verification was used for all the results presented here.

Material Properties. Results presented in this chapter are for linear-elastic behavior of both the rubbery phase and the epoxy matrix, and for elastic–plastic behavior of the matrix. For linear elasticity, the input values used for the stress-concentration studies and the toughness model are shown in Table I. The values for a typical epoxy matrix polymer are well-established. However, the values of E of the rubbery phase are far more difficult to establish; the range of values used reflects the values quoted in the literature (*18*), where the in situ polymerization has been simulated to manufacture "bulk" specimens of the rubber particles, which were subsequently used to determine the value of E from tensile stress versus strain measurements. A sensible range of values of ν was selected for the rubber. The upper value chosen for ν is very close to the maximum theoretical value of 0.5. It should be noted that the finite-element analysis package that was employed fails if $\nu = 0.5$ is used. However, the maximum value of ν used in the present work is higher than that used previously by Huang and Kinloch (*7, 8*) and Guild and Young (*13*). Such relatively high values may now be used because of improved precision of the finite-element analytical code.

The major, and most important, advantage of being able to use high values of the Poisson ratio is that this implies that the value of K of the rubber particle is high. Indeed, input values for E of 1 MPa and ν of 0.49992 imply a value of K of about 2 GPa, which is of the order expected for a rubbery polymer (*19*). Thus, the values of the bulk modulus, K, and of the shear modulus,

Table I. Elastic Material Properties Used in the Predictive Models

Phase	Input Properties		Derived Properties	
	Young's Modulus, E (GPa)	Poisson Ratio, ν	Bulk Modulus, K (GPa)	Shear Modulus, G (GPa)
Epoxy	3.0	0.35	3.333	1.111
Rubber	0.001–0.003	0.490–0.49999	0.0167–6.0	0.000333–0.001

G, are also included in Table I; these values were derived from the input properties of E and v.

The ranges of the values of K and G of the rubbery phase, which are derived from the input values of E and v, are also shown in Figure 3. As expected, of course, for a given input value of E, the value of G is relatively insensitive to the input value of v. On the other hand, for a given input value of E, the value of K is extremely sensitive to the precise value of v when v is greater than about 0.499.

The effect of simply having a void, or a hole, in the epoxy cell was also ascertained in order to determine the effect of a cavitated, or interfacially debonded, rubber particle.

The model used for the elastic–plastic behavior of the resin is shown in Figure 4. This elastic–plastic behavior used for input to the finite analysis is the true stress/true strain curve. The curve shown in Figure 4 was derived from experimental results for the compressive behavior of the resin (7). It should be noted that the material shows an intrinsic yield point followed by strain-softening, as is frequently observed in glassy thermosetting polymers. The curve for the tensile behavior was derived by assuming that the ratio of the values of yield stress in tensile and compression behavior is 0.75; this conversion has been experimentally verified (20). There is some strain-softening beyond maximum load, but the true stress/true strain curve then becomes horizontal. It is important to note that strain-hardening is not observed for strain values up to 20%, which is in agreement with results obtained by other workers (e.g., 21). The value of 20% for the true strain for tensile failure was experimentally determined (23). The elastic properties are shown in Table I.

The value of the elastic Poisson ratio used is shown in Table I. However, the analysis was carried out using the assumption that plastic deformation takes place essentially at constant volume; that is, with a Poisson ratio of 0.5. The Poisson ratio of any element is the weighted sum of the elastic Poisson ratio and 0.5, weighted via the proportion of elastic and plastic strain. However, it should be noted that this complex behavior leads to considerable difficulty regarding directional loading of the spherical model for elastic–plastic behavior.

Some linear-elastic results were obtained for direct comparison with experimental values of mechanical properties available in the literature. The materials were rubber-toughened epoxy resin (5) and epoxy resin filled with glass beads (22). The material properties used for these analyses are shown in Table II.

Linear-Elastic Results

Prediction of the Poisson Ratio.   Earlier work performed using the cylindrical-material model found good agreement between predicted and

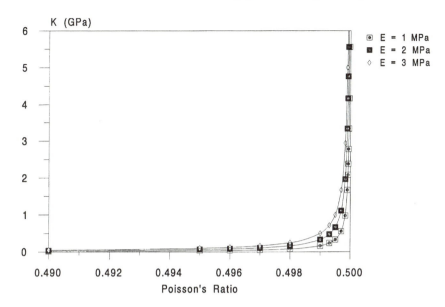

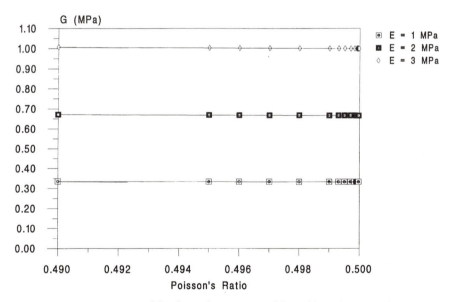

Figure 3. Variation of the derived properties of the rubber phase as a function of the input properties. Top: Variation of the bulk modulus, K, with the Poisson ratio, ν, for different values of Young's modulus, E. Bottom: Variation of the shear modulus, G, with the Poisson ratio, ν, for different values of Young's modulus, E.

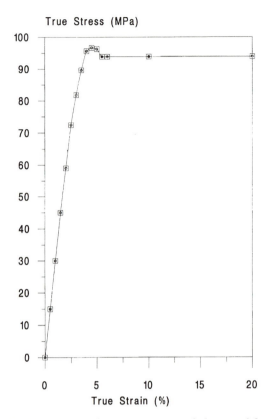

Figure 4. Model of elastic–plastic, stress–strain behavior of the epoxy resin.

Table II. Material Properties Used in the Modulus-Comparison Studies

Phase	Input Properties		Derived Properties	
	Young's Modulus, E (GPa)	Poisson Ratio, ν	Bulk Modulus, K (GPa)	Shear Modulus, G (GPa)
Rubber-Toughened Epoxy[a]				
Epoxy	2.965	0.375	3.953	1.078
Rubber	0.001	0.49992	2.083	0.000333
	0.001	0.490	0.01667	0.000336
Glass-Filled Epoxy[b]				
Epoxy	3.01	0.394	4.733	1.080
Glass	76.0	0.23	46.91	30.89

[a]From reference 5.
[b]From reference 22.

experimental values of E for both hard (*24*) and soft (*13*) particles. However, the predicted values of v for epoxy resin toughened with glass beads (*12*) were much lower than carefully measured experimental values from the literature (*22*). This discrepancy was attributed to the probable high dependence of this property on cell shape (*12*).

The predictions for the values of v for epoxy resin filled with glass beads have been repeated using the spherical-material model. The predictions from the two models are compared with experimental values in Figure 5. The values predicted using the spherical-material model are far closer to the experimental values than the values predicted using the cylindrical model. The better fit of the spherical-material model is particularly marked in the higher range of volume fraction.

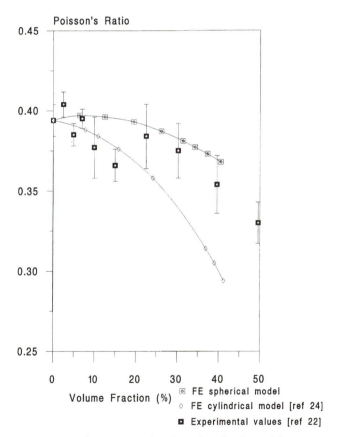

Figure 5. Comparison of experimental and predicted values of the Poisson ratio of epoxy resin filled with glass beads. Experimental values are shown with error bars, and predicted values were obtained from the two material models.

Young's Modulus of Rubber-Toughened Epoxy.   Predicted values of E for rubber-toughened epoxy were obtained using unidirectional loading of the spherical-material model. These properties were compared with values found in the literature (5). The predicted values, calculated for two values of ν of the rubbery phase, are compared with experimental values in Figure 6. The experimental values are plotted with respect to volume fraction using the amount of rubber added and the expected correction for epoxy present, in solution, within the rubbery phase (25). Also included are the predicted property bounds calculated using the analysis of Ishai and Cohen (26). These bounds are derived from the assumptions that the particles are arranged in a regular cubic packing, and that the particle and matrix are either at equal stress or strain. No effect of ν is included. As expected, the assumption of this material model increases the rate of change of stiffness with volume fraction. The experimental results lie close to the upper bound.

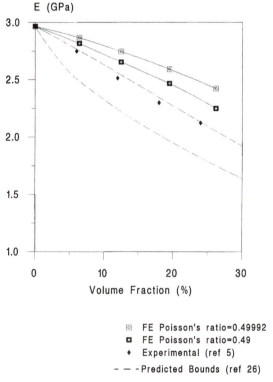

⊙	FE Poisson's ratio=0.49992
◻	FE Poisson's ratio=0.49
♦	Experimental (ref 5)
- — -	Predicted Bounds (ref 26)

Figure 6. Comparison of experimental and predicted values of Young's modulus of epoxy resin toughened with rubber particles. Finite-element predicted values were calculated using two values of the Poisson ratio of the rubber phase and the predicted bounds using the Ishai and Cohen model (26).

Figure 6 compares experimental results with finite-element predictions obtained using the spherical-material model. The predictions obtained using the lower value of v for the rubbery phase are lower than those obtained using the higher value. However, both finite-element predictions underestimate the experimentally measured reduction of modulus with volume fraction, although the difference is only about 0.2 GPa at about a 25% volume fraction. Thus, there is reasonable agreement between the predicted and the experimental results.

This small difference between the experimental results and the finite-element predictions can be accounted for by the onset of plasticity during the experimental measurements as well as by any experimental errors. However, the discrepancy could also arise from the use of axisymmetric modeling. Somewhat closer agreement was found between experimental values and values predicted using the cylindrical model (24). The extra constraints required in the cylindrical model would be expected to increase the rate of change of stiffness with volume fraction, in an analogous way to the assumption of regular cubic packing, as just described. The removal of all directional interactions in the spherical model may cause the overall effect of interactions between particles to be underestimated, which would lead to a smaller reduction in modulus with volume fraction, as found here.

Stress Distributions in Rubber-Toughened Epoxy.   The toughening mechanisms described in the introduction of this chapter are observed ahead of a crack, in a triaxial stress field. The simplest triaxial stress field that can be applied to the spherical cell is a pure hydrostatic stress. This stress field can be directly imposed by the application of stress, without prior knowledge of the material properties of the overall material. The shape of the deformed grid is automatically spherical, and this is the correct deformed shape. Pure hydrostatic stress was used for the detailed investigation of the stress distributions and the effect of rubber properties.

The overall features of the stress distribution were unchanged for the range of rubber properties employed. Application of pure hydrostatic stress to the representative cell places the rubber sphere in uniform pure hydrostatic tension. The stresses within the epoxy annulus vary, but they are radially symmetric, as expected. The maximum direct stress is at the interface, in the tangential direction, and the maximum von Mises equivalent stress is also at the interface. The radial stress is a maximum at the outer surface of the epoxy annulus, where it equals the overall imposed stress. Typical profiles through the epoxy annulus for the tangential, radial, and von Mises stresses are shown in Figure 7. These profiles are for a 20% volume fraction of rubber particles with rubber properties E = 1 MPa and v = 0.49992. The imposed hydrostatic stress was 100 MPa, which gives rise to a maximum tangential stress of about 111 MPa at the rubber–epoxy interface, and a maximum von Mises stress of about 24 MPa at this location. The maximum radial stress is at the surface of the

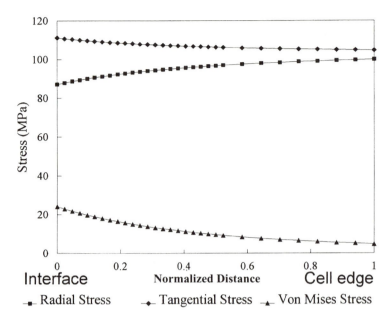

Figure 7. Stress profiles through the epoxy annulus for an applied pure hydro-static tension of 100 MPa. The volume fraction of the rubber phase is 20%, and the rubber properties are as follows: E = 1 MPa, ν = 0.49992, *and* K = 2.083 GPa.

epoxy annulus, and it has a value of 100 MPa, which is equal to the imposed applied hydrostatic stress, as required. At the rubber–epoxy interface, the radial stress is equal to the hydrostatic stress in the rubber particles and has a value of about 87 MPa.

The stress distributions for the different properties of the rubber sphere, for this pure hydrostatic applied stress, have been found to be unique functions of the bulk modulus, K, of the rubber (27). In other words, for a given volume fraction, the values of maximum stress for the different rubber properties fall on single curves when plotted as functions of the bulk modulus of the rubber. The relationships are shown for a 20% volume fraction of rubber in Figure 8; the values plotted are the hydrostatic stress in the rubber particle and the maximum von Mises stress in the epoxy, occurring at the interface. The results shown in Figure 8 demonstrate that the hydrostatic stress in the rubber sphere increases steadily with increasing values of K of the rubber, although the rate of increase is lower as the value of K rises. When the value of K of the rubber equals that of the epoxy annulus (i.e., 3.333 GPa), the model responds as an isotropic sphere and the stress state is pure hydrostatic tension. The maximum von Mises stress in the epoxy annulus decreases relatively

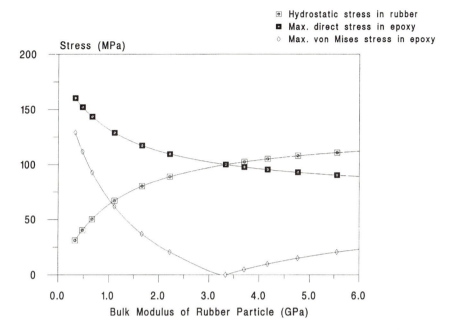

Figure 8. Variation of different types of maximum stresses with the bulk modulus, K, of the rubber particle. The applied pure hydrostatic tension is 100 MPa, and the volume fraction of rubber is 20%.

rapidly with increasing values of K of the rubber, and it attains a value of zero when the value of K of the rubber equals that of the epoxy. Since values of the von Mises stress are always positive, at higher values of K of the rubber the von Mises stress increases gradually as the value of K of the rubber continues to rise.

Figure 9 shows the variation of hydrostatic stress in the rubber particle as a function of the volume fraction of the rubber phase, for two values of K, 2.083 and 0.0167 GPa, of the rubber particle. These values of K are equivalent, for example, to the value of E of 1 MPa for both cases, and values of v of 0.49992 and 0.490 for the higher and lower values of K, respectively. The hydrostatic stresses for the higher value of K for the rubber particle are about 40 times greater than those for the lower value of K. For both values of K, the hydrostatic stress in the rubber increases slightly as the volume fraction of rubber phase increases. Thus, the hydrostatic stress is greater in the rubber particle when its bulk modulus is higher, as expected.

The variation of maximum von Mises stress in the epoxy matrix as a function of volume fraction is shown in Figure 10. The results are presented for the

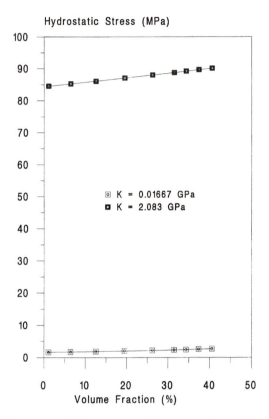

Figure 9. Variation of hydrostatic tensile stress in the rubber particle with the volume fraction of the rubber phase for different values of the bulk modulus, K, *of the rubber particle. The applied pure hydrostatic tension is 100 MPa.*

same two values of K of the rubber as were just used, and for a void replacing the particle. As just described, the maximum value of the von Mises stress occurs at the interface. The results for the lower value of K reveal relatively high values for von Mises stress in the epoxy matrix, and the stress increases significantly as the volume fraction of rubber phase increases. Further, these results are similar but not identical to the values for a hole. The values of the von Mises stress for the case when the rubber has a relatively high value of K are lower in magnitude and are almost independent of the volume fraction of the rubber phases. These effects arise from the stress concentrations being caused by the difference in properties between the rubber sphere and epoxy annulus. Hence, increasing the value of K of the rubber decreases this difference, so the von Mises stress concentration would be expected to be smaller.

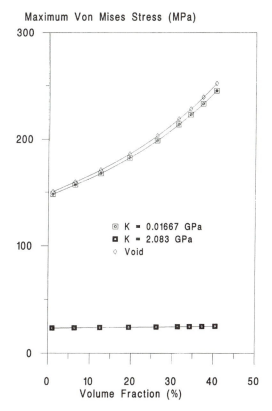

Figure 10. Variation of the maximum von Mises stress in the epoxy matrix with the volume fraction of the rubber phase, for different values of the bulk modulus, K, of the rubber particle, and a void. The applied pure hydrostatic tension is 100 MPa.

Elastic–Plastic Results

Growth of Localized Shear Zones.   The predicted growth of the localized shear zone was investigated for "voided" epoxy with a 20% volume fraction of voids. The imposed stress state was the triaxial loading used for the toughness model, described in the next section. The predicted growth of the localized shear zone is shown in Figure 11. The elements with values of von Mises stress greater than 93 MPa, that is, the plateau level on the stress–strain curve (Figure 4), are shown for three levels of applied strain. As expected, plastic yielding clearly begins at the equator, where the maximum von Mises stress occurs. Beyond the initiation, the localized shear zone grows outward

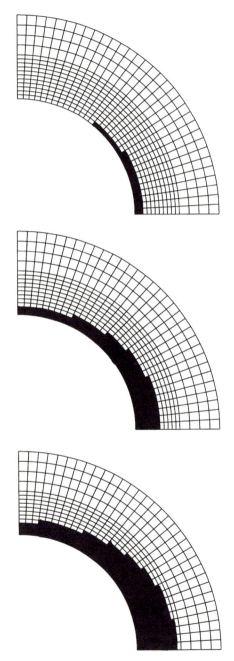

Figure 11. Growth of a localized shear zone in rubber-toughened epoxy, obtained using the spherical model. The yielded elements at the different steps are shaded. The applied strain (loading direction) is 3.0% (top), 4.0% (middle), and 5.0% (bottom).

toward the boundary of the cell, and around the interface between the void and the epoxy annulus.

It should be noted that modeling of localized shear zones performed using this spherical model is more accurate than that performed using the cylindrical-material model. The cylindrical-material model does not predict coherent growth of a localized shear zone. Although initiation occurs at the interface at the equator, secondary initiation occurs at a location remote from the particles, at the outer corner of the cylinder, separate from the original zone; this second zone must arise from the constraints imposed at the corner of the grid. Experimental evidence (6–8) clearly shows that initiation of the localized shear zone occurs at the interface, with no secondary, separate, point of initiation. Thus the growth of the localized shear zone cannot be simulated using the cylindrical-material model. The importance of using the more accurate spherical-material model is again demonstrated.

Modeling the Toughness.   It has been shown that the spherical-material model may be used to accurately model the fracture processes. The fracture processes in rubber-toughened epoxy take place ahead of the crack, in the triaxial stress field; subsequent growth of localized shear zones, and associated hole growth, must therefore be modeled using directional loading. To preserve axisymmetry, stresses were considered equal in the plane perpendicular to the applied load; the stress ratio used was 1:0.8:0.8. This triaxial stress state best describes the stress state at the crack tip while preserving the axisymmetry (28). However, directional loading of this material model is complex because the value of ν must be known, and for elastic–plastic behavior the value of ν is not constant, as described in the section "Material Properties." Therefore the changing value of ν means that this directional loading can only be achieved using an iterative process.

Analyses have been carried out assuming a cavitated particle, that is, the particle is replaced by a void (see the section "Cavitation of the Rubber Particles"). The analysis is applied to an annulus of epoxy resin. The volume fraction of the void is 20%. The elastic material properties used for the epoxy matrix are shown in Table I. The elastic–plastic material properties used are shown in Figure 4. Nonlinear geometric effects were included to take account of large deformations. Final failure of the cell was defined (23) to be the applied strain required for the maximum linear tensile strain in the resin to attain the value of 20%.

A simple global model of fracture is used. The fracture modeling is carried out using the concept of virtual crack growth (28). The crack is assumed to be surrounded by a plastic zone of radius r_y, and it is growing in a plate of thickness b. By simple geometry, if the crack grows a distance Δa, the extra volume of the plastic zone, ΔV, is given approximately by:

$$\Delta V = 2r_y b \Delta a \qquad (4)$$

The fracture energy, G_C, is related to the energy, ΔU, associated with crack growth by

$$G_C = \frac{1}{b} \frac{\Delta U}{\Delta a} \tag{5}$$

Assuming all the energy is absorbed by the crack growth, which is associated with the extra volume reaching the failure strain, the energy may be expressed in terms of the intensity, λ (the number of cells per unit volume) by

$$\Delta U = w\lambda\Delta V \tag{6}$$

where w is the plastic work performed per unit cell. Using equations 4 and 6, equation 5 may be expressed as

$$G_C = 2r_y\lambda w \tag{7}$$

The value of plastic work per unit cell, w, is calculated from the finite-element results, as described in the next section. The value of λ is calculated from the volume fraction (0.2 is used here) and the radius of the rubber particle (1.6 μm). The radius of the plastic zone, r_y, was calculated using the relationship proposed by Huang and Kinloch (8):

$$r_y/r_{yu} = k_{VM}^2 \tag{8}$$

where r_{yu} is the radius of the plastic zone for the unmodified epoxy and k_{VM} is the stress-concentration factor of von Mises stress for the toughened epoxy. The factor k_{VM} is defined as the ratio of the maximum von Mises stress within the cell to the average applied stress. The value of k_{VM} was obtained from finite-element analysis and is 1.90. The value of r_y used for the unmodified epoxy was 18 μm (8). Thus r_y was calculated to be 65.3 μm.

Calculation of w. The plastic work performed per unit cell, w, was determined from the results of the finite-element analysis. Several stepwise analyses were carried out. Results from the finite-element analysis led to stress–strain curves for the three directions; the stress–strain curve in the z-direction was identical to that for the x-direction. The curves are shown in Figure 12. The ends of the curves coincide with the value of 20% maximum linear strain. There is a slow decrease in stress after the maximum stress is obtained. The volume fraction of ellipsoids at "failure" is 24.5%; this value is in good agreement with experimental observations (8).

The curved portions of the graphs (Figure 12) were fitted to fourth-order polynomials, and the polynomials were integrated to obtain the area under the curves. The elastic area was subtracted. Thus the values of plastic work for the

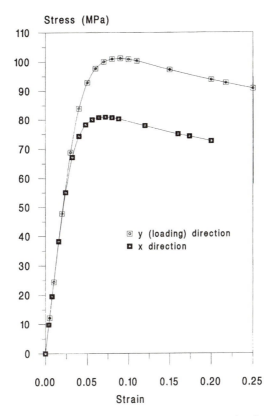

Figure 12. Predicted elastic–plastic, stress–strain curves for the epoxy annulus. The volume fraction of voids is 20%.

two directions, w_x and w_y, were calculated using the volume of the cell. The total work, w, for the cell was calculated from the sum of the work performed in the three orthogonal directions. This is the plastic work performed for the single cell assuming that the cell is strained to failure, where failure is assumed to occur at 20% linear strain.

Preliminary Value of Predicted G_C. The fracture model and the finite-element analysis lead to a predicted value of G_C of 5.5 kJ/m² for a rubber-toughened epoxy with a volume fraction of rubbery particles of 20%. In previous experimental studies (8), a rubber-toughened epoxy possessing the microstructure and mechanical properties of the epoxy matrix modeled in the present work was prepared, and G_C was measured. The measured value was 5.9 kJ/m² (8). The predicted value is in good agreement with the experimental value.

The predicted value is derived from the combined processes of plastic-hole growth and shear-band growth, assuming that the particle has cavitated. The processes of plastic-hole growth and shear-band growth cannot be separated because hole growth arises from plastic deformation of the matrix.

Discussion

Stress Distributions.   The onset of failure mechanisms in rubber-toughened epoxy is triggered by different types of stress concentrations. Initiation and growth of the shear bands are largely governed by the concentration of von Mises (deviatoric) stress in the matrix, whereas cavitation, or interfacial debonding, of the rubber particle is largely controlled by hydrostatic (dilatational) tensile stresses. The predictions presented in this chapter may be used to consider the sequence of failure events and the effect of rubber properties. It has been shown that the value of ν of the rubbery phase has a very significant effect on the values of stress concentrations (see Figures 8–10). This effect is important because it is often difficult to ascertain the value of K (or the value of ν) of the rubber phase with great precision, as discussed in the section "Material Properties."

The precise value of K (or ν) of the rubber has a major influence on the exact sequence of the toughening mechanisms. For example, Figures 9 and 10 clearly reveal that a relatively low value of K (or ν) of the rubber leads to relatively low hydrostatic stresses in the rubber but high von Mises stresses in the epoxy. These conditions tend to promote shear yielding in the epoxy in preference to cavitation (or debonding) in the rubber particle; recall that the importance of cavitation of the rubber is that such void formation may then enable subsequent plastic-hole growth in the epoxy to occur. Alternatively, a relatively high value of K (or ν) of the rubber leads to relatively high hydrostatic stresses in the rubber but low von Mises stresses in the epoxy. These conditions tend to first initiate cavitation of the rubber particle in preference to shear yielding of the epoxy matrix. Furthermore, cavitation of the rubber particle leads to a void, and in Figure 10 it is clear that, for rubber with a relatively high value of ν (or K), this void will lead to a considerable increase in the von Mises stress in the epoxy matrix.

Cavitation of the Rubber Particles.   From experimental observations on both rubber-toughened thermoplastic (*1, 29, 30*) and thermosetting plastics (*3–7, 31*), it is generally agreed that if cavitation or debonding of the rubber particle occurs, the toughness of the material is often enhanced. The improved toughness occurs because cavitation or debonding of the rubber particles leads to the formation of a series of holes or voids. This may greatly reduce the degree of triaxial tensile stresses acting in the matrix and thereby promote plastic yielding in the matrix. In the context of the present work, this behavior raises two important issues.

First, the precise value of K (or v) of the rubber particle will have a major influence on the effect of such void formation. If the value of v of the rubber particle is low (i.e., the value of K is low), any cavitation or debonding to form a void will not significantly change the von Mises stress in the epoxy matrix (see Figure 10). However, if the value of v of the rubber particle is high (i.e., the value of K is high), any cavitation or debonding to form a hole would greatly increase the von Mises stress in the epoxy matrix. In this case, cavitation or debonding would significantly promote any form of the plastic-shear yielding mechanism (see Figure 10), and so increase the value of G_C. As discussed in the section "Material Properties," the most likely values of the moduli of the rubbery particles in the rubber-toughened epoxies are Young's modulus, E, in the range of 1 to 3 MPa, and a bulk modulus, K, of about 2 GPa, which give a value of v in the range of 0.49992 to 0.49975. For these values, it is predicted that any cavitation or debonding to form a hole would greatly increase the von Mises stress in the epoxy matrix and so significantly promote further plastic-shear yielding and increase the toughness of the materials.

Second, in many formulations of rubber-toughened epoxies, cavitation of the rubber particles does indeed occur (*3–6, 31*), especially at higher test temperatures and slower test rates. Hence, in order to estimate the toughness under such conditions, it is necessary to either (1) be able to predict the cavitation event, or (2) simply assume that it occurs and base the finite-element model on an already "voided" epoxy. The first option is clearly better. However, it is not currently possible to predict the cavitation event. Obviously, we do know the stress distribution in, and around, the rubber particle, but are unable to state a satisfactory criterion for its cavitation. The problem of predicting the cavitation of rubber subjected either to a positive pressure from within or to a far-field triaxial tension has been studied by many authors (*1, 29, 32–35*). However, no fully quantitative solution has yet emerged without the size of an assumed pre-existing flaw or the initially formed void in the rubber being known. The essence of the problem is given in an elegant paper by Gent (*35*). The basic approach for predicting the cavitation of a rubber block, adhesive layer, particle, etc., has been based on fracture mechanics, and this approach assumes, of course, the existence of an initial flaw in the rubber. As discussed many years ago, and more recently explained in detail by Gent (*35*), the size of this pre-existing flaw in the rubber greatly affects the stress needed for cavitation. This is especially true when the pre-existing flaw is relatively small. In the case of our rubber-toughened epoxies, the rubber particles are typically about 1 to 5 μm in diameter, but we have no information about the size of any pre-existing flaws. However, assuming that the flaw size is at least an order of magnitude smaller than the particle diameter gives a size of about 0.1 μm or less. This dimension is in the region where a small difference in flaw size greatly affects the predicted value of the cavitation stress. More recent attempts (*30, 36*) to predict the cavitation of the rubber particles in polymeric matrices have required a knowledge of the size of the initially formed void and, because the value of this parameter is not known, they have not succeeded in accurately defining the stress needed for particle cavitation. Such work leads to the conclusion that the smaller the rubber particle, the higher the stress needed for cavitation, which is in agreement with previous work, including that of Gent (35).

In view of these difficulties, we have taken the second of the options just discussed. Namely, we assume that the particles have cavitated, which is in agreement with experimental observations, and then we base our finite-element modeling studies on examining an already "voided" epoxy matrix. This approach avoids the necessity of defining the value of the bulk modulus of the rubber with a high degree of accuracy (see the section "Material Properties").

Summary and Conclusions

Predictive modeling of rubber-toughened epoxy polymers is a powerful tool for the investigation of failure mechanisms. The new material model described in this chapter is important for modeling both the mechanical properties and the fracture processes. We have shown that the exact nature of the properties of the rubber may influence the sequence of failure events in the material.

The new spherical-material model is vital for proper modeling of the failure processes in rubber-toughened epoxy. The value for fracture energy obtained from the preliminary model shows good agreement with the experimentally measured value. It is noteworthy that this model assumes that cavitation has occurred. Further improvements to the model will include (1) exploration of any significant effects of assuming a regular distribution of particles, using a full, three-dimensional, finite-element model, (2) improvements in the modeling of yield and fracture, such as strain variation within the plastic zone, and (3) prediction of whether cavitation (and/or debonding) occurs. These improvements are the goal of current work.

Acknowledgments

We gratefully acknowledge valuable discussions with B. A. Crouch (E. I. DuPont). The results from the cylindrical-material model were obtained during F. J. Guild's work with R. J. Young, now at the Manchester Materials Science Centre, United Kingdom, and in collaboration with P. J. Davy, now at the University of Wollongong, Australia. The computer time used in the present work is supported by a grant from the Engineering and Physical Sciences Research Council.

References

1. Bucknall, C. B. *Toughened Plastics*, Applied Science Publishers: London, 1977.
2. *Toughened Composites*, Johnston, N. J., Ed. American Society for Testing and Materials: Philadelphia, PA, 1987.

3. Kinloch, A. J. In *Rubber-Toughened Plastics*, Riew, C. K., Ed. Advances in Chemistry 222; American Chemical Society: Washington, DC, 1989, pp. 67–91.
4. Kinloch, A. J.; Shaw, S. J.; Hunston, D. L. *Polymer* **1983**, *24*, 1355.
5. Yee, A. F.; Pearson, R. A. *J. Mater. Sci.* **1986**, *21*, 2462.
6. Pearson, R. A.; Yee, A. F. *J. Mater. Sci.* **1986**, *21*, 2475.
7. Huang, Y.; Kinloch, A. J. *J. Mater. Sci.* **1992**, *27*, 2753.
8. Huang, Y.; Kinloch, A. J. *J. Mater. Sci.* **1992**, *27*, 2763.
9. Huang, Y.; Kinloch, A. J. *Polymer* **1992**, *33*, 1330.
10. Broutman, L. J.; Panizza, G. *Int. J. Polym. Mater.* **1971**, *1*, 95.
11. Agarwal, B. D.; Broutman, L. J. *Fibre Sci. Technol.* **1974**, *7*, 63.
12. Davy, P. J.; Guild, F. J. *Proc. Roy. Soc. Ser.* **1988**, *A418*, 95.
13. Guild, F. J.; Young, R. J. *J. Mater. Sci.* **1989**, *24*, 2454.
14. Guild, F. J.; Young, R. J.; Lovell, P. A. *J. Mater. Sci. Lett.* **1994**, *13*, 10.
15. Guild, F. J.; Kinloch, A. J. *J. Mater. Sci. Lett.* **1992**, *11*, 484.
16. Green, D.; Guild, F. J. *Composites* **1991**, *22*, 239.
17. Margolina, A.; Wu, S. *Polymer* **1988**, *29*, 2170.
18. Kunz, S. C.; Beaumont, P. W. R. *J. Mater. Sci.* **1981**, *16*, 3141.
19. *Polymer Handbook;* Brandrup, J.; Immergut, E. H., Eds.; John Wiley: New York, 1989, Vols. 8, 9.
20. Wronski, A. S.; Pick, M. *J. Mater. Sci.* **1977**, *12*, 28.
21. Crocombe, A. D.; Richardson, G.; Smith, P. A. *J. Adhesion* **1995**, *49*, 211.
22. Smith, J. C. *J. Res. Natl. Bur. Stand. (Phys. Chem.)* **1976**, *80A*, 1.
23. Bradley, W. L. private communication, 1994.
24. Guild, F. J.; Young, R. J. *J. Mater. Sci.* **1989**, *24*, 298.
25. Kinloch, A. J.; Hunston, D. L. *J. Mater. Sci. Lett.* **1987**, *6*, 131.
26. Ishai, O.; Cohen, L. J. *Int. J. Mech. Sci.* **1967**, *9*, 539.
27. Guild, F. J.; Kinloch, A. J. *J. Mater. Sci.* **1995**, *30*, 1689.
28. Knott, J. F. *Fundamentals of Fracture Mechanics;* Butterworths: London, 1973.
29. Fukui, T.; Kikuchi, Y.; Inou, T. *Polymer* **1991**, *32*, 2367.
30. Dompass, D.; Groeninckx, G. *Polymer* **1994**, *35*, 4743.
31. Kinloch, A. J., Young, R. J. *Fracture Behaviour of Polymers;* Applied Science Publishers Ltd.: London, 1983.
32. Gent, A. N.; Lindley, P. B. *Proc. Roy. Soc. Ser. A* **1958**, *249*, 195.
33. Gent, A. N.; Tompkins, D. A. *J. Appl. Phys.* **1969**, *40*, 2520.
34. Cho, K.; Gent, A. N. *J. Mater. Sci.* **1988**, *23*, 141.
35. Gent, A. N. *Rubber Chem. Technol.* **1990**, *63*, 49.
36. Lazzeri, A.; Bucknall, C. B. *Polymer* **1993**, *28*, 6799.

2

Bimodal-Distribution Models of the Discrete Phase in Toughened Plastics

Richard A. Hall[1] and Ilene Burnstein[2]

[1]Amoco Polymers, Inc. 4500 McGinnis Ferry Road, Alpharetta, GA 30202
[2]Department of Computer Science, Illinois Institute of Technology, Chicago, IL 60616

The efficient packing of spheres that have bimodal distribution can produce a high volume of the discrete phase of a toughened plastic and a corresponding small interparticle distance. A computational method has been developed to model plastics toughened with spheres that have a bimodal distribution. The modeling method allows geometric simulation and graphic display of real materials containing bimodal distributions of reinforcing spheres. Interparticle distance and other parameters can be calculated from the geometric models. This application also facilitates the study of hypothetical materials that may be difficult or impossible to synthesize.

T HE INCORPORATION OF RUBBER SPHERES alters many physical properties of a glassy polymer. Computer modeling offers a way of visualizing the geometry of such systems. Interest in the geometric modeling of rubber-toughened plastics with graphic displays of models dates back over a decade to the work of Hobbs et al. (1), who used high-resolution graphics to visualize rubber-toughened nylons. More recently, the discrete-phase geometry of materials like high-impact polystyrene has been simulated, facilitating the calculation of interparticle-distance parameters of real and hypothetical resins (2). A previous paper explains how geometric simulations of the discrete phase are used to obtain a surface model of high-impact polystyrene (3). The work is a graphic example, showing how the discrete phase might disrupt the surface of a toughened plastic. Past work has focused on geometric modeling of plastics toughened with monosized spheres (4) and spheres having a log-normal size distribution (2), as in the rubber phase of high-impact polystyrene. The diameters of discrete spheres in toughened plastics can fit other size distributions. Bimodal systems, for example, can be made from blending two rubber-modified thermoplastics containing two distinct size distributions of rubber particles.

A potential advantage in toughness results from the more efficient packing of spheres in a bimodal system (5). However, even when the volume of the discrete phase is constant, bimodal systems may show a significantly higher fracture energy than conventional systems (6). It has been known for some time that rubber-toughened polystyrene containing small particles (<1 μm) is inferior in toughness to polystyrene reinforced with small and large particles (2–5 μm) (7). There are many examples of improvements in properties that occur when the discrete phase of the rubber-toughened resin has a bimodal particle-size distribution. Examples include preparations of high-impact polystyrene (8, 9), acrylonitrile–butadiene–styrene (10), and rubber-toughened epoxies (11, 12).

There is a need to be able to calculate interparticle distance and other parameters from geometric models of bimodal and conventional systems. This paper explains how computer programs for the placement of spheres in a three-dimensional space have been modified to accommodate bimodal distributions. Examples of models composed of simple bimodal particle distributions are presented to illustrate the technique.

Experimental Details

The general concept of the computer program for the placement of spheres in a three-dimensional space has been explained previously (2, 4). Statgraphics (13) was used to generate a file containing the desired conventional size distributions of spheres. In a typical experiment for generating a conventional system model, a file holding the diameters of spheres was read by the program until enough particles were picked to satisfy the target volume of the discrete phase. After sorting, the program placed the selected spheres in the user-specified, three-dimensional space at random while avoiding overlap with spheres already placed.

The program was modified to allow the creation of bimodal systems. The modified program required two distinct distributions of spheres, which were also generated using Statgraphics and held in separate files. In a representative experiment for creating a bimodal-system model, the user selected the boundary dimensions and the desired phase volume of each pool of spheres to be placed in a three-dimensional rectangular solid. The program acquired enough spheres from each pool to satisfy phase-volume requirements. The combined set of spheres was sorted from large to small before placement of the spheres in the three-dimensional boundary. Output files from the program were used as input to SciAn (14), a graphics-visualization and animation application.

Example of a Model Creation

This example shows how to model the discrete phase of a rubber-toughened plastic. Sphere diameters are generated at random from a distribution representative of a sphere population, and the diameters are saved in a file. The operation is repeated for a second sphere population, and the diameters are saved in another file. The distributions of sphere diameters used to generate the two files are shown in Figure 1. Then the modeling program is called with the following input information: phase volumes of the two sphere populations (2.0% and 6.2% for average diameters of 2.0 μm and 8.0 μm, respectively); the names of the distribution files; and the dimensions of the rectangular solid boundary (35.0 μm^3). After creating

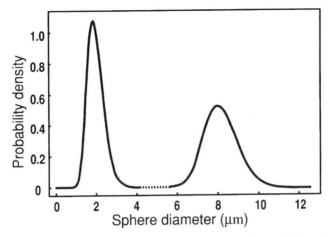

Figure 1. Log-normal distributions used to create a bimodal-distribution model of a hypothetical rubber-toughened plastic. Distribution 1: 2.0-μm average diameter, 0.4-μm standard deviation. Distribution 2: 8.0-μm average diameter, 0.8-μm standard deviation.

the model, the program calculates the interparticle distance parameters and generates an output file for SciAn that produces the model shown in Figure 2.

Results and Discussion

The program can be used in several ways to model the geometry of real materials. In the modeling of a blend of two rubber-toughened plastic components, the discrete-phase volume and rubber particle-size distribution of each blend component would have to be known. Files containing the actual diameters of discrete-phase spheres would also be required. Finally, the rubber-phase volume in each blend component would have to be multiplied by the component's weight fraction in the blend. This assumes that both components have the same continuous phase and, therefore, no volume change when blended.

This program can easily create geometric models of hypothetical blends. For example, consider two pools of spheres available for placement in a three-dimensional space. Each pool can represent a distinct size distribution, and the pools of spheres are kept separate in files available to the program for sphere placement. Figure 3 shows the results of three runs of the program. In the first run, 6.0% discrete-phase volume of spheres from one pool (2.0 μm diameter, monodispersed) and zero-discrete-phase volume from the other pool (3.0 μm diameter, monodispersed) were selected. The second model shows the placement of only the larger spheres at the 6.0% phase-volume level. Placement of

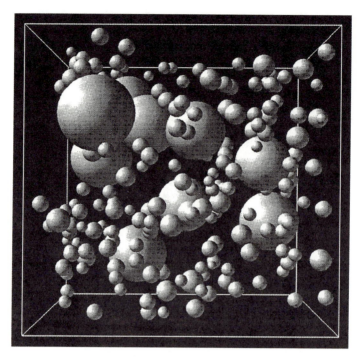

Figure 2. Bimodal-distribution model, seen in a perspective view, of sphere placement in a 35-μm³ three-dimensional space. Distribution 1: 2.0-μm average diameter, 0.4-μm standard deviation, 2.0% discrete-phase volume, 192 particles. Distribution 2: 8.0-μm average diameter, 0.8-μm standard deviation, 6.2% discrete-phase volume, 9 particles.

equal weights of materials containing spheres from each pool is shown in the third model, illustrating placement of a bimodal distribution of spheres.

Interparticle distance parameters were calculated for each model in Figure 3. The bimodal system could represent a blend of equal volumes of the two conventional systems containing 2.0- and 3.0-μm spheres. The intermediate average interparticle distance in the bimodal system shows that any advantage in toughness that the bimodal system might have over both conventional systems would not be from a small average interparticle distance. The bimodal system might allow a higher volume of the discrete phase. However, this would have to be achieved by other means besides blending two components of equal discrete-phase volume.

Some restrictions must be applied to the particle size and discrete-phase volumes supplied to the program. Monodispersed pools of spheres can be used as input, as shown in Figure 3c. Polydispersed distributions of spheres, held in separate files, can also be combined by the program to generate bi-

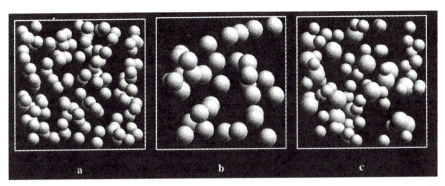

Figure 3. Orthographic-projection models of sphere placement in a 20-μm^3, three-dimensional space at 6.0% phase volume. Interparticle distance: (a) 0.69 μm, (b) 0.98 μm, and (c) 0.79 μm. Number of particles: (a) 115, (b) 34, and (c) 58 + 17.

modal systems with control over the discrete-phase volume contribution from each distribution, as illustrated in Figure 2. Models of nonoverlapping spheres comprising a bimodal distribution can be made with a relatively high total phase volume (up to 30%). This can be achieved if the two distinct sphere-population means are not too far apart, and if the distribution with the larger mean is not too broad. The reason for the restrictions is that large spheres, relative to the boundary size, have a tendency to block the available volume and restrict the placement of other spheres.

The technique has not been applied to any real resins containing bimodal distributions. A link to practicality has been established, however, by earlier work in which resin properties were correlated to computed parameters associated with spacing of the discrete-phase rubber particles (2, 3). The bimodal method described here could be applied in the same way to calculate properties dependent on the rubber particle-size distribution modality. The static, geometric models could also be used as launch points for studying the dynamic behavior of rubber-toughened materials—to model crazing in bimodal rubber-toughened polystyrene, for example (15). Finally, the bimodal model could be extended to study materials toughened with spheres that have trimodal distributions (16) or a variety of size distributions.

In summary, a method has been developed for the placement of bimodal sphere distributions within three-dimensional boundaries. The bimodal distribution is created from the combination of two sphere populations, where each population represents a distinct distribution. The efficient packing of a bimodal distribution of spheres can produce a high volume of the discrete phase in a toughened plastic and a corresponding small interparticle distance. However, combining two materials containing equal discrete-phase volumes of monosized spheres to make a bimodal system does not decrease interparticle distance in the blend. Any synergism in toughness in this system would have to be explained in other ways besides interparticle distance. The modeling method allows geometric simulation of real materials containing bimodal distributions of reinforcing spheres and facilitates the study of hypothetical materials that may be difficult or impossible to synthesize.

Acknowledgment

The authors thank Amoco for providing resources and releasing this document for publication.

References

1. Hobbs, S. Y.; Bopp, R. C.; Watkins, V. H. *Polym. Eng. Sci.* **1983,** *23,* 380.
2. Hall, R. A. *J. Mater. Sci.* **1992,** *27,* 6029.
3. Hall, R. A.; Burnstein, I. *J. Mater. Sci.* **1994,** *29,* 6523.
4. Hall, R. A. *J. Mater. Sci.* **1991,** *26,* 5631.
5. Ayer, J. E.; Soppet, F. E. J. Am. Ceram. Soc. **1965,** *48,* 180.
6. Bascom, W. D.; Hunston, D. L. In *Rubber Toughened Plastics;* Riew, C. K., Ed.; Advances in Chemistry 222; American Chemical Society: Washington, DC, 1989; pp. 135–172.
7. Turley, S. G.; Keskkula, H. *Polymer* **1980,** *21,* 466.
8. Kim, S.; Daumerie M.; Blackmon, K. P.; Sosa, J. M. Eur. Patent 620,236, 1994.
9. Dupre, C. R. U.S. Patent 4,146,589, 1979.
10. Baumgartner, E.; Hofmann, J.; Jung, R. H.; Moore, R.; Schaech, H. J. Eur. Patent 505,799, 1992.
11. Chen, T. K.; Jan, Y. H. *J. Mater. Sci.* **1992,** *27,* 111.
12. Pearson, R. A.; Yee, A. F. *J. Mater. Sci.* **1991,** *26,* 3828.
13. *Statgraphics,* Version 5.0; STSC, Inc.: Rockville, MD, 20852.
14. *SciAn,* Version 0.810; Supercomputer Computations Research Institute and Florida State University: Tallahassee, FL, 32306.
15. Okamoto, Y.; Miyagi, H.; Kakugo, M.; Takahashi, K. *Macromolecules* **1991,** *24,* 5639.
16. Henton, D. E. *U.S. Patent* 4,713,420, 1987.

Toughened Epoxy Resins: Preformed Particles as Tougheners for Adhesives and Matrices

C. Keith Riew,[1,4] A. R. Siebert,[2] R. W. Smith,[2] M. Fernando,[3] and A. J. Kinloch[3]

[1]Department of Chemical Engineering, College of Engineering, University of Akron, Akron, OH 44325–3906.
[2]BFGoodrich Company, Brecksville, OH 44141–3289
[3]Department of Mechanical Engineering, Imperial College of Science, Technology, and Medicine, Exhibition Road, London SW7 2BX, United Kingdom

Free-flowing, preformed, rigid, multilayer acrylic core–shell polymers with variable reactive-shell compositions were made by a sequential emulsion polymerization followed by spray- or freeze-drying to give powders with various particle sizes. These polymers were used to enhance the toughness of epoxy resins. The main materials studied were a blend of a solid and liquid diglycidyl ether of bisphenol A (DGEBA) epoxy resin, which was cured with 4,4'-diaminodiphenyl sulfone (DDS) and toughened with the core–shell polymers. The glass-transition temperatures of the cured epoxy materials were over 190 °C. The toughened epoxy resins were evaluated as bulk materials and as adhesives. The adhesive strength measured in the peel mode was improved by up to about fourfold by the addition of the core–shell polymers, with the lap-shear strengths being relatively unaffected. A blend of N,N,N',N',-tetraglycidyl-4,4'-diaminodiphenylmethane epoxy resin (TGDDM) and liquid DGEBA, which was cured with DDS and toughened with the core–shell polymers, was evaluated as a matrix resin for an epoxy–graphite composite. The mode II fracture toughness, G_{IIc}, was enhanced from 500 J/m^2 for the unmodified composite to 890 J/m^2 for the toughened epoxy-matrix composite.

PREFORMED PARTICLES SUCH AS THERMOPLASTIC POWDERS or core–shell polymers made from latexes are being increasingly used as tougheners for

[4]Current address: Riewkim Research and Associates, 2174 Rickel Drive, Akron, OH 44333–2917.

thermosets and thermoplastics (1). Some of the advantages of toughening plastics using preformed particles rather than reactive liquid rubbers are that it is relatively easy to form particles of different sizes and to maximize the volume fraction of the toughening phase. In general, the preformed particles enhance toughness without any tradeoff in thermomechanical properties. The conventional carboxy-terminated poly(butadiene–acrylonitrile) (CTBN) type of modifier produces the greatest degree of toughness of the existing commercial tougheners for the thermoset resin systems. Unfortunately, this enhancement is at the expense of (a) losses in the thermal or load-bearing properties, and (b) a decrease of toughness on aging, due to unsaturation in the backbone structure of the rubber.

Ricco et al. (2) have reported a quantitative assessment of the effect of structural heterogeneity of rubber particles on mechanical factors. They used a micromechanical analysis of an elementary model of a particulate-filled material, consisting of a continuous matrix phase and a discontinuous filler phase made up of discrete heterogeneous particles. They showed that distributions of the stress and strain concentrations in the surrounding elastic matrix produced by a rubber-coated, spherical, hard particle are very similar to those produced by a solid rubber particle, unless the rubbery coating on the surface of the hard particle became very thin compared with the total diameter of the particle. These results provide a theoretical explanation of what has been observed in a high-impact polystyrene having glassy polystyrene particles occluded within polybutadiene rubber particles in a polystyrene matrix. Namely, a large volume fraction of occluded glassy polystyrene may exist within the rubber particle without impairing its toughening efficiency.

This observation means that a hard particle with a thin rubber coating can behave like a solid rubber particle in toughening efficiency. The toughening effect of the thin film of rubber is mainly due to energy-absorbing plastic flow induced around such particles in the vicinity of the crack tip. In order to test Ricco et al.'s analytical approaches experimentally, we designed three-layer, preformed, particulate tougheners that consist of a large plastic core (60% by weight), a thin inner elastomeric shell (20% by weight), and a thin outer plastic shell (20% by weight), which may behave like core–shell polymers with 80% by weight of rubber core. Two-layer polymeric particles with a plastic core and a thin elastomeric shell (which may behave like 100% rubber particles), may have been better for the test program; but it would have been difficult to keep the particles separate and free-flowing. In industry, most of the manufacturers of commercial core–shell tougheners prefer to produce higher rubbery-core contents, for example, more than 50%, to have higher toughening efficiency.

For the outer-shell compositions, the choice of a monomer alters the reactive functional group(s) on the surface of the particle (e.g., epoxy, carboxy, mercaptan, etc.). Such groups will enable the toughener particles to compatabilize with, and possibly chemically bond with, the matrix resin. Thus for the outer shell, methyl methacrylate (M) and ethyl acrylate (E) were copolymerized with a minor amount of a monomer with a reactive functional group, such as glycidyl acrylate (MEG, where G is glycidyl acrylate), and carboxy groups in acrylic acid (MEAc, where Ac is acrylic acid) or itaconic acid (MAnI, where An is acrylonitrile and I is itaconic acid). Another outer shell was produced by using copolymers of methyl methacrylate with acrylonitrile and caprolactonediol acrylate (MAnC, where C is caprolactonediol acrylate) for better compatibility.

Experimental Details

Adhesive A.   One-part epoxy adhesive formulations are listed in Table I. The particular tougheners evaluated all have the same composition; namely, they are three-layer, core–shell polymers with an outer shell containing polymethyl methacrylate–ethyl acrylate–acrylic acid (MEAc), made by emulsion polymerization with reactive groups. They are processed to make free-flowing powders. All tougheners were made by spray-drying the emulsion. The one exception was for the emulsion that produced the particle size of 5 to 45 μm. This emulsion was co-agulated, filtered, washed, dried, and pulverized. Particle sizes of the three MEAc tougheners were 5 to 45 μm (T-A), 17.8 μm (T-B), and 2.3 μm (T-C).

Adhesive formulations were mixed by hand, then cured for 1 h at 170 °C. They were then tested using cold-rolled steel and electrogalvanized steel sub-strates according to the ASTM procedures D1002 and D1876 for lap-shear strength and T-peel resistance, respectively. Five specimens were tested in lap-shear and three in T-peel tests. The metal coupons were $25.4 \times 101.6 \times 0.76$ mm. A 12.7-mm overlap was used for lap-shear tests and a 76.2-mm bond length was used for the T-peel tests. The adhesive bond thickness was 0.48 mm and was con-trolled using wire spacers.

Adhesive B.   Bulk materials and adhesive film were made with a blend of diglycidyl ether of bisphenol A (DGEBA) epoxy resin in a solid form (epoxy equiv-alent weight 530), 25 parts, and liquid form (epoxy equivalent weight 190), 75 parts, and were cured with 4,4′-diaminodiphenyl sulfone (DDS, amine equivalent weight 64), 28 parts, and toughened with the core–shell polymers, 0 or 15 parts (Table II).

Bulk samples were cured at 125 °C for 1 h, followed by 177 °C for 2 h, and fi-nally 220 °C for 1 h. Fracture-toughness tests on bulk samples were performed ac-cording to the ASTM E399–83 procedure using compact-tension specimens with a thickness of 6.35 mm. Tensile tests were performed following ASTM D638–86. Both tests were run at a 50-mm/min jaw speed. Adhesive formulations were blend-

Table I. Adhesive A Formulation

Material	Parts
DGEBA resin	100
Tabular alumina	40
Cab-O-Sil	3.5
Dicyandiamide	6
Omnicure 94[a]	2
Core–shell toughener	0, 2.5, 7.5, 10, and 15

[a] 1-phenyl-3,3-dimethylurea.

Table II. Adhesive B Formulation

Material	Parts
DGEBA resin (epoxy equivalent weight 530)	25
DGEBA resin (epoxy equivalent weight 190)	75
4,4'-diaminodiphenyl sulfone (amine equivalent weight 64)	28
Core–shell toughener	0, 5

ed using a three-roll ink-mill after three passes. The blend was then heated to 60–70 °C and poured onto a running release paper over a hot platen adjusted to 60–80 °C. The thickness of the resin film was adjusted by passing the film through a Gardner knife and then flanking it by a layer of nylon scrim and release polyfilm. The glass-transition temperature of the adhesives after cure was over 190 °C.

Aluminum panels, which had a thickness of 0.61 and 1.63 mm, were etched with chromic acid. ASTM procedure D3167–76 (reapproved 1981) was followed for 135° peel tests. The adhesive film was placed between the aluminum panels and press-laminated at a temperature of 177 °C for 1 h at a pressure of approximately 5 psi (34.5 × 10^3 Pa). The temperature was then increased to 220 °C, and the joints were kept under pressure for another hour. The heaters in the hydraulic press were then switched off and the platens air-cooled and then water-cooled until the platen temperature was down to 100 °C. The bonded panels were cut into 12.7-mm-wide joints and tested at a rate of 20 mm/min and at a peel angle of 135°.

Following ASTM D1002–72 (1983), the lap shear tests were undertaken. Aluminum alloy sheets that had a thickness of 1.626 mm were bonded and press-laminated as described above for the peel tests. The panels were then cut into 25.4-mm panels and tested at a rate of 5 mm/min.

Fiber-Composite Materials.   For the epoxy–graphite composites, a blend of 60 parts of N,N,N',N',-tetraglycidyl-4,4'-diaminodiphenylmethane epoxy resin (TGDDM) and 40 parts of liquid DGEBA was used as the matrix. The blend was cured with DDS at 42 parts per hundred parts of epoxy resin (phr) by weight (3). It was toughened using the MEAc-based toughener at a 10-phr level, which had a range of particle sizes of 5 to 45 μm. Processing was accomplished using two impregnation steps followed by an autoclave curing process to consolidate the final laminate, according to the procedure described by Lee et al. (4). The final multilayer laminate structure contained layers of matrix resin with reinforcing carbon fibers separated by thin layers of matrix resin with the particulate toughener. Mode II interlaminar fracture toughness tests were performed using the end-notch-flexure specimen (5).

Results and Discussion

Adhesive A.   For the Adhesive A system, the tougheners were evaluated at levels of 0, 2.5, 7.5, 10, and 15 phr, as shown in Table I for the MEAc-based core–shell toughener with a particle-size range of 5 to 45 μm (T-A). The toughener T-A showed the best improvement at a toughener level of 7.5 phr and higher (Table III). Figure 1 shows the lap-shear strength and T-peel strength of the unmodified adhesive (CT), the three tougheners at the 15 phr

Table III. T-Peel and Lap-Shear Properties of Adhesive Joints of Adhesive A Formulations at Various Levels of MEAc Toughener

Test Type	Toughener Level				
	0	*2.5*	*7.5*	*10*	*15*
Lap-shear strength on oily cold rolled steel (psi)	1160	1100	1380	1390	1440
Lap-shear strength on oily electrogalvanized steel (psi)	1180	1330	1430	1510	1460
T-peel strength on oily cold rolled steel (pli)	13.5	13.1	18.3	21.6	23.9
T-peel strength on oily electrogalvanized steel (pli)	13.5	18.6	26.4	24.9	25.5

NOTE: The particle-size range for the MEAc toughener is 5–45 μm (T-A).

level (T-A, T-B, and T-C), and CTBN (L13). The figure shows the effect of particle size on the lap-shear and T-peel strengths at 15 phr of the toughener. The core–shell polymers of T-C did not improve the adhesive properties at all, at any levels. It is noteworthy that the T-A toughener, which had the largest particle sizes, showed the best improvements in T-peel adhesive strength. However, in spite of the chemical reactivity of particulate tougheners, the improvement in toughness obtained using the core–shell tougheners is lower than that observed for the reactive liquid rubber, CTBN. One reason for this difference is the terminal difunctionality of CTBN, which can chain-extend and reduce the cross-link density of the epoxy adhesive. However, the potential advantage of saturated, particulate, core–shell tougheners is better ultraviolet-light and thermo-oxidative stability than can be obtained using CTBN.

Adhesive B. The fracture toughness measured using compact-tension specimens and the tensile strength are shown in Table IV. All the systems were cured with DDS, and the glass-transition temperatures of all the samples were above 190 °C. In general, it is extremely difficult to toughen such brittle epoxy resins. The toughness enhancement shown in Table IV is, therefore, reasonably encouraging for the three different types of particulate tougheners, considering the brittleness of the base resin. Note that the mode I fracture toughness, the tensile fracture stress, and the failure strain are all lowered as particle size increases in the MEAc-based tougheners.

Among the various types of tougheners, the MAnC-based core–shell polymer gave the best enhancement in toughness. The caprolactone segments, along with the acrylonitrile groups, most probably increased the compatibility with the epoxy-matrix system. The carboxylic functional groups of MEAc or MAnI should have given a better adhesion because of reactivity between the

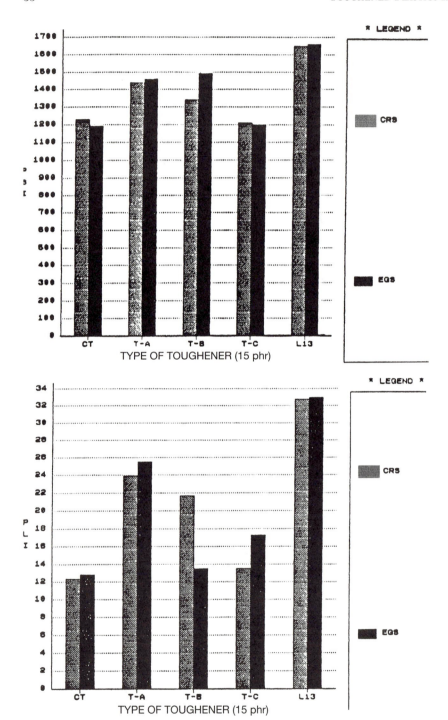

**Table IV. Bulk Properties of Adhesive B Formulations
Containing 5 phr of Toughener**

Type of Outer Shell[a]	Particle Size (μm)	Tensile Property			Fracture Toughness, G_{Ic} (J/m^2)
		Elongation (%)	Stress (MPa)	Modulus (GPa)	
None		3.9	76.5	3.29	107
MAnC	9.9	3.3	69.7	3.01	266
MAnI	10.4	2.4	64.12	3.45	161
MEAc	2.3	3.1	60.0	4.01	220
MEAc	17.8	2.6	58.4	3.65	175
MEAc	5–45	2.6	56.4	3.58	141

[a]M, methyl methacrylate; An, acrylonitrile; C, caprolactonediol acrylate; I, itaconic acid; E, ethyl acrylate; Ac, acrylic acid.

domains and matrix resin. However, they may have poorer compatibility than the MAnC-based tougheners. The MAnC toughener also gave the highest peel strength, as shown in Table V. The peel strength was improved fourfold with lap-shear strength being retained, compared with the adhesives based on un-modified epoxy resin.

Fiber-Composite Materials.   Interlaminar fracture energies of epoxy–graphite composite were measured in mode II fracture (Table VI). The mode II fracture toughness, G_{IIc}, was increased from 500 J/m^2 for the unmod-ified resin to 890 J/m^2 for the toughened epoxy resin (3). Particle sizes of this particular MEAc in Table VI ranged from 5 to 45 μm and averaged about 18 μm. In mode I fracture, the toughener particle sizes, being relatively large, may act as flaws, and the G_{Ic} values are actually lower for the cured epoxy ma-terials toughened with core–shell polymers, as shown in Table VI.

Fractographic Studies.   Visual examination of the fracture surfaces from the various tests revealed no signs of significant plastic deformation but did show that stress-whitening had occurred in some of the toughened epoxy resins. An exception was the MEAc core–shell toughener with a particle-size range of 5 to 45 μm. There, no signs of stress-whitening were observed.

The fracture surfaces were then examined using scanning electron mi-croscopy (SEM). In the case of the MEAc core–shell toughener with a parti-

Figure 1. Lap-shear strength (top) and T-peel strength (bottom) of one-part epoxy adhesives (Adhesive A formulations). Key: CRS, oily cold-rolled steel; EGS, electrogalvanized steel; CT, control (unmodified resin); T-A, T-B, and T-C refer to particle sizes of MEAc tougheners: 5 to 45 μm (T-A), 17.8 μm (T-B), and 2.3 μm (T-C); and L13, Hycar CTBN (1300×13).

Table V. Peel and Lap-Shear Properties of Adhesive Joints of Adhesive B
Formulations Containing 15 phr of Toughener

Type of Outer Shell	Particle Size (μm)	135° Peel Strength (lb/in.)	Lap-Shear Strength (psi)
None		1.9	2,300
MAnC	9.9	6.8	2,030
MEAc	2.3	2.5	1,600
MEAc	17.8	1.8	2,180
MEAc	5–45.0	3.0	1,740
MAnI	10.4	2.1	1,600

NOTE: An aluminum-alloy substrate was used.

cle-size range of 5 to 45 μm, good adhesion was found between irregularly shaped particles and the epoxy resin. (Recall that the irregularly shaped particles arise from the pulverization process used to form this type of toughener.) Fractographic features associated with crack-pinning were also discernible (Figure 2). Such fractures are normally observed when inorganic particles are added to a brittle matrix (6, 7). Thus, for these materials some increase in toughness appears to arise from a crack-tip pinning mechanism.

SEM studies of the toughened formulations based on core–shell polymers that were spherical in shape revealed the presence of debonding around the particles. The greater the toughness of the formulation, the more extensive the debonding, and a more intense stress-whitening of the fracture surfaces was also seen visually. The SEM images are shown in Figures 3 and 4. Especially in Figure 4 (bottom), the debonding and associated plastic dilatation of the epoxy resin may be clearly seen. The parabolic, or clam-shaped, features in Figures 3 and 4 (top) have been previously seen (8), and the feature usually has a particle as its focus.

Summary and Conclusions

We made free-flowing, preformed, rigid, three-layer acrylic core–shell polymers that consisted of a large plastic core, a thin elastomeric inner shell, and a

Table VI. Interlaminar Fracture Energies of Epoxy–Graphite Composites:
TGDDM–DGEBA–DDS System Toughened with MEAc Polymer

Resin	Fracture Energy	
	G_{Ic} (J/m²)	G_{IIc} (J/m²)
Unmodified epoxy resin	131	500
Epoxy resin modified with MEAc	50	890

NOTE: The particle-size range for the MEAc toughener is 5–45 μm (T-A).

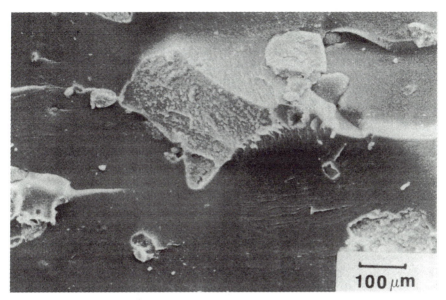

Figure 2. Micrograph of Adhesive B showing good adhesion between the pulverized MEAc toughener and the matrix resin, with crack-pinning features discernible. MEAc toughener was added at a 15-phr level and has a particle-size range of 5 to 45 μm (average 18 μm).

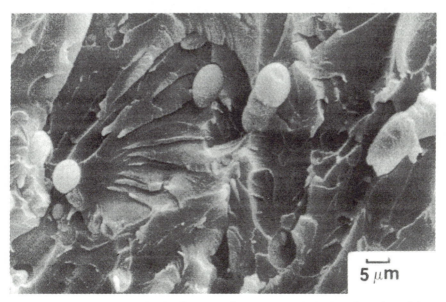

Figure 3. Micrograph of Adhesive B showing a paraboloid or clam-shaped fracture surface in the left half. The paraboloid is normally focused by a toughener particle. MAnC toughener was added at a 5-phr level and has an average particle size of 9.9 μm.

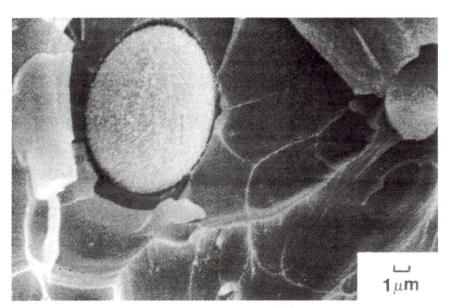

Figure 4. Top: Micrograph of Adhesive B similar to Figure 3, showing parabo-loids or clam-shaped fracture surfaces (center to the right half). MAnC toughener was added at a 5-phr level and has an average particle size of 9.9 μm. Bottom: Higher-magnification micrograph showing debonding and dilatation around the egg-shaped toughener particle, with a few petaloids (center to the right half).

thin plastic outer shell. The reactive outer shells had variable compositions. The preformed particles were used to enhance the toughness of epoxy resins. The glass-transition temperatures of the cured epoxy resins ranged from 120 °C (Adhesive A formulation) to 190 °C (Adhesive B).

The thin inner shell of the core–shell polymers contained less than 20% rubber by weight, with 60% by weight of a plastic core in the particles. It is unusual to enhance the toughness of the normally brittle epoxy-resin system at such a low level of rubber content. We believe that the unique design, the use of a multilayer toughener with a thin, elastomeric inner layer, and the large plastic core together produced behavior like that of solid rubber particles and, hence, enhanced crack resistance.

The acrylic core–shell polymers are considered to offer superior ultraviolet-light and thermal-oxidative aging properties than does the more conventional reactive liquid polymeric toughener, CTBN. Hence, there is current interest in the use of acrylic core–shell polymers as tougheners for adhesives and composite matrices that possess a relatively high glass-transition temperature.

In the case of the adhesive formulation with a glass-transition temperature of 120 °C (i.e., Adhesive A), the MEAc (T-A) core–shell polymer, which had a particle-size range of 5 to 45 μm, gave a significant improvement in peel strength and lap-shear strength. However, the conventional CTBN liquid-rubber toughener was more effective. The toughening mechanisms involved were a combination of increased plastic shear-yielding induced in the epoxy, and pinning of the advancing crack tip by the irregularly shaped, but well-bonded, toughener particles.

In the adhesive formulation with a glass-transition temperature of 190 °C (i.e., Adhesive B), the MAnC core–shell polymer, with a particle size of 9.9 μm, gave about a fourfold improvement in peel strength with no major loss in lap-shear strength. The toughening mechanisms involved were a combination of increased plastic shear yielding induced in the epoxy and debonding of the relatively poorly bonded spherical particles, enabling plastic dilatation of the epoxy to occur.

Finally, the basic Adhesive B formulation was toughened using the MEAc core–shell polymer and used as the matrix for an epoxy–graphite fiber composite. The mode II interlaminar fracture toughness, G_{IIc}, was increased from 500 J/m^2 to 890 J/m^2 by the presence of the acrylic core–shell toughener.

References

1. *Toughened Plastics: Science and Engineering;* Riew, C. K.; Kinloch, A. J., Eds.; Advances in Chemistry 232; American Chemical Society: Washington, DC, 1993; and references therein.
2. Ricco, T.; Pavan, A.; Danusso, F. *Polym. Eng. Sci.* **1978,** *18(10)*.
3. These data were kindly provided by Mr. Mark A. Hoisington, who was a graduate student under Professor J. C. Seferis, University of Washington, Seattle, WA.
4. Lee, W. L.; Seferis, J. C.; Bonner, C. D. *SAMPE Quarterly* **1986,** *17,* 58.

5. Carlson, L. A.; Pipes, R. B. *Experimental Characterization of Advanced Composite Materials*; Prentice-Hall: Englewood Cliffs, NJ, 1987; pp. 160–170.
6. Kinloch, A. J.; Young, R. J. *Fracture Behavior of Polymers*; Applied Science: London, 1983.
7. Lange, F. F.; Radford, K. C. *J. Mater. Sci.* **1971,** 6, 1197.
8. Riew, C. K.; Rowe, E. H., Siebert, A. R. In *Toughness and Brittleness of Plastics*; Deanin, R. D.; Crugnola, A. M., Eds.; *Advances in Chemistry* 154; American Chemical Society: Washington, DC, 1976; pp. 326–343.

4

Influence of Adhesive Strength upon Crack Trapping in Brittle Heterogeneous Solids

Todd M. Mower[1] and Ali S. Argon[2]

[1]Room D–450, Massachusetts Institute of Technology Lincoln Laboratory, P.O. Box 73, Lexington, MA 02173
[2]Room 1–306, Department of Mechanical Engineering, Massachusetts Institute of Technology, Cambridge, MA 02139

An experimental technique was devised in which a crack could be propagated past macroscopic obstacles in a transparent brittle epoxy under stable conditions, to determine the effective toughening resulting from discrete, isolated, crack-trapping processes in the absence of all other toughening mechanisms. The results indicate that toughness may be enhanced by a factor of nearly 2 through the crack-trapping mechanism alone, with particle equivalent volume fractions of just 0.06. To produce this level of toughening, a high adhesive strength was found to be required between the matrix and tough spherical particles. With long cylindrical trapping obstacles, the adhesive strength does not exert such a critical role.

TOUGHENING OF BRITTLE SOLIDS by the inclusion of tough particles has often been attributed to pinning of the crack front. By analogy with dislocations bowing around impenetrable obstacles, it has been assumed that the local crack front bows out between such tough particles until a critical breakaway configuration is reached (1, 2). Recent numerical simulations of crack-front interactions with arrays of tough particles predict enhancement of global fracture toughnesses ranging from a factor of approximately 2 (3, 4) to a factor of nearly 3 (5) for volume fractions of particles in the range of 0.1 to 0.25. These numerical predictions and other analytical models (6, 7) assume that the impeding particles are "perfectly bonded" to the host matrix, a condition seldom achieved in practice.

A precise understanding of the adhesive strength necessary to promote crack trapping by impenetrable obstacles in brittle solids has still not been reached, despite numerous investigations. In one of the earliest studies, Broutman and Sahu used tapered double-cantilever beam (TDCB) specimens to measure the fracture energies of epoxies containing glass spheres that had various surface treatments to modify their adhesive strengths (8). They credited the crack-trapping mechanism for a maximum measured enhancement of fracture energy, by a factor of approximately 3, in specimens containing particles with either unmodified or adhesion-enhanced surfaces. Further increases in fracture energy, to a level about five times greater than in the neat resin, were measured with composites prepared from spheres that had reduced adhesion. This augmented fracture energy was attributed to the greater fracture surface area resulting from rougher surfaces containing debonded hemispheres. We suggest that other factors were responsible instead (such as crack-tip shielding via decohesion of particles above and below the crack plane), because recent statistical analyses of toughening due to increased surface areas suggest that, at best, such increases can be expected to account for only a 15% rise in fracture energy (9).

Other investigations of toughening by glass spheres in brittle epoxies have concluded that little improvement of toughness (K_{Ic}) of the composite results from increased strength of adhesion of the particles to the matrix (10–12). The primary mechanisms suggested to contribute to overall toughness are crack trapping and crack-tip blunting, where the total toughness results from a competition between these two mechanisms modulated by the degree of adhesion (13–15). Lack of significant variation in measured critical stress intensities with apparent adhesive strengths (in particular composites) is not significant evidence that crack trapping is unaffected by adhesion levels. Rather, with decreasing adhesion, other mechanisms may become operative while crack trapping becomes less effective.

The role that adhesion plays in toughening by the process of crack trapping has remained unclear for two distinct reasons. First, no prior particulate-toughening research (that the authors are aware of) has included a quantitative study of adhesive strengths; rather, they have relied on fractographic information for qualitative rankings only. In addition, previous studies have used measurements of toughness in composite materials with microconstituents only. With that methodology it is always difficult, if not impossible, to separate the effects of fracture mechanisms and deformation processes that may be operating in unison.

In the study of crack trapping reported here, an experimental technique was devised in which a crack could be propagated past tough macroscopic obstacles in a transparent brittle epoxy under stable conditions. The shape and motion of the crack front were recorded cinematographically, with increasing crack driving force, to determine the effective toughening resulting only from discrete, isolated, crack-trapping events *in the absence of all other toughening mechanisms*. A parametric approach was adopted to study the importance of inclusion spacing, interfacial adhesion, and residual thermal stresses on the crack-front behavior and enhanced stress intensity required to propagate the cracks past obstacles. This chapter briefly details the experimental procedures, presents typical crack-front images, and discusses the trends observed in the measured toughening due to crack trapping. The influence of adhesion on these results is emphasized.

Experimental Details

Specimen Fabrication. The crack-trapping experiments reported here were performed with double cantilever-beam (DCB) fracture specimens having a square cross section of 38 × 38 mm and a height of 89 mm. The specimens were cast from a transparent epoxy bearing tough, 3.17-mm-diameter rods or spheres to act as obstacles to crack propagation. Epoxy was chosen for the matrix material because it lacks other distracting damage and energy-absorption mechanisms, such as crazing and microcracking, and can be formulated to a brittle state.

A standard DGEBA epoxy (Shell 815), cured with a bifunctional polyamide (V-40), was employed so that partial curing at room temperature would occur, in order to position obstacles in the desired locations within the specimens. A resin-rich composition of 3 parts resin to 1 part curing agent was employed to create an epoxy solid that was as brittle as possible using this particular curing agent. An elevated postcuring temperature (135 °C) activated the secondary amines and completed the cross-linking of epoxy groups, creating a matrix material with a glass-transition temperature, T_g, of 80 °C and a room-temperature fracture-initiation toughness as low as $K_{Ic} \sim 1.0$ MPa $\sqrt{m}$, $G_{Ic} \sim 350$ J/m^2, as measured with tapered DCB specimens at displacement rates of 0.1 to 10 mm/min.

The materials chosen for obstacles, polycarbonate (PC) and nylon-6, had elastic properties very similar to those of the epoxy matrix, so that stress concentrations would not be present and the analysis and interpretation of data would be simplified. All rods and spheres were solvent-cleaned and then thoroughly dried prior to their inclusion in the specimens to prevent plasticization of adjacent epoxy by diffusion of water from the rods during curing of the epoxy. Interparticle spacings (R/L, as defined in Figure 1) of 0.125, 0.187, and 0.250 were used, corresponding to equivalent volume fractions of approximately 0.06, 0.14, and 0.27.

Specimens were milled to the dimensions just indicated, and a semicircular groove was machined across the bottom of each specimen, parallel to the crack plane, to accommodate a pin that provided vertical support during testing. To promote planarity of crack growth, side grooves were cut with a 0.5-mm-thick slitting saw to a depth of 6.3 mm, and a reverse-chevron-tipped starter crack was cut to a

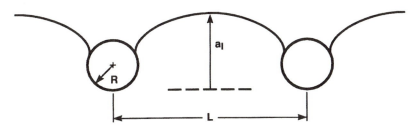

Figure 1. Geometry of particle spacing and local crack advance. Reproduced with permission from reference 26.

length of 43 mm. The specimens were then finish-sanded and polished with alumina paste. Photographs of a typical specimen are shown in Figure 2.

Constituent Material Properties.   Fracture of most epoxies at room temperature occurs in an unstable, jerky manner, which results from blunting of stationary cracks, followed by their re-initiation as sharp cracks and energetically "supercharged" extensions, and then blunting again upon arrest (*16, 17*). This mode of fracture gives way to stable, continuous fracture as the yield stress is increased, as a result of either increased loading rates or reduced testing temperatures (*18*). In the epoxy used here, stable fracture was consistently found to occur at a temperature of –60 °C. Because stable fracture was required both to photograph the crack front as it interacted with the obstacles and to accurately measure the resulting increases in toughness, the experiments were performed at –60 °C.

Relevant mechanical properties at this temperature, such as Young's modulus, E, yield strength, σ_y, and fracture toughness, K_{Ic}, were determined for the individual constituent materials and are listed in Table I. Glass-transition tempera-

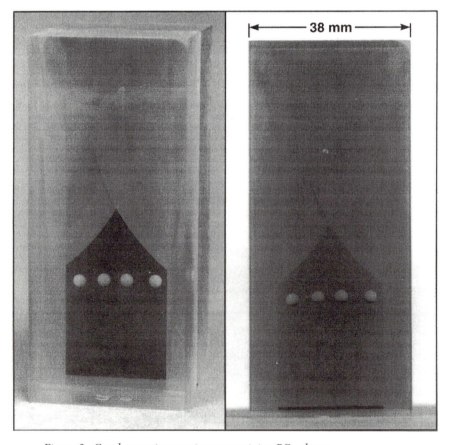

Figure 2. Crack-trapping specimen containing PC spheres.

Table I. Physical Properties of Constituent Materials at –60 °C

Material	T_g (°C)	E (GPa)	σ_y (MPa)	K_{Ic} (MPa$\sqrt{m}$)
Epoxy	80	3.5	145	0.5
PC	145	3.0	90 (20)	1.9 (20, 21)
Nylon	50	3.6	140 (22)	3.3 (22, 23)

ture, T_g, was determined both with differential scanning calorimetry and thermomechanical analysis. The values of T_g are also listed in the table. The uniaxial modulus and yield strength of the epoxy were determined by compression experiments. Fracture toughness of the epoxy was measured with TDCB specimens at displacement rates ranging from 0.1 to 10 mm/min. Tensile moduli of the PC and nylon rods were measured with the aid of a strain-gage extensometer. More details on the conduct of these tests can be found elsewhere (*19*). Included in Table I are yield-stress and fracture-toughness data for PC and nylon, derived from the literature.

Adhesive Strengths. The strengths of the adhesive bonds between the inclusion materials and the epoxy were determined with an experimental technique described in detail elsewhere (*24*). The technique involves embedding a sphere of the candidate material in the contoured-neck portion of a cylindrical epoxy bar. The decohesion of the sample from the epoxy is then observed as the bar is strained in tension. The radial stress at decohesion is then determined by means of a numerical solution of the deformation problem. Because no stress singularities exist at the interface prior to debonding, as is often the case with the usual popular tests, the technique used here provides a measure of the "true" adhesive strength for the bimaterial systems employed.

The specimens tested contained either nylon or PC spheres and were made using the same epoxy formulation as was employed in the crack-trapping experiments. All spheres were solvent-cleaned and dried prior to inclusion in test specimens. Analysis of the test results indicates that the nylon–epoxy system exhibits an adhesive strength (σ_a) of approximately 31 ± 5 MPa. Various attempts to enhance the adhesion through acid etching or use of primers did not result in adhesive strengths outside the indicated range. Polycarbonate–epoxy systems consistently exhibited higher adhesive strengths, such that σ_a = 53 ± 2 MPa. Modification of the PC surface with one coat of release agent lowered the strength to 33 ± 4 MPa; a second coat resulted in σ_a = 25 ± 3 MPa.

Residual Thermal Stresses. Residual thermal stresses in the model specimens at the surface of inclusions (*i.e.*, *maximum matrix stresses*) have been calculated (*25, 26*) using appropriate equations from the literature and thermal strains measured with a thermomechanical analyzer (TA 2940). The actual thermal stresses present in the model specimens were equal to the sum of the stresses generated during the initial cooling from the T_g of the epoxy to room temperature, reduced by some unknown amount of inelastic relaxation, plus the stresses generated during cooling from room temperature to –60 °C. Because experiments with a similar epoxy have demonstrated a relaxation time of several years at room temperature (*27*), the residual stresses during the crack-trapping experiments are taken to be the sum of those generated during the two cooling stages, as indicated in Table II.

Table II. Computed Matrix Thermal Stress (MPa) at Inclusion Surfaces

Inclusion	80 to 20 °C		20 to –60 °C		Final	
Spheres	σ_r	σ_θ	σ_r	σ_θ	σ_r	σ_θ
Nylon	1.8	–0.9	2.2	–1.1	4.0	–2.0
PC	–6.2	3.1	–2.2	1.1	–8.0	4.0
Rods	σ_r	σ_z	σ_r	σ_z	σ_r	σ_z
Nylon	–1.0	– 0.1	–0.4	–0.3	–1.4	–0.4
PC	–3.7	0.55	–1.0	0.15	–4.7	0.7

Testing Procedures.   The crack-trapping experiments were performed at –60 °C in a temperature-controlled chamber. Cooling was accomplished with forced convection of liquid nitrogen boil-off vapor, and was maintained to within ±2 °C with a controller, utilizing a thermocouple placed adjacent to the specimens. Testing was executed with a screw-driven (displacement-controlled) machine operated at a constant crosshead velocity of 2.5 mm/min. Loading of the specimens was performed with a polished, stainless steel, 30° wedge, prying apart the DCB specimens. Specimens were supported with a cylindrical steel pin positioned parallel to the fracture plane.

The load was directly recorded by a load cell, while the wedge displacement was measured with a linear potentiometer sensing the relative motion between the wedge and the load-cell platen. To prevent sticking of the actuator, a resistive film heater was mounted on the potentiometer housing. Both load and displacement data were digitally recorded and stored.

Images of the crack front and its interaction with obstacles were recorded on 16-mm film at a frame rate of 24 frames per second. Two cameras were used: one with a normal view (looking along the axis of the rods) to image the crack front, and one looking down on the specimens at an angle of approximately 25° in order to record possible debonding along the rod–epoxy interfaces. Lighting for the crack-front images was provided by light from a tungsten lamp reflected by a heated mirror (to increase the spot size) and passed through a diffuser plate. Lighting for the debonding images was provided by a fiber-optic light source aimed directly at the surface of the obstacles where crack intersection was anticipated. The elapsed time from crack initiation to specimen failure was typically about 40 s, so approximately 1000 photographic images were obtained during crack growth in each specimen.

Results

Load-Displacement History.   A typical load *versus* displacement (P–δ) trace obtained from testing a specimen containing no obstacles is shown in Figure 3. As the crack extends through the chevron starter section, its width increases, resulting in a nonlinear rise in load. The maximum load is reached when the crack reaches the full width of the specimen (end of the chevron), and is followed by a steady drop in load during further crack extension. A precipitous drop in load results as the crack approaches the end of the DCB and final specimen cleavage occurs. In crack-trapping specimens, the obstacles

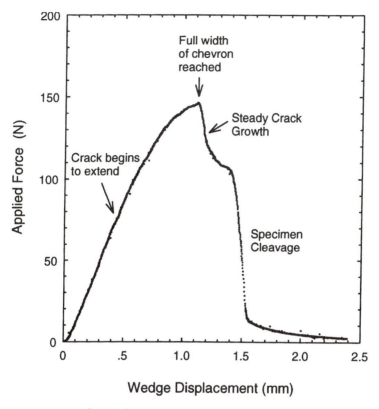

Figure 3. Typical P–δ plot obtained from testing a neat specimen. Reproduced with permission from reference 26.

were positioned such that the crack front encountered them during stable extension.

Typical *P*–δ traces from specimens containing two PC spheres are shown in Figure 4. The slight reduction in load at the initiation of fracture (pop-in) can be seen in the data from the specimen with full adhesive strength (> 54 MPa). When the crack front reaches the obstacles and becomes "trapped," the applied load (and hence the stress intensity) rises as the crack front begins to advance past the obstacles and assumes locally increasingly bowed configurations. A further load increase causes the local crack front to continue to bow between the obstacles until the breakaway configuration is reached. At this point, one or more of the obstacles are left behind, and the previously independent crack fronts coalesce into one and continue to grow. If the impeding obstacle was tough enough to be left intact (as was the case here with PC), and if it is sufficiently well-bonded to the matrix, then bridging of the crack flanks takes place. In specimens containing PC spheres with adhesive strength re-

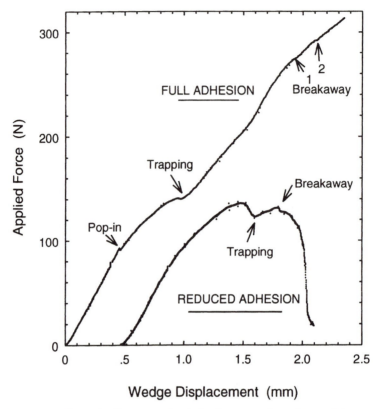

Figure 4. Effect of adhesion on load–displacement data obtained from testing two specimens containing PC spheres: one with full (> 54 MPa) and one with re-duced (22 MPa) adhesive strength. R/L = 0.125. Reproduced with permission from reference 26.

duced to about 22 MPa, crack trapping was still exhibited, but the cracks were able to bypass the obstacles by fracturing along the interfaces. As indicated by the data in Figure 4, such behavior resulted in a greatly reduced maximum ap-plied loading during the crack-trapping process.

Analysis of Crack-Front Shapes. The cinematographically recorded images provide clear histories of the crack-front interactions with the obstacles in each of the model specimens. Digitized and enhanced reproduc-tions of frames recorded during the testing of a typical specimen with PC spheres are shown in Figure 5. These sequential images give direct evidence for the complete evolution of the crack-trapping mechanism, from the initial "pinning" of the crack faces to the final breakaway configuration and transition to bridging, once the crack front has gone around the obstacles and has left

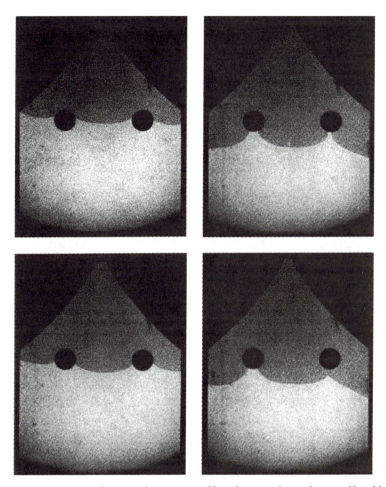

Figure 5. Digitized images from 16-mm film, showing the evolution of local bow-ing of a brittle matrix crack past tough PC spheres, R/L = 0.125.

them behind. Registration of the movie films with the applied loading provides the magnitude of applied loading (and corresponding toughness enhance-ment) that results from forcing the cracks into the configurations recorded in each image.

Determination of Stress Intensity. The accurate determination of the elevation of stress intensity due to the crack-path obstacles in the specif-ic DCB specimens requires taking note of the special features of the specimen and its mode of separation. The simple beam theory solution does not account for shear deformations or the compliance of the "built-in" cantilever ends. A more exact analysis considering these problems was developed by Kanninen using the model of a beam resting on an elastic foundation (28). Kanninen's derivation was for crack extension under conditions of constant opening displacement, and yields the Mode I stress intensity as:

$$K_I = \left(\frac{Pa}{1 - v^2}\right)\left(\frac{12}{B^2 h^3}\right)^{1/2} \Psi \tag{1}$$

where the correction term, Ψ, is a function of the ratio of ligament length to beam height and is not reproduced here; nonetheless, its minor influence in these experiments was recognized. Following convention, P represents the applied load, B and h are the beam width and height, a is the crack length, and v is Poisson's ratio.

Although the application of equation 1 to data obtained from fracture in the neat portions of the DCB specimens results in K_{Ic} values that are in exact agreement with the value obtained from our TDCB fracture tests (0.5 MPa√m, at –60 °C), a procedure was adopted to eliminate the effects of minor specimen-to-specimen differences. The toughness enhancements (effective stress intensity (K_I^∞), remotely measured, normalized by matrix toughness) due to trapping of the crack front by tough particles were determined using the ratio given by

$$\left(\frac{K_I^\infty}{K_{Ic}^{matr}}\right) = \frac{Pa\Psi}{P_o a_o \Psi_o}, \tag{2}$$

where the numerator is evaluated throughout the crack-growth experiment, and the denominator is evaluated at the point where the crack just reaches the obstacles. The effect of residual stresses was accounted for by a superposition procedure, using remote loadings measured from testing of specimens containing no particles.

Discussion

Influence of Residual Stresses. Though the magnitude of residual thermal stresses in the vicinity of the obstacles in the model specimens was small in comparison to crack-tip stresses, their effect on crack growth was noticeable and qualitatively predictable. The compressive longitudinal stress (–0.4 MPa) in the epoxy adjacent to the nylon rods caused a slight retardation in the rate of crack growth as the rods were approached. Conversely, the longitudinal tensile stress (0.7 MPa) near PC rods caused slight accelerations in crack growth. In the case of spheres, the effect of residual stresses was more dramatic. With PC spheres, the tensile, tangential matrix stress (4 MPa) attracted approaching cracks to the sphere equators. In the matrix near the surface of nylon spheres, the combination of compressive tangential (−2 MPa) and tensile radial (4 MPa) stresses caused approaching cracks to spall away from the obstacles, so that no crack trapping was induced in these specimens.

Influence of Adhesive Strengths. The effective toughening due to crack trapping in specimens containing either nylon or PC rods is plotted in Figure 6 as a function of the local bow-out amplitude of the crack, along with a numerical prediction provided by Bower and Ortiz (5, 29). The particle spacing of these data ($R/L = 0.125$) corresponds to an equivalent particle volume fraction of ~0.06. The data indicate that, at a given applied stress intensity, the local crack fronts in specimens with nylon rods were able to advance slightly farther than the numerical model predicts. This advance was accompanied by limited, albeit steady, debonding (fracture) of the nylon–epoxy interface,

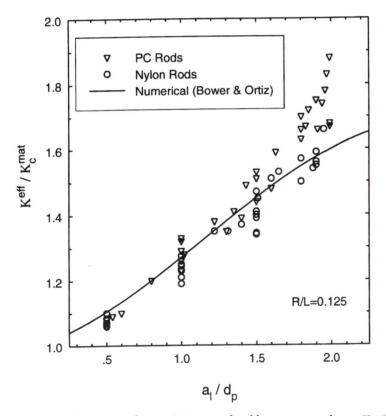

Figure 6. *Effective toughening (K_I^∞, normalized by matrix toughness, K_{Ic}^{matr}) resulting from crack trapping as a function of the local crack advance, normalized by particle diameter (d_p). R/L = 0.125. Numerical results, shown by the curve, are from Bower and Ortiz (5, 29).*

recorded by the upper camera as a bright area growing from the point of crack contact, along the interface to a distance of approximately one rod diameter. Though the crack faces remain "pinned" at the trailing edges of the nylon rods, the action of the partial debonding is to reduce the effective R/L and thus diminish the toughening from the levels that would be achieved with "perfect" bonding.

Apparently perfect bonding was exhibited by the specimens containing PC rods. No debonding was ever observed in these specimens, not even in those with adhesive strengths reduced by a surface coating of release agent. The effective toughening levels corresponding to local crack advances in all the specimens containing PC rods or spheres compare well with numerical predictions (which assume perfect bonding) up to the stages at which the cracks approach the final breakaway configuration. At this point, the toughening observed in the model specimens begins to exceed the predictions by as much as 15%. The probable cause of this discrepancy is not clear, but it is likely related to some finite size effect manifested in the experiments at the highest load levels.

The behavior of the specimens containing nylon rods, when compared with the behavior of those containing PC rods, is consistent with the measured adhesive strengths of 31 and >54 MPa, respectively. An unexpected result was that reduction of the PC–epoxy adhesive strength to about 28 MPa, accomplished by treating the rod surfaces with a release agent, did not result in any debonding. Further reduction, with a second coat, to about 22 MPa still did not result in any debonding or decrease in trapping-induced toughness. Evidently, the *toughness* of the PC–epoxy interfaces remained high enough to prevent the probing cracks from prying open the interface. In contrast, the nylon–epoxy interfacial toughness must be sufficiently low to enable the limited debonding that was observed in these specimens. This observation is supported by determinations of interfacial toughness (*30*) obtained from analysis of adhesive-strength tests performed on specimens fabricated from these material pairs (*24*). In those experiments, the toughness of release-agent-coated PC–epoxy interfaces was found to be about 20% greater than the toughness of nylon–epoxy interfaces, even though the two systems exhibited similar adhesive strengths.

Reduction of the adhesive strength of spherical particles had a much more dramatic effect on their ability to trap cracks, as illustrated by Figure 4. When the adhesive strength of PC spheres was reduced from above 54 MPa to about 22 MPa, crack trapping was still exhibited, but the cracks were able to bypass the obstacles by fracturing along the interfaces. As a result, the maximum levels of toughness enhancement achieved in such specimens was only on the order of 10–15% of the toughening possible with "perfect" adhesion.

Maximum Toughness Enhancement Achievable by Crack Trapping. In specimens containing well-bonded obstacles at an equivalent volume fraction of approximately 0.06 (R/L = 0.125), the maximum measured toughness enhancement consistently approached 80–90% prior to crack-front breakaway. Increased toughening was measured as a function of decreased particle spacing, such that at an equivalent particle volume fraction of about 0.27 (R/L = 0.25), the measured crack-trapping toughening was approximately 270% relative to the neat epoxy. It is emphasized here that the model matrix material does not craze or produce microcracks. It was completely brittle at the test temperature of –60 °C: the calculated plane-strain plastic-zone size (radius, r_p, approximated by $[K_{Ic}/\sigma_y]^2/6\pi$) is less than one μm. Consequently, toughening by local plastic flow or any mechanism other than crack trapping was inconsequential in the model specimens.

Summary

Experiments were performed to definitively determine the magnitude of toughening generated by crack trapping in the absence of all other toughening mechanisms. Model composite specimens consisting of a transparent brittle matrix containing two or more tough inclusions in a row perpendicular to the crack front were fractured under conditions of stable, controlled crack growth. Crack–particle interactions were recorded cinematographically both to document the evolution of the crack-trapping mechanism and to enable identification of various stages of crack growth with the applied stress intensity.

The sample images presented here reveal the evolution of crack-front trapping from the initial pinning to the fully bowed, breakaway configuration and finally the transition to crack-flank bridging by inclusions in model fracture specimens. The quantitative results indicate that toughness enhancement by nearly a factor of 2, relative to neat matrix values, may be achieved through the crack-trapping mechanism alone, with particle equivalent volume fractions of just 0.06. To produce this level of toughening, it was found that a high adhesive strength (of the order of one-third of the matrix flow stress) is required between the matrix and tough spherical particles. With cylindrical trapping obstacles, the requirement on the adhesive strength is not as critical because the crack front cannot readily escape from the pinning configuration.

Acknowledgments

Support for this work was provided by the U.S. Department of the Air Force and is gratefully acknowledged. Opinions, interpretations, conclusions, and recommendations are those of the authors and are not necessarily endorsed by the U.S. Air Force. We thank Professors A. F. Bower and M. Ortiz of Brown University for providing results of specific numerical simulations relating to our experimental conditions, and acknowledge several helpful discussions with them. The skilled assistance of Michael Imbeault (photography) and Chester Beals (computer graphics), both at the Massachusetts Institute of Technology

Lincoln Laboratory, was instrumental in this work and is gratefully appreciated. The generous provision of epoxy by the Shell Chemical Co. is acknowledged.

References

1. Lange, F. F. *Philos. Mag.* **1970,** *22,* 983–992.
2. Evans, A. G. *Philos. Mag.* **1972,** *26,* 1327–1344.
3. Fares, N. *J. Appl. Mech.* **1989,** *56,* 837–843.
4. Gao, H.; Rice, J. R. *J. Appl. Mech.* **1989,** *56,* 828–836.
5. Bower, A. F.; Ortiz, M. *J. Mech. Phys. Solids* **1991,** *39(6),* 815–858.
6. Rice, J. R. *ASME AMD* **1988,** *91,* 175–184.
7. Rose, L. R. F. *Mech. Mater.* **1987,** *6,* 11–15.
8. Broutman, L. J.; Sahu, S. *Mater. Sci. Eng.* **1971,** *8,* 98–107.
9. Faber, K. T.; Evans, A. G.; Drory, M. D. In *Fracture Mechanics of Ceramics;* Bradt, R. C.; Evans, A. G.; Hasselman, D. P. H.; Lange, F. F., Eds.; Plenum: New York, 1983; pp. 77–91.
10. Spanoudakis, J.; Young, R. J. *J. Mater. Sci.* **1984,** *19,* 487–496.
11. Kinloch, A. J.; Maxwell, D. L.; Young, R. J. *J. Mater. Sci.* **1985,** *20,* 4169–4184.
12. Moloney, A. C.; Kausch, H. H.; Stieger, H. R. *J. Mater. Sci.* **1984,** *19,* 1125–1130.
13. Moloney, A. C.; et al. *J. Mater. Sci.* **1987,** *22,* 381–393.
14. Cantwell, W. J.; et al. *J. Mater. Sci.* **1990,** *25,* 633–648.
15. Amdoudi, N.; et al. *J. Mater. Sci.* **1990,** *25,* 1435–1443.
16. Kinloch, A. J.; Williams, J. G. *J. Mater. Sci.* **1980,** *15,* 987–996.
17. Kinloch, A. J. *Adv. Polym. Sci.* **1985,** *72,* 45–67.
18. Scott, J. M.; Wells, G. M.; Phillips, D. C. *J. Mater. Sci.* **1980,** *15,* 1436–1448.
19. Mower, T. M.; Argon, A. S. *J. Mater. Sci.* to be submitted.
20. Pitman, G. L.; Ward, I. M. *Polymer* **1979,** *20,* 895–902.
21. Brown, H. R. *J. Mater. Sci.* **1982,** *17,* 469–476.
22. Mai, Y.; Williams, J. G. *J. Mater. Sci.* **1977,** *12,* 1376–1382.
23. Kinloch, A. J.; Young, R. J. *Fracture Behavior of Polymers;* Elsevier: Amsterdam, Netherlands, 1988; p. 344.
24. Mower, T. M.; Argon, A. S. *J. Mater. Sci.* **1996,** *31,* 1585–1594.
25. Mower, T. M. Ph.D. Thesis, Department of Mechanical Engineering, Massachusetts Institute of Technology, 1993.
26. Mower, T. M.; Argon, A. S. *Mech. Mater.* **1995,** *19(4),* 343–364.
27. Theocaris, P. S. *Rheol. Acta* **1967,** *6(2),* 246–251.
28. Kanninen, M. F. *Int. J. Fract.* **1973,** *9(1),* 83–92.
29. Bower, A. F.; Ortiz, M. Brown University, personal communication, 1992.
30. Mower, T. M.; Argon, A. S. *J. Adhes.* to be submitted.

5

Rheological Monitoring of Phase Separation Induced by Chemical Reaction in Thermoplastic-Modified Epoxy

C. Vinh-Tung[1], G. Lachenal[1], B. Chabert[1], and J.-P. Pascault[2]

[1]Laboratoire d'Etudes des Matériaux Plastiques et Biomatériaux, Centre National de la Recherche Scientifique–Unité Mixte de Recherches no. 5627, Université Claude Bernard Lyon 1, 43 bd du 11 novembre 1918, F-69622 Villeurbanne Cedex, France
[2]Laboratoire des Matériaux Macromoléculaires et Composites, Centre National de la Recherche Scientifique–Unité Mixte de Recherches no. 5627, Institut National des Sciences Appliquées de Lyon, 20 av Albert Einstein, F-69621 Villeurbanne Cedex, France

Rheological measurements were made to monitor in situ the phase separation induced by chemical reaction in blends of tetraglycidyl-diaminodiphenylmethane epoxy resin with an aromatic diamine hardener and a thermoplastic having a high glass-transition temperature. The final morphologies were observed by optical and electron microscopy. We showed that the type of morphology can be identified during phase separation by means of viscoelastic measurements. Gelation and vitrifications of the two phases can also be monitored with this technique. The strain applied during the experiments had a strong influence on the final morphology.

DEVELOPING BIPHASIC MATERIALS in order to improve the fracture toughness of thermoset resins is now a common practice. Thermoplastics that have a high glass-transition temperature (T_g) are used as tougheners in preference to low-T_g elastomers because of their insignificant effect on the thermal and modulus properties.

Phase separation in initially miscible blends of thermoplastic (TP) and epoxy-diamine occurs during the reaction because the increase in the molecular weight of epoxy–amine copolymers diminishes the entropy of the blend. We can consider our systems as pseudobinary blends of TP and copolymer epoxy–amine for ease of understanding. The Flory–Huggins theory can be used in this case (*1, 2*). The miscibility during the reaction is controlled by the temperature, molar masses, and proportions in the blend, and the final morphology results from the competition between the kinetics of reaction and phase separation.

Rheological measurements are usually performed on polymer blends to characterize their behavior during processing (3). Miscible blends have been studied using several theories (4–6). The characterization of miscible blends exhibiting phase separation is now receiving more interest, and a connection between thermodynamics and rheology has been established. Until now, rheology has been performed to study physical phase separation in nonreacting blends. The most studied systems are blends of polystyrene and poly(vinyl methyl ether) because their miscibility behavior is well known and the temperature of phase separation is accessible (7–9). With these systems, the viscoelastic response is sensitive to phase separation. The appearance of the biphasic co-continuous morphology is characterized by an increase of shear moduli in the low-frequency range. This effect is attributed to the large differences between the moduli of the two phases. In blends of polystyrene and poly(methyl methacrylate) (10–11), the morphologies are made of dispersed particles, and an emulsion model is used to describe the viscoelastic behavior of the blends. To our knowledge, studies of phase separation induced by chemical reaction have not been conducted in modified thermosets. Experimentally, it is difficult to make measurements over a wide frequency range because in such blends, phase separation occurs in a short time. In epoxies modified with a high-T_g, thermoplastic, we expect to observe wide variations in viscosity during phase separation, as the properties of the phases are expected to be very different.

Experimental Details

Materials. The epoxy resin was tetraglycidyldiaminodiphenylmethane (TGDDM) from Ciba-Geigy (MY721). The hardener was crystalline aromatic diamine with a melting point of 88 °C. The modifier was an amorphous, nonreactive TP with T_g around 200 °C, weight-average molecular weight, $\overline{M}_w$, of 50,000 g/mol, and number-average molecular weight, $\overline{M}_n$, of 26,000 g/mol. Fine powder of TP (average particle diameter < 160 μm) was used to be easily dissolved in the resin. For neat resin mixtures, the epoxy resin and the hardener were mixed in stoichiometric ratio at 90 °C for 20 min (until the mixture was clear and homogeneous).

A different procedure was used for ternary blends. The TP was dried for 4 h at 150 °C and then dissolved in the epoxy resin at 140 °C for 20 min. The blend was cooled to 100 °C and the hardener added. Stirring for 20 min at 90 °C was necessary to homogenize the blend. The reaction during this stage was negligible. This process is interesting because no solvent is used.

Techniques. The common technique for studying phase separation is the cloud point from light transmission. The cloud-point experiment has already shown its reliability in many studies (1, 2, 12). The turbidity of the blend is followed by measuring the intensity of the transmitted white-light beam with a photosensor. To detect the beginning of phase separation as precisely as possible, we started the measurement with a very weak light intensity and a high sensitivity of detection. The cloud point was taken as the onset of the loss of light intensity. By not taking into consideration the variations of the refractive index, we could assume that the phase separation was completed when the turbidity remained constant. So we increased the light intensity to determine the end of the phase separation. This experimental procedure was necessary so as not to underestimate the time during which phase separation took place.

Rheological measurements were performed with a Rheometrics RDA700 instrument equipped with parallel plates (diameter of plates = 40 mm). Unreacted resin was placed between the preheated plates, and isothermal measurements were made at 160 °C under a nitrogen flow until the measured values stabilized, reflecting the end of the reaction. The temperature of the experiments was measured by a thermocouple near the plates. Frequencies were chosen to reduce the time between two measurements (10 to 200 rad/s), because phase separation occured quickly. Lower frequencies were therefore not used. It took approximately 1 min to scan all the frequencies. Small shear strains (less than 2%) were used to minimize the shear effect on the miscibility and morphology. This effect was studied by using a larger shear strain (more than 20%); the strain was reduced to a low value as the resin approached gelation.

Experiments were carried out on blends of different compositions (from 0 to 20.9 wt% TP) under isothermal conditions.

Simultaneous cloud-point and rheological measurements were performed by driving the light beam through glass plates via an optical fiber. A relatively low temperature (130 °C) was chosen to slow down the phase-separation kinetics and to precisely correlate the responses of each technique.

The cured samples were microtomed at room temperature, and morphologies were identified by transmission electron microscopy (TEM) and optical microscopy.

Results and Discussion

Neat Resin. For the neat resin, the evolution of the complex viscosity, the tangent of the loss angle (tan δ), and the loss modulus are shown in Figures 1 and 2. The absence of measurements at the beginning of the reaction is due to the lack of sensitivity of the apparatus.

Chemical reaction in TGDDM–diamine systems leads to the formation of a tridimensional network, which is characteristic of thermosets. The appearance of a tridimensional macromolecule of infinite size is defined as the gelation of the system. At this moment, the viscosity suddenly increases to high values. Gelation in TGDDM–diamine systems can be detected by viscoelasticity, and the gel point corresponds to the time when tan δ is independent of frequency. This criterion describes the passage from a liquid to a viscoelastic material and is characteristic of many other systems (*13–16*).

When the T_g of the reacting system becomes equal to the cure temperature, the system passes from a gel to a glassy state. This transition occurs with a mechanical relaxation that depends on the frequency, and it is identifiable by a peak on the loss modulus curve (G''). This relaxation is attributed to the generalized motion of molecular chains in the polymer (*17*).

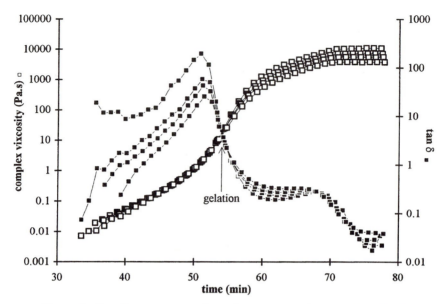

Figure 1. Complex viscosity and tan δ versus time for unmodified TGDDM–diamine at 160 °C (100–300 rad/s).

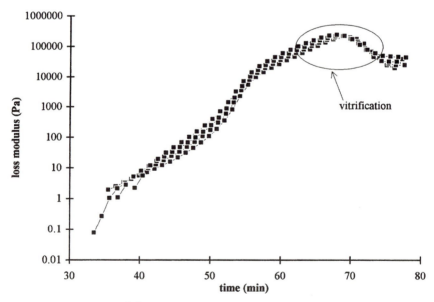

Figure 2. Loss modulus versus time for unmodified TGDDM–diamine at 160 °C (100–300 rad/s).

Sensitivity of the Techniques with Respect to Phase Separation. Simultaneous cloud-point and viscoelastic measurements were made at 130 °C. Figure 3 shows that phase separation is not detected at the same time by these two techniques. Complex viscosity is more sensitive to the beginning of phase separation. Cloud-point sensitivity is a function of the wavelength of the light. For our experiment, the use of white light determines the sensitivity at approximately 0.1 μm. However, in the last step of phase separation (when the transmitted light intensity still varies), the increase of viscosity due to gelation of the resin hides the end of the phenomenon.

Chen et al. (*12*) showed that phase separation in elastomer-modified epoxies can also be detected with much more sensitivity using small-angle X-ray scattering (SAXS). Then, the choice of the cloud point as a criterion for detecting the beginning of phase separation can be discussed. We have not performed SAXS studies on our systems because phase separation is much faster in thermoplastic-modified epoxy and the scanning time is too long.

Final Morphologies. Micrographs of morphologies obtained in the fully cured material are shown in Figure 4. They reveal that a phase separation takes place during the cure. For convenience, we call the α-phase the epoxy-amine-rich phase and the β-phase the TP-rich phase. We observed three types of morphologies. Type 1 morphology (Figure 4a) is made of a continuous α-phase with small particles of β-phase (≈1 μm). Type 2 morphology (Figure 4b), has irreg-

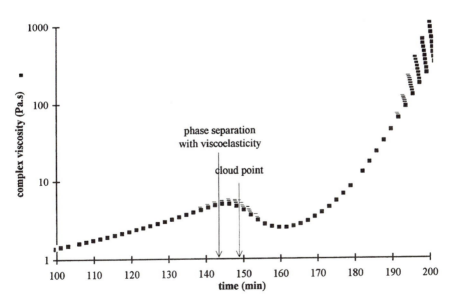

Figure 3. Complex-viscosity and cloud-point curves in simultaneous experiments for a TGDDM–diamine blend containing 10.5 wt% TP at 130 °C (10–200 rad/s).

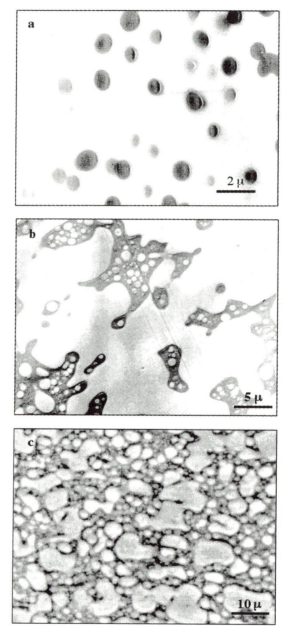

Figure 4. Three morphologies observed by TEM on TGDDM–diamine blends containing various amounts of TP and cured at 160 °C. (a): Type 1 morphology (10.5 wt% TP). (b) Type 2 morphology (15.0 wt% TP). (c) Type 3 morphology (20.9 wt% TP). Parts a and b are TEM images; part c is an optical microscopic image.

ular large β-phase domains with subinclusions of α-phase. Type 3 morphology (Figure 4c) shows a complete phase inversion where the β-phase is continuous and the α-phase is dispersed in domains of different size (0.1–10 μm). For a cure temperature of 160 °C, the different types of morphology appear for different initial concentrations of TP (Figure 5). From lower to higher initial concentrations of TP in the blends, we have type 1 to type 3 morphologies. Those morphologies are typical in epoxy blends (*18, 19*) and can be related to the phase diagrams calculated with the Flory–Huggins theory. The main problem with this theory is that it does not include the polydispersity of the polymers, the specific interactions, and the variations of the interaction parameter with composition. For these reasons, the model sometimes shows poor agreement with reality (*1, 2*). The next section shows that viscoelastic measurements can be an excellent tool for identifying the type of morphology during its formation.

Phase-Separation Monitoring. Viscoelastic behavior during phase separation is different for the three types of final morphology. For type 1 morphology (Figure 6a), complex viscosity is weakly dependent on the frequency, reflecting a quasi-Newtonian behavior. For type 2 and 3 morphologies (Figures 6b and 6c), complex viscosity strongly depends on frequency, which suggests a viscoelastic behavior.

Before trying to explain the variations in viscosity during phase separation, it is useful to think about what happens in each phase. When phase separation begins, the TP demixes from the α-phase and the copolymer epoxy–amine demixes from the TP β-phase. Therefore, the viscosity of the α-phase diminishes and that of the β-phase increases. As the TP's T_g is higher than the cure temperature, it is evident that the β-phase has a high viscosity and that it may vitrify if it is sufficiently rich in TP. The second effect is due to the reticulation of the resin. The polycondensation between the epoxy and the amine induces an increase in the viscosity of the resin, as in the neat resin experiment. Therefore, the viscosity of the α-phase increases as well. The viscosity of the β-phase is lightly affected by this effect because the reaction is slower in this phase as it is extremely viscous and poor in epoxy–amine. The third effect controlling the viscosity is morphological. In a biphasic material, the whole viscosity is strongly influenced by the geometry of the phases. For example, in a system with dispersed particles (an emulsion), the viscosity mainly depends on the viscosity of the continuous phase. After the geometry is known, the variable volume frac-

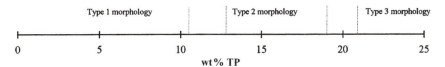

Figure 5. Final morphologies versus initial concentrations (wt%) of TP in TGDDM–diamine blends cured at 160 °C.

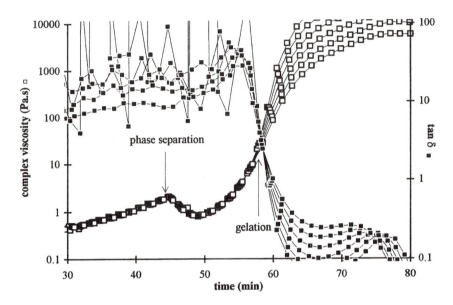

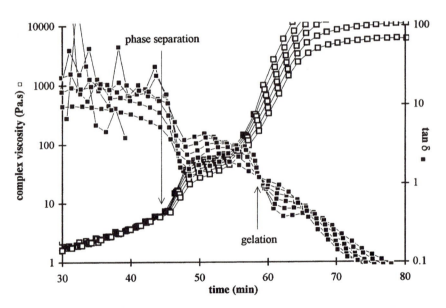

Figure 6. *Complex viscosity and tan δ versus time during cure at 160 °C, for TGDDM–diamine blends containing various amounts of TP. Frequencies were 10 to 200 rad/s for parts a and b, 10 to 300 rad/s for part c. (a) Type 1 morphology (10.5 wt% TP). (b) Type 2 morphology (15.0 wt% TP).*

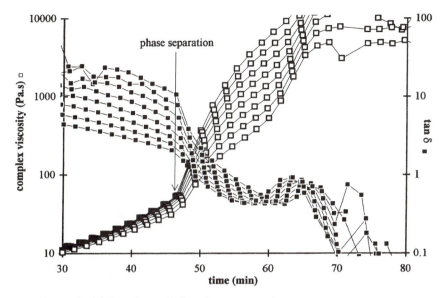

Figure 6. (c) Type 3 morphology (20.9 wt% TP).

tion of the phases constitutes one more parameter. At this level of the study, we are not able to describe the evolution of the morphology and of the volume fraction. So we discuss only the effects of variations in the composition of the phases and reticulation of the resin.

Before phase separation, the viscosity evolves as expected for a classical epoxy–amine system. The viscosity increases because of the polycondensation of the epoxy with the diamine. When phase separation occurs, the behavior becomes characteristic of the final morphology.

For type 1 phase separation (blends from 5.6 to 10.5 wt% TP), it is reasonable to suppose that the quasi-Newtonian behavior reflects the fact that the α-phase is continuous during the entire phase separation (Figure 6a). If it were not, the behavior would be non-Newtonian, as it is for types 2 and 3. For this reason, we can affirm that phase separation occurs by nucleation and growth for type 1 morphology. The decrease in viscosity is due to the demixing of the TP from the α-phase. When the resin approaches gelation, the reticulation effect is emphasized and the viscosity increases again. As the α-phase is continuous, we can also detect its gelation by a crossover of the multifrequential curves of tan δ. The vitrification of the α-phase is seen in the shear loss modulus G'' curves (Figure 7).

For type 2 phase separation (blends from 12.8 to 19.0 wt% TP), the viscoelastic behavior shows that the β-phase is continuous because at this moment which is before the gelation of the resin, it is the only phase that can be

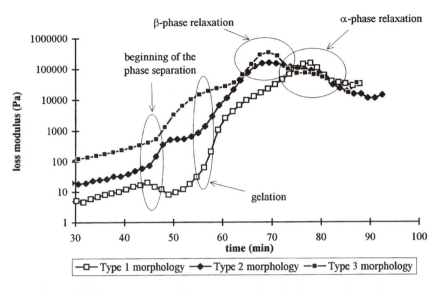

Figure 7. Shear viscous modulus curve at 10 rad/s during polycondensation at 160 °C. Blends of TGDDM–diamine contain 10.5 wt% (type 1), 15.0 wt% (type 2), and 20.9 wt% (type 3) of TP.

non-Newtonian (Figure 6b). The complex viscosity values initially increase faster for all frequencies. This increase is due to the contribution of the β-phase, which becomes richer in TP. The following shoulder on the complex viscosity curves is due to the influence of the α-phase, and finally, the increase of the viscosity of both phases brings the global viscosity to high values. In addition, gelation is detected for this kind of morphology. This feature indicates that the α-phase is continuous and, therefore, that the mechanism of the phase separation is probably a spinodal decomposition with co-continuous phases.

When type 3 phase separation occurs (blends above 20.9 wt% TP), the evolution of complex viscosity is predominated by the β-phase as it increases more rapidly from the beginning of phase separation (Figure 6c). The difference in this behavior is that the response of the α-phase is hidden by that of the β-phase. Obviously, neither its contribution to the decrease in the global viscosity nor the gelation criterion is apparent. It is reasonable to say that the α-phase is dispersed in this case from the beginning of phase separation, and we propose a nucleation and growth mechanism with a phase inversion for this morphology.

This interpretation shows that the viscoelastic measurements can characterize the phase-separation mechanism in TP-modified epoxy systems. If we apply the Flory–Huggins theory to this pseudobinary blend (20), the calculated phase diagrams are in very good agreement with the morphologies encoun-

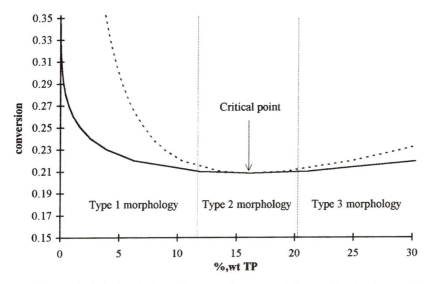

Figure 8. Calculated phase diagram for a pseudobinary blend of TGDDM–diamine and TP during cure at 160 °C.

tered. Figure 8 shows the phase diagram at 160 °C for a pseudobinary blend of TGDDM–diamine with TP, in the plane conversion/weight percentage of TP. The critical concentration of TP is localized at 16.1 wt%, which corresponds to type 2 morphology. The interpretation of this diagram is that the phase-separation mechanisms depend on the region where the system enters. Near the critical point, the binodal region is small and the system can easily enter into the spinodal region, where the mechanism is spinodal decomposition. With concentrations far from the critical point, the system probably stays in the binodal region during the entire phase separation, and the nucleation and growth mechanism occurs. The possibility of entering the spinodal region is determined by the phase diagram as well as competition between the kinetics of phase separation and polycondensation of the resin. So it is not possible to predict the final morphology exclusively with the theory. In addition, the viscoelasticity provides a wealth of information about the morphology and also about the cure of the thermoset.

Figure 7 shows the shear loss modulus G'' curves at 10 rad/s for the three types of blends during the cure at 160 °C. The interesting feature of these curves concerns the vitrifications of the phases, which are each characterized by a peak. In type 1 morphology, the peak observed around 75 min corresponds to the α-phase relaxation because this phase is continuous and logically brings the main contribution to the global response. In type 2 morphology, a second relaxation appears before the α-phase relaxation. This transition is attributed to the vitrification of the β-phase. When the β-phase is continuous (type 3), its relaxation peak is more pronounced and the α-phase vitrification is hidden. We have already said that if the T_g of the TP gets higher than the cure temperature, the β-phase vitrifies. In the β-phase, the composition and the epoxy–amine reaction evolve from the beginning of the phase separation until the molecular motions are frozen by vitrification. Vitrification appears relatively late, meaning that these evolutions are slow because of the extremely high viscosity in the β-phase.

Influence of the Strain Rate on Viscoelastic Response. In high-shear-strain experiments, we observe significant differences only with type 2 and type 3 blends (Figures 9b and 9c). The complex viscosity still depends on frequency, but it decreases as phase separation begins. Moreover, a crossover of the tan δ curves appears at this moment. The differences reported in this kind of experiment can be found at two levels:

- The high-strain experiments do not verify the viscoelastic linearity.
- The shear applied to the blends implies some modifications in the morphology.

Before phase separation, viscoelastic linearity is respected because the viscoelastic values are the same in the low-strain and high-strain experiments. The difference begins to appear at the beginning of phase separation. For

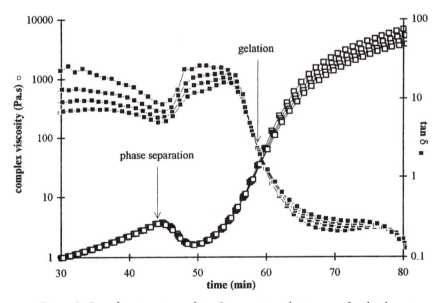

Figure 9. Complex viscosity and tan δ versus time during cure for the three types of phase separation, under high shear strain. TGDDM–diamine blends contain various amounts of TP at 160 °C. (a) Type 1, 30% strain (10.5 wt% TP, frequencies 100–300 rad/s).

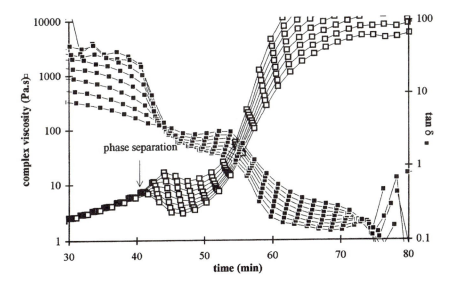

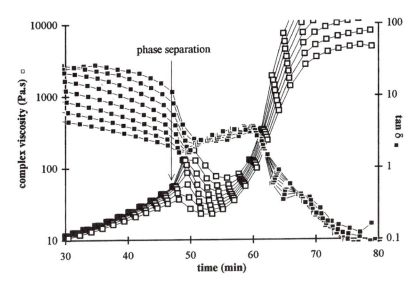

Figure 9. (b) Type 2, 30% strain (15.0 wt% TP, frequencies 10–300 rad/s). (c) Type 3, 20% strain (20.9 wt% TP, 10–300 rad/s).

high-strain experiments, shear strain is reduced as the resin approaches gelation. No discontinuity in the viscoelastic curves exists at this moment, showing that the experiments take place in the linear viscoelastic range. Finally, the differences result from an exclusively morphological aspect. Microscopic observations made on a type 3 sample that has been sheared with a 20% strain (Figure 10) clearly show that large domains of the β-phase with inclusions of the α-phase have been dispersed in a continuous α-phase.

Viscosity curves obtained with constant-shear viscosimetry are shown in Figure 11. In this kind of measurement, phase separation is monitored, but the same behavior is obtained with the three types of phase separation. As in the high-strain dynamic measurements, the constant shear applied during phase separation forbids the formation of a continuous β-phase, which would normally exist in type 2 and 3 blends. On the contrary, because it is liquid, the α-phase can be continuous, and in all cases the viscosity of this phase governs the viscosity of the blend. The fast increase in viscosity is therefore characteristic of the gelation of the α-phase.

The effect of the shearing on miscibility in polymer blends has been reported to give a modification of the phase diagram (21–24). In our case, we cannot discuss this point because the differences measured in phase-separation times can also be attributed to thermal problems.

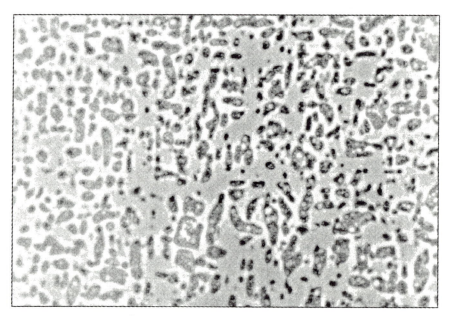

Figure 10. Optical microscopic image of a type 3 blend of TGDDM–diamine and 20.9 wt% TP, cured at 160 °C under a 20% shear strain.

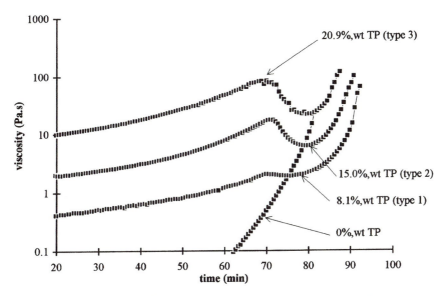

Figure 11. Constant-shear measurements of viscosity versus time during cure at 150 °C for TGDDM–diamine blends containing various amounts of TP.

Conclusions

In TP-modified epoxy, phase separation generates three types of morphologies according to the initial concentration of TP: an uninverted, a partially inverted, and a totally inverted morphology. Phase separation, which is often studied with light transmission, was monitored with rheology. Dynamic measurements show three types of characteristic behavior corresponding to the three types of final morphology. This powerful technique gives a lot of information and even allows us to determine the phase-separation mechanisms. In agreement with the phase diagrams, a nucleation and growth mechanism takes place for type 1 and 3 morphologies, and a spinodal decomposition occurs for type 2 phase separation. Shear strain has a strong influence on the final morphology. For example, the continuous β-phase of the type 3 morphology becomes dispersed when a 20% strain is applied during phase separation. The gelation of the resin and the vitrifications of the phases can also be monitored with dynamic measurements. With constant-shear viscosimetry, the morphologies cannot be identified because of the breakdown of the continuous β-phase by shearing, but detection of the gelation of the resin is always possible.

References

1. Verchere, D.; Sautereau, H.; Pascault, J. P.; Moschiar, S. M.; Riccardi, C. C.; Williams, R. J. *J. Polymer* **1989,** *30,* 107–115.

2. Ruseckaite, R. A.; Williams, R. J. *J. Polym. Int.* **1993**, *30*, 11–16.
3. Utracki, L. A. In *Multiphase Polymers: Blends and Ionomers*, American Chemical Society: Washington DC, 1989, pp 153–210.
4. Han, C. D.; Kim J. K. *Macromolecules* **1989**, *22*, 1914–1921.
5. Roovers, J.; Toporowski, P. M. *Macromolecules* **1992**, *25*, 1096–1102.
6. Han, C. D., Kim, J. K. *Macromolecules* **1989**, *22*, 4292–4302.
7. Ajji, A.; Choplin, L.; Prudhomme, R. E. *J. Polyrn. Sci. Polym. Phys,* **1988**, *26*, 2279–2289.
8. Mani, S.; Malone, M. F.; Winter, H. H. *J. Rheol.* **1992**, *36*, 1625–1649.
9. Ajji, A.; Choplin, L.; Prudhomme, R. E. *J. Poly. Sci. Polym. Phys.* **1991**, *29*, 1573–1578.
10. Graebling, D.; Muller, R.; Palierne, J. F. *Macromolecules* **1993**, *26*, 320–329.
11. Bousmina, M.; Muller, R. *J. Rheol.* **1993**, *37*, 663–679.
12. Chen, D.; Pascault, J. P.; Sautereau, H.; Vigier, G. *Poly. Int.* **1993**, *32*, 369–379.
13. Winter, H. H.; Chambon, E. *J. Rheol.* **1986**, *30*, 367.
14. Feve, M. *Makromol. Chem. Macromol. Symp.* **1989**, *30*, 95–107.
15. Matejka, L. *Polym. Bull.* **1991**, *26*, 109–116.
16. Lairez, D.; Adam, M.; Emery, J. R.; Durand, D. *Macromolecules* **1992**, *25*, 286–289.
17. Boiteux, G.; Dublineau P.; Feve, M.; Mathieu, C.; Seytre, G.; Ulanski, J. *Polym. Bull.* **1993**, *30*, 441–447.
18. Cho, J. B.; Hwang, J. W.; Cho, K.; An, J. H.; Park, C. E. *Polymer* **1993**, *34*, 4832–4836.
19. Bucknall, C. B.; Gomez, C. M.; Quintard, I. *Polymer* **1994**, *35*, 353–359.
20. Vinh-Tung, C.; Lachenal, G.; Chabert, B.; Pascault, J.-P. (to be submitted).
21. Mani, S.; Malone, M. F.; Winter, H. H. *Macromolecules* **1992**, *25*, 5671–5676.
22. Wu, R.; Shaw, M. T.; Weiss, B. A. *J. Rheol.* **1992**, *36*, 1605-1623.
23. Muniz, E. C.; Nunes, S. P.; Wolf, B. A. *Macromol. Chem. Phys.* **1994**, *195*, 1257–1271,
24. Fernandez, M. L.; Higgins, J. S.; Richardson, S. M. *Chem. Eng. Res. Des.* **1993**, *71*, 239–244.

6

Compatibilized, Segmented Liquid Rubbers as Epoxy-Toughening Agents

R. Mülhaupt and U. Buchholz

Freiburger Materialforschungszentrum und Institut für Makromolekulare Chemie der Albert-Ludwigs Universität, Stefan-Meier-Strasse 31, D-79104 Freiburg im Breisgau, Germany

Novel hybrid composites and advanced structural materials, such as structural adhesives and fiber–metal laminates that are resistant to high-velocity impact, are based on epoxy resins toughened with compatibilized, segmented, reactive liquid rubbers. Upon cure, interpenetrating networks containing dispersed rubber micro- and nanoparticles are formed. Morphological and mechanical properties are controlled by liquid-rubber molecular architectures, especially by the balance of segments that are either compatible or incompatible with the epoxy matrix. Novel poly(ε-caprolactone)-block-poly(dimethylsiloxane)-block-poly(ε-caprolactone) liquid rubbers improve the toughness of silica-filled epoxy resins without sacrificing stiffness. Blends of compatibilized liquid rubbers, such as bisphenol-terminated segmented polyetherurethanes with epoxy-terminated nitrile rubber, possess rubber-blend microphases that account for unusual property synergisms, namely, substantially improved static and high-velocity impact resistance, high T-peel strength combined with high lap shear strength, excellent adhesion, and improved fatigue resistance.

HIGHLY CROSS-LINKED EPOXY RESINS combine high strength; stiffness; thermal, chemical, and environmental stability; adhesion; low weight; processability; excellent creep resistance; and favorable economics. These resins are widely applied as coatings, casting resins, structural adhesives, and matrix resins of advanced composite materials. The broad spectrum of applications ranges from the automotive and aerospace industries to corrosion protection and microelectronics.

As a result of their high cross-link densities, structural epoxy materials are inherently brittle and can fail prematurely when exposed to mechanical stresses. Especially at high strain rates or during high-velocity impact, most conventional toughened epoxy materials undergo brittle failure. Moreover, when fillers or fibers are added as matrix reinforcement, improved strength and stiffness are frequently accompanied by drastically lower toughness. For epoxy resins to compete with metals and metal alloys in structural material applications, resistance to mechanical stresses, deformation, and crack propagation must be improved. An important target is the development of novel toughened highly filled or fiber-reinforced epoxy resins that overcome the tradeoff between toughness and stiffness to exhibit improved toughness at static and high-velocity impact.

Polymer fracture behavior depends on many parameters relating to polymer architecture, molecular weight, phase transition, and morphology, as well as on test conditions, for example, sample geometry, test temperature, and stress mode (1, 2). In principle, two different approaches to improving impact resistance are possible. First, producing entirely homogeneous materials without stress-concentrating inhomogeneities would eliminate potential crack initiators. In an industrial environment, however, such defect-free materials would be extremely difficult to manufacture. The second, much more viable approach is to produce heterogeneous materials with a large content of stress-concentrating microphases, such as dispersed rubbers, crystalline or amorphous domains, fibers, fillers, other polymers, and even microbubbles. Such microphases can induce localized plastic deformation, crazing, or shear yielding as impact-energy-dissipating processes. In rubber microphases, rubber cavitation under stress represents an important mechanism for impact-energy dissipation. Here, in contrast to homogeneous materials, energy dissipation is not restricted to the extremely small region near the crack tip but involves a much larger sample volume. The controlled formation of multiphase epoxy materials, especially epoxy resins containing dispersed, discrete rubber microphases, is proving to be the method of choice for developing impact-resistant, epoxy-based structural materials. Preferably, such rubber microphases are produced when reactive liquid rubbers phase-separate during cure.

Polymer compatibility plays an important role in such materials. On the one hand, liquid rubbers must be miscible with the epoxy resin–hardener mixture. On the other hand, they must be incompatible in order to phase-separate during cure. Compatibility also influences dispersion of the phase-separated rubber and interfacial adhesion at the epoxy–rubber interface, both of which are prime requirements for functional toughening agents. These requirements prompt challenges for designing liquid rubbers with tailor-made polymer architectures (3). A family of carboxy-terminated poly(butadiene-co-acrylonitrile) liquid rubbers, abbreviated as CTBN, was pioneered by BFGoodrich and introduced commercially as versatile epoxy-toughening agents (4, 5). Compatibility between CTBN and epoxy resin is matched by incorporating polar acrylonitrile into a nonpolar poly(butadiene) backbone and also by modifying end groups, for example, by converting carboxy into epoxy end groups via adduct formation.

In contrast to liquid rubbers based on random copoymers, much less is known about segmented liquid rubbers compatibilized by linking together highly incompatible segments with compatible segments. Our research is aimed at a better understanding of the basic structure–property relationship of segmented reactive liquid rubbers as epoxy-toughening agents and the controlled formation of micro- and nanoparticle dispersions. This chapter gives an overview of novel, compatibilized, segmented silicone rubbers and blends of nitrile liquid rubbers with segmented polyetherurethanes, especially in view of applications in toughened hybrid composites and structure materials that have high-velocity impact resistance.

Experimental Details

Linear and Branched Poly(ε-caprolactone)-block-poly(dimethylsiloxane)-block-poly(ε-caprolactone). The linear diol-terminated HO–PCL–PDMS–PCL–OH (Structure **1**) was prepared at 80 °C in toluene via Pt-catalyzed coupling of bis-(HSiMe$_2$)-terminated oligo(dimethylsiloxane) with vinyl-terminated oligo(ε-caprolactone), which was readily available by transesterification of undecen-1-ol with ε-caprolactone in the presence of catalytic amounts of Bu$_2$Sn(OOC–C$_{11}$H$_{23}$)$_2$ at 220 °C. For preparation of branched, segmented, tetrol-functional silicone rubbers (HO–PCL)$_2$–PDMS–(PCL–OH)$_2$ (Structure **1**), O-allylglycidyl ether was added to bis-(HSiMe$_2$)-terminated oligo(dimethylsiloxane) to produce tetrol-terminated oligo(dimethylsiloxane), which was then reacted with ε-caprolactone, as described for undecen-1-ol, to grow ε-caprolactone onto the hydroxy end groups. These silicone liquid rubbers were purified either by repeated precipitation or, preferably, by thin-film evaporation of volatiles under oil-pump vacuum at 80 °C. Detailed synthetic procedures were reported elsewhere (6–8). The hybrid composites were prepared using the linear HO–PCL–PDMS–PCL–OH with number average molecular weight (M_n), as determined by end-group titration, of 6500 g/mol, poly(caprolactone) content of 46.1 wt%, oligo(dimethylsiloxane) average segment length of M_n 3800 g/mol, and poly(ε-caprolactone) segment length of approximately M_n 1500 g/mol.

Epoxy-Terminated Poly(butadiene-co-acrylonitrile) (ETBN). A mixture of 730 g of bisphenol A diglycidyl ether (5.4 mol epoxy/kg), 200 g of carboxy-terminated poly(butadiene-co-acrylonitrile) (Hycar 1300 × 13 from BFGoodrich, 26 wt% acrylonitrile content, acid number of 32 mg KOH/g), 64 g of bisphenol A, and 5 g triphenylphosphin was heated at 130 °C for 3 h to yield ETBN with a viscosity of 130,000 mPa·s (40 °C) and 3.3 mol epoxy/kg.

Bisphenol-Terminated Segmented Polyetherurethane (PPU). Under dry nitrogen, a mixture of 354 g of dry poly(propylene oxide) with M_n 2000 g/mol, 1.8 g of trimethylolpropane, and 0.1 ml of Bu$_2$Sn(OOCC$_{11}$H$_{23}$)$_2$ was slowly added to 54.4 g of hexamethylene-1,6-diisocyanate at 100 °C. After stirring for 2 h at 100 °C, the isocyanate content was 4 wt%. Then the isocyanate-terminated prepolymer was added to 135 g of dry 3,3′-diallylbisphenol A at 80 °C and heated for 2.5 h at 80 °C and 30 min at 100 °C. The viscous isocyanate-free PPU resin was obtained with a viscosity of 128,600 mPa·s (40 °C) and a phenol content of 2.5 mol/kg. The end-capped polyetherurethanes were prepared accordingly by substituting 3,3′-diallylbisphenol A with glycidol, caprolactam, 2-thioethan-1-ol, 4-aminophenol, and bisphenol A, or by substituting poly(propylene oxide) with the equivalent molar amount of diol-terminated oligo(tetrahydrofuran) with M_n 2000 g/mol.

Hybrid Composites. In a stirred resin kettle, 100 g of bisphenol A diglycidyl ether (GY250, Ciba-Geigy, 5.4 mol epoxy/kg) was mixed with 5 wt% of HO–PCL–PDMS–PCL–OH (M_n 6500 g/mol, 46.1 wt% PCL) at 80 °C until a clear solution was obtained. Then half of the calculated amount of SiO_2 filler ("Quarzmehl W12" of Quarzwerke Frechen, average particle size of 17 μm) was dispersed using an Ultrathorax disperser operating at 12,000 rpm. In order to prevent filler sedimentation, especially at low SiO_2 contents of 15 and 25 vol%, 3 wt% of fine calcium carbonate powder (Calofort S) with an average particle size of 0.1 μm was added as a thioxotropic agent. After achieving a fine dispersion, 71 g of hexahydrophthalic anhydride (HT907, Ciba-Geigy) and the second batch of SiO_2 filler were added and dispersed as described. Then the SiO_2-dispersion was degassed for 1 h under oil-pump vacuum and stirring at 100 rpm with an anker-type stirrer. The SiO_2 content had values of 31 g (15 vol%), 62 g (25.5 vol%), and 124 g (42 vol%). Prior to casting, 1 wt% *N,N*-dimethylbenzylamine accelerator was added. The SiO_2-filled epoxy resin was cured according to the following cure schedule: 1 h at 80 °C, 3 h at 150 °C, and 1 h at 180 °C.

Rubber-Blend-Toughened Epoxy Resins. Formulations listed in Tables I and II were mixed together on a three-roll mill and applied to aluminum and steel sheets 1.5 mm thick that were degreased with acetone and sand-blasted. Cure was performed for 1 h at 180 °C. For mechanical testing, especially K_{Ic} measurements, epoxy plates (150 × 60 × 4 mm) were prepared using the following formulation: 200 g of bisphenol A diglycidyl ether (5.4 mol epoxy/kg), 10 g of butane-1,4-dioldiglycidyl ether (9.2 mol epoxy/kg), 22.8 g of dicyandiamide, 1.0 g of chlorotolurone accelerator, and 2.0 g of pyrogenic silica (Aerosil 380). Cure was performed for 2 h at 140 °C and 1 h at 160 °C. A faster cure could have caused charring because of the exothermic cure reaction!

Table I. Rubber-Blend-Toughened, Structural Epoxy Adhesives

				Lap Shear Strength[b]			
Run	ETBN[a] (wt%)	PPU[a] (wt%)	T_g (°C)	Al (MPa)	Steel (MPa)	T-Peel Strength[b] (N/mm)	Cohesive Failure (%)
A	27	4.5	104	29	24	<0.1	0
B	25	12.5	97	30	26	2.3	0
C	24	16	91	32	26	4.0	0
D	22	22	87	28	26	7.8	100
E	19	32	72	28	23	6.3	100
F	17	43	61	20	18	5.0	100
G	—	10	107	26	n.d.[c]	<0.1	0
H	—	30	91	20	n.d.	<0.1	0
I	—	50	n.d.	11	n.d.	<0.1	0
K	44	—	120	26	23	<0.1	0

[a]Weight percent with respect to total amount of ETBN, PPU, and epoxy resins.
[b]Model formulation: 70 parts of bisphenol A diglycidyl ether (5.4 mol epoxy/kg), 5 parts of butane-1,4-dioldiglycidyl ether, 0.1 part of glycidyloxypropyltrimethoxysilane, 30 parts of Wollastonit P1 filler, 9.8 parts of dicyandiamide, 0.5 part of chlorotolurone accelerator, 0.1 part of Aerosil 380; ETBN contents were 5, 15, 20, 30, and 70 parts; cure: 1 h at 80 °C.
[c]Not determined.

Table II. Influence of Polyurethane End Groups on Structural Adhesive Properties

Run	Polyurethane End Group	Polyol[a]	Lap Shear Strength[b] Al (MPa)	Steel (MPa)	T-Peel Strength[c] (N/mm)
A		PPG2000	18	17	1.4
B		PPG2000	23	25	2.8
C	-O-CH₂-CH₂-SH	PPG2000	20	20	2.3
D		PPG2000	28	22	6.0
E		PPG2000	27	26	5.6
F		PPG2000	30	23	4.5
G		PTHF2000	32	27	5.7
H		PTHF2000	33	24	7.4
I		PPG2000	28	27	7.8

NOTE: In all cases, the isocyanate was HMDI, hexamethylene-1,6-diisocyanate.
[a]PPG2000, diol-terminated poly(propylene glycol) with M_n 2000 g/mol; PTHF2000, diol-terminated poly(tetrahydrofuran) with M_n 2000 g/mol.
[b]Model formulation: 70 parts of bisphenol A diglycidyl ether (5.4 mol epoxy/kg), 5 parts of butane-1,4-dioldiglycidyl ether, 0.1 part of glycidyloxypropyltrimethoxysilane, 30 parts of Wollastonit P1 filler, 9.8 parts of dicyandiamide, 0.5 part of chlorotolurone accelerator, 7 parts of Aerosil P380, 30 parts of ETBN, 30 parts of polyurethane prepolymer with functional end groups; cure: 1 h at 180 °C.

Polymer Characterization. Glass-transition temperatures were determined as the maximum of the loss factor (tan δ)–temperature curve, which was recorded using the Rheometrix RSA II solid analyzer with dual-cantilever geometry, 1-Hz frequency, 0.1% amplitude, and 5-K/min heating rate. Mechanical properties such as tensile strength, flexural strength, and Young's modulus were measured according to standardized tests, for example, stress–strain measurement (DIN 53455) on an

Instron 4202 machine at a 5-mm/min crosshead speed. Flexural strength was determined using a three-point-bending test (DIN 53452). The stress intensity factor K_{Id} was determined using an instrumented Charpy impact pendulum. Cracks were initiated with razor blade cuts. For measurement of the stress intensity factor, K_{Ic}, compact tension specimens were used. More detailed descriptions of dynamic and static K_I measurements are published elsewhere (9, 10). Transmission electron microscopy was performed on a Zeiss CEM 902, 80-kV instrument, using ultrathin cuts prepared with a diamond knife Ultracut E supplied by Reichert-Jung. Si-specific imaging was carried out at 120 eV. Fractography was performed using a scanning electron microscope and sample surfaces sputtered with Au.

Segmented Silicone Liquid Rubbers and Hybrid Composites

Silicone liquid rubbers offer several advantages over conventional CTBN or ETBN toughening agents. First, poly(dimethylsiloxane)s are much less polar and exhibit much lower glass-transition temperatures, about –125 °C as compared with –40 °C for poly(butadiene-*co*-acrylonitrile). Second, silicones do not contain olefinic groups, which adversely affect thermooxidative stabilities. Third, the viscosity of most silicone liquid rubbers is much lower than that of CTBN with equivalent molecular weight. Consequently, silicone-toughened epoxies have better low-temperature toughness, low water uptake, low dielectric constants, and good weatherability. However, a major drawback is associated with the chemical nature of silicone rubbers, namely, most silicones, including low-molecular-weight liquid rubbers, are highly incompatible with many other polymers and uncured polar epoxy resin–hardener mixtures. As with CTBN development, traditional approaches have improved silicone compatibility with epoxy resins by randomly incorporating more polar groups into the silicone backbone or by end-group variation or adduct formation (11, 12).

In our research, we have coupled two plasticizing miscible segments, especially oligo(ε-caprolactone), with an immiscible silicone center segment to form triblock copolymers. Structure 1 shows linear and branched diol- and tetrol-functional segmented silicones compatibilized with oligo(caprolactone) segments. Typically, M_n values of oligo(dimethylsiloxane) segments vary between 2200 and 3800 g/mol, and poly(ε-caprolactone) weight fractions vary between 45 and 65 wt%, corresponding to molecular weights of the oligo(caprolactone) segments of $600 < M_n < 2000$ g/mol. Such segmented liquid rubbers are readily obtained by using bis(hydridodimethylsilane)-terminated oligo(dimethylsiloxane)s, resulting from the equilibration of tetramethyldisiloxane with octamethylcyclotetrasiloxane, as reactive intermediates. In linear HO–PCL–PDMS–PCL–OH, ε-caprolactone is polymerized onto undecen-1-ol. Subsequently, this vinyl-terminated oligo(ε-caprolactone) is coupled via Pt-catalyzed SiH-addition with the vinyl group, producing diol-terminated triblock silicone liquid rubber. Similarly branched, segmented, silicone liquid rubbers are produced when ε-caprolactone is polymerized onto tetrol-terminated oligo(dimethylsiloxane), which is formed by the addition of *O*-allylgly-cidyl ether to Si–H end groups (*7, 8*).

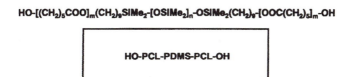

While oligo(ε-caprolactone)s are well known as efficient epoxy plasticizers that drastically lower glass-transition temperatures, stiffness, and strength of the epoxy matrix, the same amounts of oligo(ε-caprolactone) segments coupled with oligo(dimethylsiloxane) center segments do not impair thermal and mechanical properties at a total weight fraction of less than 10 wt% segmented silicone. In fact, the oligo(ε-caprolactone) segments greatly improve miscibility of the segmented silicone liquid rubber with epoxy–hardener mixtures. As reported previously, blends of epoxy resin, hardener, and the novel branched liquid rubbers (HO–PCL)$_2$–PDMS–(PCL–OH)$_2$ exhibit unusually low viscosities. This observation has been attributed to the formation of colloidal dispersions (*7, 8*). In contrast to most other difunctional oligo(dimethylsiloxane) rubbers, which form dispersed silicone microphases, segmented rubbers afford transparent toughened epoxy resins with much smaller dispersed silicone nanoparticles (10–20 nm average diameter). Typical morphologies of nanophase-separated epoxy resins modified with linear and branched segmented silicone rubbers are shown in Figure 1. Adding 5 wt% compatibilized segmented silicone rubber to epoxy resins cured with hexahydrophthalic anhydride accounted for a fivefold increase in impact strength.

At a constant level of 5 wt% HO–PCL–PDMS–PCL–OH (M_n 6500 g/mol, 46.1 wt% oligo(ε-caprolactone) content), SiO$_2$ filler with an average particle size of 17 μm was added to bisphenol A diglycidyl ether cured with hexahydrophthalic anhydride. The filler content had values of 0, 15, 25, and 40 vol%, corresponding to 0, 26, 42, and 60 wt%. Such liquid-rubber-toughened composites are often referred to as hybrid composites. CTBN-based hybrid composites have been studied extensively (*13–18*). In glass beads, toughness increased at low filler contents of 10 vol% and decreased with increasing filler volume fraction.

 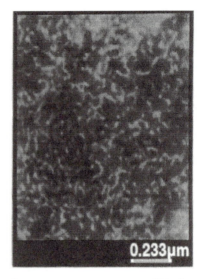

Figure 1. Element-specific transmission electron microscopic images of bisphenol A diglycidyl ether cured with hexahydrophthalic anhydride and toughened with 10 wt% (HO–PCL)$_2$–PDMS–(PCL–OH)$_2$ (left), and HO–PCL–PDMS–PCL–OH (right). Both parts contain PDMS segments with M$_n$ 2200 g/mol and 48 wt% oligo(ε-caprolactone).

Figures 2 and 3 illustrate the influence of SiO$_2$ filler volume fraction on rheological and mechanical properties. As with the silicone-modified neat resin, the viscosity of the SiO$_2$-filled epoxy resin depended primarily on SiO$_2$ volume fraction and was not affected by 5 wt% segmented silicone liquid rubber. When the SiO$_2$ content was raised from 0 to 40 vol%, Young's modulus increased from 2990 to 11,000 MPa, tensile strength from 69 to 82 MPa, and flexural strength from 114 to 140 MPa. Linear correlations with SiO$_2$ volume fraction were found for both flexural and tensile strength. The glass-transition temperature of the matrix, as determined by means of dynamic mechanical analysis, increased slightly from 120 to 125 °C with increasing SiO$_2$ volume fraction. In the presence of silicone rubber, a second glass transition was detected near −130 °C and was attributed to phase-separated silicone nanophases. Because of the nanoscale of the silicone phases, such structures were not imaged when carrying out scanning electron microscopic analyses of fracture surfaces. A much more pronounced impact of silicone rubbers was found for static and dynamic stress intensity factors, as evidenced in Figure 3. In modified and unmodified SiO$_2$-filled epoxy, K_I increased linearly with SiO$_2$ volume fraction. Whereas both K_{Id} and K_{Ic} were identical in the absence of silicone, both K_{Id} and K_{Ic} of the silicone-modified system were significantly larger than those of the unmodified system. Furthermore, K_{Ic} was larger than K_{Id} in the

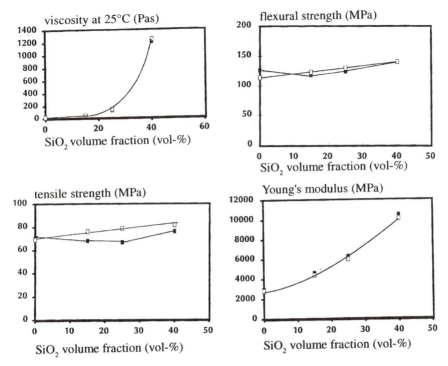

Figure 2. Influence of SiO₂ volume fraction on resin viscosity (top left), flexural strength (top right), tensile strength (bottom left), and Young's modulus (bottom right) in the absence (■) and presence (□) of 5 wt% HO–PCL–PDMS–PCL–OH.

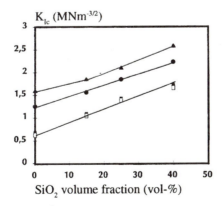

Figure 3. Influence of SiO₂ volume fraction on stress intensity factors K_{Ic} and K_{Id}. Shown are K_{Ic} (■) and K_{Id} (□) of silicone-free SiO₂-filled epoxy, and K_{Ic} (▲) and K_{Id} (●) of SiO₂-filled epoxy modified with HO–PCL–PDMS–PCL–OH.

silicone-modified system. The addition of 5 wt% HO–PCL–PDMS–PCL–OH to neat resin increased K_{Ic} from 0.67 to 1.70 MNm$^{-3/2}$. When 40 vol% SiO$_2$ was dispersed in the silicone-toughened epoxy matrix, a further improvement was achieved, with K_{Ic} = 2.59 MNm$^{-3/2}$.

Most likely, HO–PCL–PDMS–PCL–OH phase-separates during cure to form discrete silicone nanophases and simultaneously, as a result of its am-phiphilic nature, to accumulate at the filler–epoxy interface. Although such structures are too small to be imaged by scanning electron microscopy (SEM), segmented silicones promote both filler dispersion and interfacial adhesion. This in situ steric stabilization of SiO$_2$ dispersion prevents premature agglom-eration of filler particles, which could initiate cracks at comparatively low stresses. Because of the presence of nanoscale structures, it is not surprising that SEM fractography of silicone-based hybrid composites is very similar to that of silicone-free composites. In fact, in accord with earlier observations by Cantwell and co-workers (16, 19, 20), SEM analysis of fracture surfaces re-vealed that fast crack propagation took place exclusively in the epoxy matrix. As is apparent in Figure 4, the SiO$_2$ particles of 17 μm average diameter are well embedded in the epoxy matrix and show no sign of debonding. In the re-gion of slow crack propagation, however, we also found a few isolated SiO$_2$ par-ticles debonded from the epoxy matrix. This debonding was accompanied by intense stress-whitening. Immediately after cracks were initiated, they propa-gated throughout the epoxy matrix.

This investigation of silicone-modified hybrid composites demonstrates that compatibilized, segmented liquid rubbers can be tailored to promote the formation of colloidal rubber dispersions, which enhance the toughness of the epoxy matrix. Furthermore, such segmented liquid rubbers can in situ–modify filler surfaces to form core–shell particles with a hard filler core and a thin

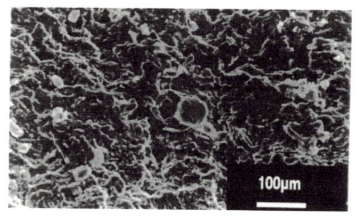

Figure 4. SEM fractography of fracture surfaces of compact tension samples.

elastomeric surface shell that is attached to the polymer matrix via covalent bonds between rubber end groups and anhydride hardener.

Compatibilized Liquid-Rubber Blends and Structural Adhesives

In addition to matching compatibility between polymer matrix and dispersed rubber micro- and nanophases, it is possible to use blends of two separate liquid rubbers, which are compatibilized with each other as well as the epoxy matrix, to achieve unusual blend property synergisms, especially much improved static and dynamic toughness, fatigue resistance, and tolerance of fillers and fibers. The first example of such rubber-blend systems was discovered and systematically investigated by Mülhaupt, Powell, and co-workers during the late 1980s (*21–28*). As illustrated in Structure **2**, key blend components are ETBN, which was prepared by adduct formation of CTBN with bisphenol A diglycidyl ether. ETBN contains 20 wt% CTBN. The second blend component, PPU containing poly(alkylene oxide) segments was prepared by reacting poly(alkylene oxide)s, such as poly(propylene oxide) or poly(tetrahydrofuran), with excess diisocyanate to yield isocyanate-terminated polyetherurethane prepolymers. Such preprolymers are end-capped with bisphenols, aminophenols, caprolactam, or glycidol. Preferably, aliphatic diisocyanates such as hexamethylene-1,6-diisocyanate or isophorone diisocyanate are used because of the better thermal stability of aliphatic urethane groups. Branches are introduced by adding triols, for example, trimethylolpropane, as chain extenders.

When the individual liquid rubbers were added separately to dicyandiamide-cured epoxy structural adhesives, flexible polyetherurethanes proved to be efficient matrix flexibilizers, causing substantial decreases of lap shear strength, Young's modulus, and glass-transition temperature with increasing PPU content. In contrast, both CTBN and ETBN were much better toughening agents, which preserved high lap shear strength and glass-transition temperature even when nitrile rubber content increased. Blending PPU and ETBN gave rise to unusual blend-property synergisms, which were not expected on the basis of the PPU/ETBN mixing ratio.

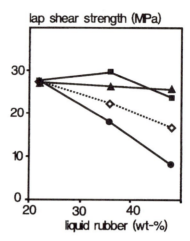

Figure 5. Influence of type of liquid rubber and weight fraction on the lap shear strength of bonded aluminum sheets. Key: ▲, *ETBN;* ●, *PPU;* ◇, *PPU/ETBN (50/50) calculated; and* ■, *PPU/ETBN (50/50) found.*

Figure 5 shows the lap shear strength of bonded aluminum sheets plotted against the liquid rubber weight fraction for different rubbers. When ETBN containing 20 wt% CTBN was blended with PPU, the resulting lap shear strength was much larger than that of the individual components and the blend calculated taking into account the PPU/CTBN mixing ratio. The influence of the PPU/ETBN mixing ratio and total blend weight fraction on adhesive properties is shown in Table I. The amount of ETBN or CTBN was kept constant, whereas PPU content was raised from 30 to 60 wt%, corresponding to PPU/ETBN ranging from 0.17 to 2.5 and PPU/CTBN ranging from 0.6 to 20. When approaching PPU/ETBN = 1 (Runs D, E in Table I), which corresponds to PPU/CTBN = 5, lap shear strength, adhesion, and also T-peel strength exhibited extraordinary improvements in bonding of aluminum and steel sheets. In fact, using the formulation described in Table I, both PPU and ETBN added separately gave no adhesion on steel. As a result, it was not possible to determine T-peel strength because cracks propagated extremely rapidly along the steel–epoxy phase boundary. As soon as PPU was added together with ETBN, adhesion improved drastically. As a consequence, resistance to crack propagation was much higher because crack propagation occurred exclusively in the toughened epoxy matrix, as evidenced by 100% cohesive mechanical failure. Moreover, fatigue resistance of the rubber-blend-toughened adhesives was also improved substantially. Although conventional ETBN-toughened adhesives sustained 10^4 to 10^5 cycles, ETBN/PPU-blend-toughened adhesives required 10^6 to 10^7 cycles to fail. In fatigue tests, adhesive failure was frequently accompanied by mechanical failure of the steel parts of the test equipment.

Obviously, adhesive performance was closely associated with compatibility between the PPU/ETBN blend and epoxy matrix as well as between PPU and ETBN. As in CTBN-modified epoxy adhesives (Figure 6), the nitrile content of CTBN played an important role. When PPU containing a poly(propylene oxide) segment with M_n 2000 g/mol was blended with ETBN, both lap shear strength and T-peel strength increased with increasing nitrile content. The best results were obtained with ETBN derived from CTBN containing 26 wt% acrylonitrile.

In addition to the compatibility of ETBN and polyetherurethane, especially the nitrile rubber segment and the poly(alkylene oxide) segment, the nature of the rubber end groups was an important factor. The end groups of nitrile rubber and polyetherurethane should preferably be co-reactive. In Table II, the poly(alkylene oxide) segment and end-group compositions are varied. Isocyanate-terminated polyetherurethanes were end-capped with glycidyl, caprolactam, 2-thioethan-1-ol, p-cresol, 4-aminophenol, bisphenol A, and 3,3'-diallylbisphenol. In the absence of polyetherurethane, complete adhesive failure of bonded steel prevented measurement of T-peel strength. All polyetherurethane compositions gave markedly improved adhesion and T-peel strength. Best results were obtained with phenolic polyetherurethanes, especially polyetherurethanes end-capped with cresol, aminophenol, and bisphenols.

The basic reaction scheme is represented in Figure 7. Upon heating at temperatures above 100 °C, preferably above 150 °C, the phenolic end groups are split off to form free phenolics and isocyanate-terminated polyetherurethane. Under those reaction conditions, nucleophilic attack of phenolics at the epoxy groups results in hydroxy-functional ethers that react with iso-

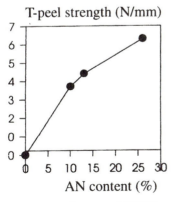
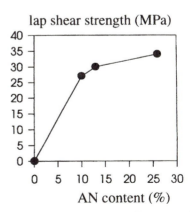

Figure 6. Influence of CTBN acrylonitrile (AN) content on adhesive performance. Left: T-peel strength of bonded steel sheets. Right: Lap shear strength of bonded aluminum sheets. A formulation equivalent to Run D in Table I was used.

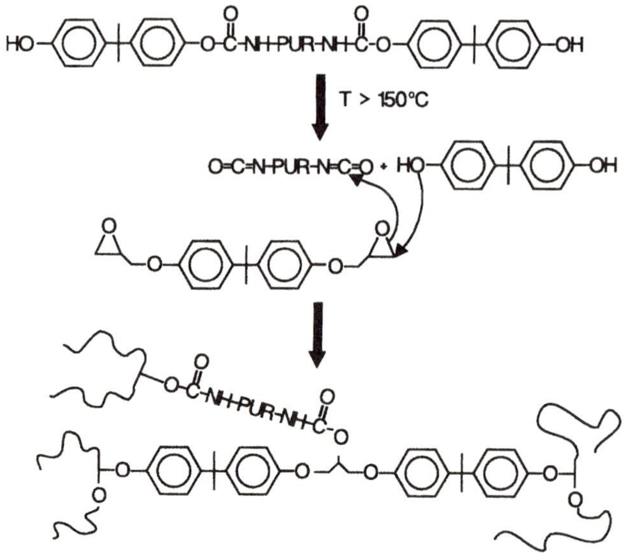

Figure 7. Formation of microphase-separated interpenetrating networks.

cyanates to form urethanes. In the bisphenols, bisphenol and bisepoxides advance to yield in situ higher-molecular-weight polyhydroxy ethers, which are well known to improve both adhesion and toughness. The result of advancement reaction and cross-linking between isocyanate and polyhydroxy ether is the formation of interpenetrating networks (IPNs). Cross-linking of epoxy matrix and liquid-rubber components promotes both phase separation and excellent interfacial adhesion of the rubber component. Simultaneously, ETBN and PPU can co-cure to form rubber IPNs as a dispersed rubber phase. Most likely, bisphenols may also migrate to the epoxy–metal phase boundary and enhance adhesion and corrosion protection.

The morphology of rubber-blend-toughened epoxy resins was investigated by means of transmission electron microscopy (TEM) on OsO_4-stained samples. The typical morphology of ETBN-toughened epoxies (Figure 8, left) disappeared when PPU was blended with ETBN. In fact, with small PPU contents the average size of phase-separated nitrile rubber particles decreased with increasing PPU content. At the synergistic blend composition of PPU/ETBN = 2 (Figure 8, right), the typical nitrile rubber microparticles were absent or too small to be detected with TEM. Moreover, OsO_4 staining revealed the presence of nitrile rubber in the much larger dispersed polyetherurethane phases. This observation supports the formation of multiphase interpenetrating networks as proposed in the foregoing discussion.

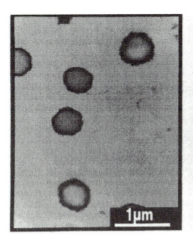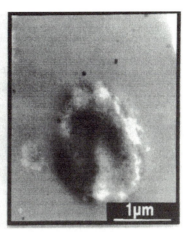

Figure 8. TEM image of dicyandiamide-cured bisphenol A diglycidyl ether containing 28 wt% ETBN (left) and a blend of 14 wt% ETBN and 14 wt% PPU (right).

Morphological studies are in excellent accord with dynamic mechanical analyses performed on ETBN-, PPU-, and PPU/ETBN-toughened dicyandiamide-cured epoxy resins in the absence of fillers. When 28 wt% ETBN (equivalent to 4.5 wt% CTBN rubber) was added to the epoxy resin (Figure 9, dotted curve), the matrix glass-transition temperature of 122 °C was marginally reduced and a second phase transition at –34 °C revealed the presence of phase-separated nitrile-rubber microphases, which were also detected by TEM. Addition of 28 wt% PPU (Figure 9, dotted-dashed curve) produced phase-separated polyetherurethane rubber microphases with a much lower glass-transition temperature of –58 °C and a markedly lower matrix glass-transition temperature of 87 °C. When both 28 wt% ETBN and 22 wt% PPU were blended, the resulting rubber-blend-toughened epoxy did not exhibit nitrile-rubber-type, low-temperature phase transitions. Because glass transition of the rubber-blend and polyetherurethane microphases fell in the range of the epoxy low-temperature relaxation (see the dashed curve in Figure 9, which corresponds to unmodified epoxy), it was not possible to detect a glass temperature depression of a few degrees expected for the rubber-blend phase with respect to the polyetherurethane phase. The depression of the epoxy–matrix glass transition at 94 °C indicated that the rubber blend had flexibilized the epoxy matrix to a certain extent. In contrast to PPU addition, however, PPU/ETBN-blend addition gave considerably less matrix flexibilization. When image processing was used to quantify phase-separated rubber, approximately 20–30 wt% of the rubber was unaccounted for. Therefore, it is reasonable to assume that phase separation, combined with epoxy-matrix flexibilization and

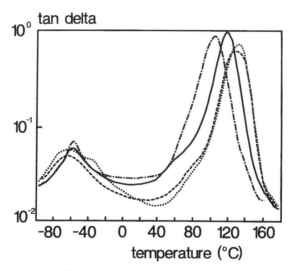

Figure 9. Results of dynamic mechanical analysis, showing loss factor as a function of temperature for unmodified epoxy (dashed curve), epoxy containing 28 wt% ETBN (dotted curve), epoxy containing 28 wt% PPU (dotted-dashed curve), and epoxy containing a blend of 14 wt% ETBN and 14 wt% PPU (solid curve). Epoxy formulation: 100 parts of bisphenol A diglycidyl ether (5.4 mol epoxy/kg), 5 parts of butane-1,4-dioldiglycidyl ether, 5 parts of dicyandiamide, 0.5 part of chlorotolurone accelerator, and 1 part of pyrogenic silica (Aeorosil 380). Cure: 1 h at 140 °C, 1 h at 160 °C.

entanglement of matrix and rubber IPNs, contributed to blend synergisms and especially to the excellent adhesion at interfaces of epoxy–rubber and epoxy–metal. Because of the high nitrogen content of the matrix, which resulted from the dicyandiamide curing agent and chlorotolurone accelerator, element-specific TEM was not capable of detecting nitrile rubber nanophases. More research is required to study the nature of interfacial regions, localize rubber components in the epoxy matrix, and identify nanostructures likely to be present.

Extraordinary high-velocity impact resistance was another striking feature of bisphenol A diglycidyl ether that had been toughened with PPU/ETBN rubber blend and cured with dicyandiamide. As shown in Figure 10 (dotted curve), conventional ETBN-toughened epoxy exhibited high T-peel strength in the bonding of thin steel sheets only under static test conditions, at very low strain rates of 1 mm/s. With increasing strain rates, T-peel strength sharply decreased; at 1 m/s (equivalent to 3.6 km/h), it was very similar to that of the unmodified epoxy. In contrast, rubber-blend-toughened structural epoxy adhesives, T-peel strength increased drastically with increasing strain rates. At 10 m/s, the T-peel strength exceeded 20 N/mm, and the steel sheets fractured,

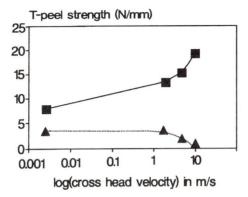

Figure 10. T-peel strength of bonded thin steel sheets as a function of strain rate of epoxy adhesives containing 22 wt% ETBN (▲) and a blend of 22 wt% ETBN and 22 wt% PPU (■). The formulation is given in Run D in Table I.

but the adhesive bond-line was not destroyed. In fact, large deformation of the steel substrate accounted for this extraordinary T-peel strength. Evaluation of bulk material fracture toughness of neat dicyandiamide-cured epoxy resins as well as hybrid composites, especially filler- and fiber-reinforced epoxy, revealed similar improvements of static and high-velocity impact resistance. In Figure 11, the influence of strain rate on K_{Ic} is compared for dicyandiamide-

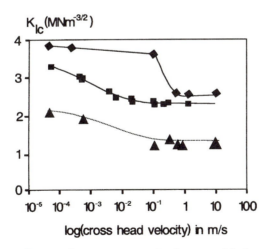

Figure 11. Influence of strain rate on K_{Ic} for unmodified epoxy, (▲), rubber-blend-modified epoxy containing 28 wt% rubber (14 wt% ETBN/14 wt% PPU) (■), and 44 wt% rubber (22 wt% ETBN/22 wt% PPU) (◆). Epoxy formulation and cure are described in Figure 8.

cured bisphenol A diglycidyl ether containing 28 wt% ETBN, 28 wt% ETBN/PPU (50/50), and 44 wt% ETBN/PPU (50/50). As is the case with structural adhesives, rubber-blend-toughened epoxy exhibits much improved fracture toughness at low as well as high strain rates.

Applications of Toughened Epoxy Adhesives

New generations of rubber-blend-toughened epoxy adhesives and hybrid composites, combining strength, stiffness, and adhesion with high impact resistance and fatigue resistance, are the key to modern advanced materials and new applications of epoxy-based materials. In the automotive industry, rubber-blend-toughened structural adhesives join various load-bearing structural materials. Especially in developing structural aluminum parts, for example, car frames and sandwich structures, such adhesives offer attractive potential.

Another important application is fiber–metal laminates for structural aerospace materials, which offer significant weight savings and potential manufacturing cost reduction compared with aluminum and especially conventional fiber-reinforced composites (22). In fiber–metal laminates, thin aluminum alloy sheets are bonded together by fiber-reinforced, rubber-blend-toughened adhesives. As a result of adhesive technology, the expensive processing typical of polymer composites is not required. The toughened epoxy interlayers account for superior impact resistance, as evidenced by 10- to 100-fold slower crack growth with respect to aluminum, while the aluminum interlayers improve stiffness and strength. Through epoxy formulation, it is also possible to improve fire retardancy and corrosion resistance. Such materials are commercially available, for example, from Structural Laminates Company in Delft, Netherlands. Because of their unique spectrum of properties, fiber–metal laminates find applications in wing structures, fuselage and cargo sections, and stabilizer structures, competing successfully with metal alloys as well as conventional composites.

Conclusion

The design of compatibilized, segmented liquid rubbers and liquid-rubber blends represents an attractive route to novel structural materials with unique combinations of properties, especially stiffness, strength, and creep resistance combined with fatigue, impact, and corrosion resistance. As a result of recent breakthroughs, resistance of epoxy-based materials, especially hybrid composites, to low- and high-velocity impacts has been improved significantly without sacrificing other useful properties. This improvement has enhanced the competitiveness of epoxy-based structural materials with respect to toughened, less-creep-resistant thermoplastics. Polymer compatibility and controlled formation of multiphase epoxy materials and blends will continue to play a key role in the development of advanced structural materials and novel technologies.

Acknowledgments

The authors thank Ciba-Geigy and the Deutsche Forschungsgemeinschaft for supporting this research. We also thank J. H. Powell and J. A. Bishopp at Ciba Duxford, U.K., and K. Jud, W. Gabriel, W. Rüfenacht, J. Etter, and M. Struny at Ciba-Geigy Marly, Switzerland, for their contributions. We gratefully acknowledge W. Böhme at the Freiburg Fraunhofer Institut für Werkstoffmechanik for allowing us to measure impact strength with his instrumented Charpy impact testing equipment, and Structural Laminates Company in Delft for supplying, and granting permission to quote, information on their Glare fiber–metal laminates.

References

1. Kausch, H. H. In *Polymer Fracture;* Springer: Berlin, 1987.
2. Kausch, H. H. *Makromol. Chem., Macromol. Symp.* **1991,** *48/49,* 155.
3. Mülhaupt, R. *Chimia,* **1990,** *44,* 43.
4. Rowe, E. H.; Siebert, A. R.; Drake, R. S. *Mod. Plast.* **1970,** *47,* 110.
5. Riew, C. K.; Rowe, E. H.; Siebert, A. R. *Am. Chem. Soc. Adv. Chem. Ser.* **1976,** *154,* 326.
6. Buchholz, U. *Ph.D. thesis,* Freiburg, Germany, 1992.
7. Buchholz, U.; Mülhaupt, R. *Am. Chem. Soc. Polym. Chem. Div. Polym. Prepr.* **1992,** *33(1),* 205.
8. Buchholz, U.; Mülhaupt, R. *Polymer,* submitted.
9. Könczöl, L.; Döll, W.; Buchholz, U.; Mülhaupt, R. *J. Appl. Polym. Sci.,* in print.
10. Döll, W.; Könczöl, L. *Adv. Polym. Sci.* **1990,** *91/92,* 137.
11. Yorkgitis, E. M.; Eiss, N. S., Jr.; Tran C.; Wilkes, G. L.; McGrath, J. E. *Adv. Polym. Sci.* **1985,** *72,* 791.
12. Saito, N.; Nakajima, N.; Ikushima, T.; Kanagawa, S. *Polym. Mater. Sci. Eng.* **1987,** *57,* 558.
13. Kinloch A. J.; Maxwell, D. L.; Young, R. J. *J. Mater. Sci.* **1985,** *20,* 4169.
14. Young, R. J.; Maxwell, D. L.; Kinloch, A. J. *J. Mater. Sci.* **1986,** *21,* 380.
15. Young, R. J.; Maxwell, D. L.; Kinloch, A. J. *J. Mater. Sci. Lett.* **1985,** *4,* 1276.
16. Cantwell, W. J.; Smith, J. W.; Kausch, H. H.; Kaiser, T. *J. Mater. Sci.* **1990,** *25,* 633.
17. Smith, J. W.; Kaiser, T.; Roulin-Moloney, A. C. *J. Mater. Sci.* **1988,** *23,* 3833.
18. Smith, J. W. *Ph.D. thesis,* Lausanne, Switzerland, 1989.
19. Cantwell, W. J.; Roulin-Moloney, A. C.; Kaiser, T. *J. Mater. Sci.* **1988,** *29,* 1615.
20. Cantwell, W. J. In *Fractography and Failure Mechanisms of Polymers and Composites,* Roulin-Moloney, A. C., Ed.; Springer: Berlin, 1990, p.233.
21. Mülhaupt, R.; Powell, J. H. In *Adhesion 90;* Plastics and Rubber Institute: London, 1990.
22. Bishopp, J. A. *Int. J. Adhesion Adhesives* **1992,** *12(3),* 178.
23. Mülhaupt, R.; Powell, J. H.; Adderley, C. S.; Rüfenacht W. Eur. Patent 308 664, 1988, assigned to Ciba-Geigy AG.
24. Mülhaupt, R.; Rüfenacht, W. Eur. Patent 353 190, 1989, assigned to Ciba-Geigy AG.

25. Mülhaupt, R.; Rüfenacht, W.; Powell, J. H.; Parrinello, G. Eur. Patent 338 985, 1989, assigned to Ciba-Geigy AG.
26. Mülhaupt, R.; Rüfenacht W.; Powell, J.H.; Mechera, K. Eur. Patent 358 603, 1989, assigned to Ciba-Geigy AG.
27. Mülhaupt, R.; Rüfenacht, W. Eur. Patent 381 625, 1990, assigned to Ciba-Geigy AG.
28. Mülhaupt, R.; Powell, J.H.; Reischmann, F.-J. Eur. Patent 307 666, 1988, assigned to Ciba-Geigy AG.

Phase Separation of Two-Phase Epoxy Thermosets That Contain Epoxidized Triglyceride Oils

Stoil Dirlikov[1], Isabelle Frischinger[2], and Zhao Chen[3]

[1]MEZA Polymers, Inc., 2430 Draper Avenue, Ypsilanti, MI 48197
[2]Ciba-Geigy Corporation, Basel, Switzerland
[3]Eastern Michigan University, Coatings Research Institute, Ypsilanti, MI 48197

Homogeneous formulations of diglycidyl ether of bisphenol A (DGEBA) epoxy resin and a commercial diamine that contain liquid rubber, based on an epoxidized triglyceride oil and diamine, form two-phase epoxy thermosets under certain conditions. The phase-separation process of these thermosets depends on the miscibility of their two phases (DGEBA and triglyceride). The effect of two diamines (4,4'-diaminodiphenyl sulfone and 4,4'-diaminodiphenylmethane) and four epoxidized triglyceride oils (vernonia, soybean, linseed, and crambe oils) has been evaluated. Particle size and phase inversion can be regulated in a broad range by the nature of the epoxidized triglyceride oil and the diamine used for thermoset preparation.

T WO-PHASE EPOXY THERMOSETS ARE OBTAINED, under certain conditions, from homogeneous formulations of diglycidyl ether of bisphenol A (DGEBA) epoxy resin and a commercial diamine that contain liquid rubber; the liquid rubber is based on an epoxidized triglyceride oil and diamine. These two-phase thermosets have excellent toughness and other physicomechanical properties (*1, 2*). Epoxidized soybean oil (ESO) is industrially produced in large volume and is available at a low price, about $0.50/lb. Its two-phase thermosets have better properties than thermosets based on commercial liquid-rubber adduct tougheners, which presently cost more (>$2.50/lb) (*1*). The epoxies modified with ESO are therefore attractive for many commercial applications, including coatings, sealants, and composites (*2*).

In the present study, two commercial diamines that have similar molecular structures but different polarities were used to cure DGEBA. 4,4'-Diaminodiphenylmethane (DDM) is the less polar diamine (see Figure 1)

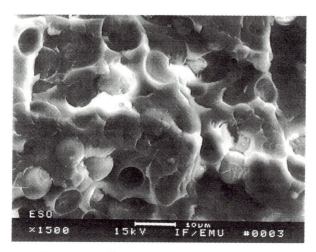

Figure 1. SEM image of the fracture surface of a DGEBA/DDS/ESR(25) thermoset.

whereas 4,4′-diaminodiphenyl sulfone (DDS) is the more polar diamine (see Figure 2).

Two epoxidized triglyceride oils, vernonia oil (VO) and ESO, which have similar molecular structures and molecular weights but different polarities and functionalities, were used to prepare two liquid rubbers: vernonia rubber (VR) and epoxidized soybean rubber (ESR). Both rubbers were prepared by pre-

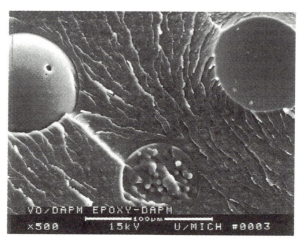

Figure 2. SEM image of the fracture surface of a DGEBA/DDM/VR(20) thermoset.

polymerization below the gel point of mixtures of the corresponding oil and DDM. VO is a naturally epoxidized triglyceride oil with a relatively low average epoxy functionality (2.4). Its structure is given by:

$$CH_3(CH_2)_4CH-CHCH_2CH=CH(CH_2)_7COOCH_2$$
$$\diagdown\diagup$$
$$O$$

$$CH_3(CH_2)_4CH-CHCH_2CH=CH(CH_2)_7COOCH$$
$$\diagdown\diagup$$
$$O$$

$$CH_3(CH_2)_4CH-CHCH_2CH=CH(CH_2)_7COOCH_2$$
$$\diagdown\diagup$$
$$O$$

Rubbers of VO with diamines, therefore, have a lower cross-linked density and polarity. ESO consists of a mixture of different triglycerides and has a higher average epoxy functionality (about 4.5) than VO:

$$CH_3(CH_2)_4CH-CHCH_2CH-CH(CH_2)_7COOCH_2$$
$$\diagdown\diagup \quad \diagdown\diagup$$
$$O \qquad O$$

$$CH_3(CH_2)_4CH-CHCH_2CH-CH(CS_2)_7COOCH$$
$$\diagdown\diagup \quad \diagdown\diagup$$
$$O \qquad O$$

$$CH_3(CH_2)_7CH-CH(CH_2)_7COOCH_2$$
$$\diagdown\diagup$$
$$O$$

The fatty acid residues of ESO triglycerides are flexible, and steric hindrance has little or no effect on the reactivity of the epoxy groups. ESO rubbers with diamines are therefore characterized by a higher cross-link density and polarity.

The effect of the nature of the diamine and the epoxidized triglyceride oil (or liquid rubber) on the phase separation of DGEBA epoxy thermosets was studied by evaluating four formulations: DGEBA/DDM/VR(x) (3), DGEBA/DDS/VR(x) (4), DGEBA/DDS/ESR(x) (5), and DGEBA/DDM/ESR(x) (6). The DGEBA/DDM/VR(x) thermosets were obtained from an initial homogeneous stoichiometric mixture of DGEBA and DDM modified by VR. The content of the liquid rubber (x) was varied between 0 and 100 wt%: x = 0 corresponds to unmodified DGEBA/DDM thermoset, whereas x = 100 corresponds to a pure cross-linked VR. Analogous abbreviations are used for the remaining three formulations.

At the end of this chapter, the effect of two other epoxidized triglyceride oils—epoxidized linseed oil (ELO) and epoxidized crambe oil (ECO)—on the phase separation of DGEBA thermosets is discussed as well.

Experimental Details

The preparation and characterization of the VR and ESR liquid rubbers and of the two-phase epoxy thermosets were described in detail previously (3–6). DGEBA is a solid epoxy resin (Epon 825) from Shell Chemical Company. ESO and ELO are commercial products from Atochem. Samples of VO and crambe oil were obtained from the U.S. Department of Agriculture. All other reagents and solvents were purchased from Aldrich Chemical Co. and Fisher Scientific Co. and used without additional purification.

Fully epoxidized ECO was obtained by epoxidation of crambe oil with m-chloroperbenzoic acid following a procedure described by Chang (7). The iodine values of the initial (nonepoxidized) crambe oil and the ECO were 91.27 and 0.31, respectively (8). The oxirane value of the ECO corresponds to 366.67 molecular weight per epoxy equivalent.

Epoxidized Soybean and Vernonia Liquid Rubbers.
ESR was obtained by prepolymerization of a mixture of 100 g of ESO and 22.8 g (a stoichiometric amount) of DDM, at 135 °C under nitrogen for 37–40 h. Similarly, VR was obtained by prepolymerization of 100 g of VO and 25.63 g (twice the stoichiometric amount) of DDM under nitrogen at 180 °C for 37–40 h. The time of prepolymerization varied slightly from batch to batch, depending on the functionality of the oil.

Both ESR and VR consist of soluble oligomers with a broad molecular-weight distribution (1,000–10,000) as determined by gel permeation chromatography. The exact conversion of the rubbers is not known, but further heating for several hours leads to gelation. Obviously, the liquid rubbers contain unreacted epoxy and amine groups that cure at elevated temperature (~150 °C) and lead to cross-linking. This curing reaction, however, proceeds at a low rate at ambient temperature; the gelation phenomenon does not take place at room temperature, and cross-linking does not occur in storage within several years.

Preparation of Two-Phase Epoxy Thermosets.
Two-phase epoxy thermosets were prepared from homogeneous stoichiometric mixtures of DGEBA and diamine (DDM or DDS), which contained varying amounts of liquid rubber (ESR or VR). The rubber was dissolved first in DGEBA at 70 °C. Then the diamine (DDM or DDS) was added and the mixture was stirred at 70 °C until the diamine dissolved (about 15 min). The homogeneous (one-phase) transparent mixture was then degassed before it reached its cloud point (phase separation). The formulations cross-linked with DDS were cured at 150 °C for 2 h, whereas the formulations cross-linked with DDM were cured first at 75 °C for 4 h and then at 150 °C for 2 h.

Morphology.
The morphology of the fracture surface of the two-phase epoxy thermosets was examined by scanning electron microscopy (SEM, Amray model 1000B). SEM specimens were sputter-coated with a thin film of gold.

Results

The direct introduction of ESO and VO to DGEBA/DDM and DGEBA/DDS formulations results in one-phase (homogeneous) epoxy thermosets (*1*). Homogeneous DGEBA/DDM or DGEBA/DDS formulations that contain VR or ESR, however, form two-phase epoxy thermosets. In contrast to the pure oils, the liquid rubbers have a higher molecular weight and a lower miscibility with the DGEBA phase (due to a lower entropy of mixing), and they undergo easier phase separation. Therefore, only DGEBA formulations that contain VR or ESR were evaluated in this study.

Although the exact conversion of VR and ESR is not known, both liquid rubbers have an advanced degree of prepolymerization close to their gel points. Because VO has a lower epoxy functionality than ESO, the gel point of VR is reached at a higher conversion than ESR gelation requires. VR, therefore, is believed to have a higher average molecular weight than ESR, which probably leads to a better (i.e., more complete) VR phase separation.

DDM and DDS diamines are much more reactive with DGEBA than with ESR and VR; they react first with DGEBA and form the rigid epoxy phase at elevated temperature: 75 °C for the DGEBA/DDM formulations and 150 °C for the DGEBA/DDS formulations (see "Experimental Details" section). The liquid rubber phase separates at the initial stage of curing; then, the unreacted epoxy groups of the liquid rubber cure with the remaining unreacted amine groups at 150 °C and form the cross-linked rubbery phase. The morphology, particle-size distribution, and phase inversion in DGEBA/DDM/VR(x), DGEBA/DDS/VR(x), DGEBA/DDS/ESR(x), and DGEBA/DDM/ESR(x) two-phase thermosets are summarized in Tables I through IV.

At a low rubber content, these four formulations form two-phase thermosets with a continuous rigid DGEBA phase (matrix) and randomly distributed, small spherical rubbery particles (triglyceride) that are several micrometers in diameter and have a unimodal particle-size distribution (Figure 1).

At a certain content of rubber, which is different for each of the four types of formulations, the formulations exhibit a different morphology. With the exception of DGEBA/DDM/ESR, this morphology is characterized by a bimodal particle-size distribution with the formation of small (several micrometers) and large (several hundred micrometers) rubbery particles dispersed in the DGEBA matrix. This morphology indicates the beginning of a phase-inversion phenomenon. The small particles are formed by unmodified cross-linked triglyceride rubber. The larger rubbery particles are a result of a local phase inversion (Figure 2). They contain occlusions of many small, rigid DGEBA particles of several micrometers.

At a higher rubber content, small (several micrometers) and large (several hundred micrometers) DGEBA particles are dispersed in the continuous rubbery triglyceride phase. The large particles probably contain occlusions of much smaller rubbery particles. This is the end of the phase inversion.

**Table I. Morphology and Particle-Size Distribution
for DGEBA/DDM/VR(x) Formulations**

Thermoset Formulation	Morphology	Dispersed Phase	Particle Size (μm)
DGEBA/DDM[a]	Homogeneous		
DGEBA/DDM/VR(5)	Two-phase	Vernonia	0.5–2
DGEBA/DDM/VR(10)	Two-phase	Vernonia	1–4
DGEBA/DDM/VR(15)	Two-phase	Vernonia	1–5
DGEBA/DDM/VR(20)	Two-phase PI[b] start	Vernonia	0.5–2, 20–100
DGEBA/DDM/VR(30)	Two-phase PI[b] end	DGEBA	2–4, 20–250
DGEBA/DDM/VR(50)	Two-phase	DGEBA	0.5–2
DGEBA/DDM/VR(60)	Two-phase	DGEBA	c
DGEBA/DDM/VR(70)	Two-phase	DGEBA	c
VR(100)[d]	Homogeneous		

[a]Pure epoxy resin.
[b]PI, phase inversion.
[c]Data not available.
[d]Pure cross-linked vernonia rubber.

**Table II. Morphology and Particle-Size Distribution
for DGEBA/DDS/VR(x) Formulations**

Thermoset Formulation	Morphology	Dispersed Phase	Particle Size (μm)
DGEBA/DDS[a]	Homogeneous		
DGEBA/DDS/VR(5)	Two-phase	Vernonia	3–10
DGEBA/DDS/VR(10)	Two-phase	Vernonia	3–25
DGEBA/DDS/VR(15)	Two-phase PI[b] start	Vernonia	3–25, 25–100
DGEBA/DDS/VR(20)	Two-phase PI[b]	Vernonia	3–25, 25–350
DGEBA/DDS/VR(30)	Two-phase PI[b] end	DGEBA	2–5, 20–400
DGEBA/DDS/VR(50)	Two-phase	DGEBA	0.5–4
DGEBA/DDS/VR(60)	Two-phase	DGEBA	c
DGEBA/DDS/VR(70)	Two-phase	DGEBA	c
VR(100)[d]	Homogeneous		

[a]Pure epoxy resin.
[b]PI, phase inversion.
[c]Data not available.
[d]Pure cross-linked vernonia rubber.

**Table III. Morphology and Particle-Size Distribution
for DGEBA/DDM/ESR(x) Formulations**

Thermoset Formulation	Morphology	Dispersed Phase	Particle Size (μm)
DGEBA/DDM[a]	Homogeneous		
DGEBA/DDM/ESR(10)	Homogeneous		
DGEBA/DDM/ESR(15)	Two-phase	Soybean	0.1–0.5
DGEBA/DDM/ESR(20)	Two-phase	Soybean	0.1–0.4
DGEBA/DDM/ESR(30)	Two-phase PI[b]	Soybean	0.1–0.8
DGEBA/DDM/ESR(35)	Two-phase	DGEBA	3–5
DGEBA/DDM/ESR(50)	Two-phase	DGEBA	1–2
DGEBA/DDM/ESR(60)	Two-phase	DGEBA	0.5–1
DGEBA/DDM/ESR(70)	Homogeneous		
ESR(100)[c]	Homogeneous		

[a]Pure epoxy resin.
[b]PI, phase inversion.
[c]Pure cross-linked soybean rubber.

The DGEBA/DDM/ESR phase inversion proceeds in a narrow composition range. Formation of larger particles and transitional states with bimodal particle-size distribution found in the other three formulations have not been observed for DGEBA/DDM/ESR(x).

A further increase in the rubber content leads to a complete phase inversion in all four formulations, with the formation of two-phase thermosets

**Table IV. Morphology and Particle-Size Distribution
for DGEBA/DDS/ESR(x) Formulations**

Thermoset Formulation	Morphology	Dispersed Phase	Particle Size (μm)
DGEBA/DDS[a]	Homogeneous		
DGEBA/DDS/ESR(10)	Two-phase	Soybean	1–2
DGEBA/DDS/ESR(20)	Two-phase	Soybean	1–5
DGEBA/DDS/ESR(25)	Two-phase	Soybean	2–10
DGEBA/DDS/ESR(30)	Two-phase PI[b] start	Soybean	2–10, 10–100
DGEBA/DDS/ESR(35)	Two-phase PI[b] end	DGEBA	3–5, 10–100
DGEBA/DDS/ESR(50)	Two-phase	DGEBA	1–3
DGEBA/DDS/ESR(60)	Two-phase	DGEBA	c
DGEBA/DDS/ESR(70)	Homogeneous		
ESR(100)[d]	Homogeneous		

[a]Pure epoxy resin.
[b]PI, phase inversion.
[c]Data not available.
[d]Pure cross-linked soybean rubber.

having a continuous rubbery phase and randomly distributed, small, rigid DGEBA particles (Figure 3).

In all four formulations, however, there is also a pronounced dependence of the phase-separation process on the nature of the diamine and the epoxidized triglyceride rubber used for their preparation. As mentioned earlier, the DGEBA/DDS matrix is more polar than the DGEBA/DDM matrix because DDS is more polar than the DDM diamine. ESR is more polar than VR because the epoxy functionality of the initial ESO is twice as high as the epoxy functionality of VO. Both the particle-size distribution and the rubber content at which phase inversion occurs are determined by the miscibility of the two phases of the four types of formulations discussed.

Discussion

Particle-Size Distribution.   The dependence of particle-size distribution on the composition was elucidated by comparison of all four formulations at 10% rubber content (Table V).

The two phases in the DGEBA/DDS/VR thermosets—highly polar DGEBA/DDS matrix and highly nonpolar VR—are characterized by the lowest miscibility. As a result, VR has the lowest solubility in the matrix; it undergoes phase separation at the lowest rubber content and forms the largest particles (3–25 μm) in DGEBA/DDS/VR(10). VR has better miscibility and solubility in the less polar DGEBA/DDM matrix than in the DGEBA/DDS matrix and forms smaller particles (1–4 μm) in DGEBA/DDM/VR (10).

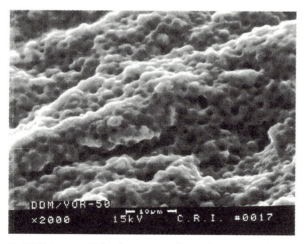

Figure 3. SEM image of the fracture surface of a DGEBA/DDM/VR(50) thermoset.

**Table V. Particle-Size Distribution in Different Rubber-Modified
DGEBA Thermosets at 10 wt% Rubber Content**

Thermoset Formulation	*Particle Size (μm)*
DGEBA/DDS/VR(10)	3–25
DGEBA/DDM/VR(10)	1–4
DGEBA/DDS/ESR(10)	1–2
DGEBA/DDM/ESR(10)	Homogeneous

The ESR has better miscibility in both DGEBA matrixes because it is more polar and perhaps because it has a lower average molecular weight than VR. The DGEBA/DDS/ESR(10), therefore, undergoes phase separation at a later stage of curing and forms smaller particles (1–2 μm) than both types of vernonia formulations.

The ESR has better solubility in the less polar DGEBA/DDM than in the DGEBA/DDS matrix. In fact, the two phases of the DGEBA/DDM/ESR are characterized by the highest miscibility of all four types of formulations. As a result, DGEBA/DDM/ESR(10) does not phase-separate at all. The saturation of the DGEBA/DDM matrix with ESR and the phase separation of the DGEBA/DDM/ESR thermosets occur only at a higher content of ESR.

Similar observations can be made by comparing all four types of formulations with 20% rubber (Tables I–IV).

The results obtained for particle-size distribution lead to four conclusions:

1. The miscibility of the two phases of the formulations decreases in the order of DGEBA/DDM/ESR, DGEBA/DDS/ESR, DGEBA/DDM/VR, and DGEBA/DDS/VR. A lower miscibility results in the formation of larger rubbery particles, and a higher miscibility results in the formation of smaller rubbery particles.

2. Below phase inversion, the size of the rubbery particles of all four types of formulations gradually increases with increasing rubber content. Obviously, at a higher rubber content the saturation of the DGEBA matrices of the formulations is reached at an earlier curing stage. An "early" phase separation results in the formation of larger rubbery particles.

3. An analogous situation is observed above phase inversion for formulations with a high rubber content (>50%). The highly polar DGEBA/DDS phase of the DGEBA/DDS/VR formulations easily undergoes phase separation from the nonpolar VR rubber, and small rigid particles (1–2 μm) are formed even at a high rubber content in DGEBA/DDS/VR(10). The less polar DGEBA/DDM phase of the DGEBA/DDM/ESR formulations has better solubility in the more polar ESR rubber, and phase

separation does not occur in DGEBA/DDM/ESR(70) or even in DGEBA/DDM/ESR(60). The two phases of the remaining two types of formulations—DGEBA/DDM/VR and DGEBA/DDS/ESR—have miscibility characteristics that are intermediate between the two extreme formulations, and they show intermediate types of morphology.

4. DGEBA particle size for all four formulations decreases gradually with increasing rubber content (>50%). The formulations that contain smaller amounts of DGEBA resin reach their saturation point at a later curing stage. A "late" phase separation results in the formation of smaller DGEBA particles.

Phase Inversion.   The rubber content required for phase inversion increases gradually for the formulations with increasing miscibility of their two phases (Table VI). In this regard, the four types of formulations follow exactly the same order as discussed previously for particle-size distribution. This relationship between rubber content, miscibility, and phase inversion is probably due to a different DGEBA partition in the two phases of the formulations. A tentative mechanism for the phase-inversion process is proposed in this chapter.

DGEBA/DDS/VR formulations start phase inversion at the lowest rubber content (15%). Under the curing conditions, the initial homogeneous formulations form a highly polar DGEBA/DDS epoxy matrix, and its polarity increases gradually with cure time because of the formation of additional hydroxyl groups by the reaction of DGEBA epoxy groups with DDS. Even more important for phase separation, the molecular weight of the DGEBA/DDS resin increases, which lowers the entropy of mixing. At a certain point, the formulations reach their saturation point, and phase separation of the nonpolar VR rubber starts.

At this point, the unreacted DGEBA monomer has a different partition in the two phases. DGEBA is a nonpolar compound and has a relatively high partition in the nonpolar VR and a low partition in the highly polar DGEBA/DDS phase. Thus, vernonia particles contain a relatively large amount of dissolved, unreacted DGEBA. This DGEBA is believed to cure at a later stage and form

**Table VI. Percentage of Rubber Content
at Which Phase Inversion Occurs in Different
Rubber-Modified DGEBA Thermosets**

Thermoset Formulation	Rubber (wt%)
DGEBA/DDS/VR	15–30
DGEBA/DDM/VR	20–30
DGEBA/DDS/ESR	30–35
DGEBA/DDM/ESR	~33 (sharp)

the small rigid occlusions (several micrometers) observed within the large rubbery particles (100–200 μm). The molecular weight and polarity of the DGEBA/DDS phase gradually increase further with curing time as the solubility of the VR rubber decreases in the matrix. This mechanism leads to phase separation of an additional amount of rubber at a later stage of curing and the formation of small particles of pure vernonia rubber (several micrometers) because most of the DGEBA has already been consumed. A bimodal particle-size distribution with the formation of small and large vernonia particles is observed for DGEBA/DDS/VR formulations at phase inversion. The apparent volume fraction of the rubbery phase is much larger than the actual vernonia volume fraction as a result of DGEBA occlusion in the large vernonia particles, which triggers phase inversion in DGEBA/DDS/VR formulations at a lower rubber content.

The partition of DGEBA monomer in the less polar DGEBA/DDM matrix of the DGEBA/DDM/VR formulations is significantly higher than in the highly polar DGEBA/DDS matrix of DGEBA/DDS/VR. As a result, DGEBA partition in the vernonia phase and the apparent rubber volume fraction of DGEBA/DDM/VR formulations are lower than in DGEBA/DDS/VR at the same rubber content. Phase inversion in DGEBA/DDM/VR, therefore, requires a higher rubber content (20%).

ESR is more polar than VR. DGEBA partition in soybean particles is therefore lower than in vernonia particles, and the apparent volume fraction of soybean rubbery particles is smaller than the vernonia volume in the corresponding formulations with the same rubber content. As a result, phase inversion of ESR formulations requires a much higher rubber content. Indeed, phase inversion in DGEBA/DDS/ESR formulations starts at a 30% rubber content.

Of all four types of formulations, the unreacted DGEBA of the DGEBA/DDM/ESR formulations is expected to have the highest partition in the (DGEBA/DDM) matrix and the lowest partition in the (soybean) rubber, which would reduce the formation of large soybean particles with DGEBA occlusions. In fact, DGEBA/DDM/ESR formulations are the only ones for which formation of larger particles and transitional states with bimodal particle-size distribution are not observed at all during phase inversion. The morphology of a continuous rigid DGEBA matrix and rubbery particles at 30% rubber changes in a narrow composition range into a continuous rubbery phase with rigid DGEBA particles at 35% rubber content. The intermediate stages of phase inversion found in the other three formulations have not been observed here. Obviously, the phase inversion in DGEBA/DDM/ESR follows a different mechanism than the other three types of formulations. The phase inversion proceeds practically without DGEBA partition in the rubbery particles.

Phase inversion in all four types of formulations occurs at about a 35% apparent rubber volume fraction. This volume fraction is achieved in the DGEBA/DDS/VR formulations at about 15% VR rubber by DGEBA occlusion in the large rubbery particles. DGEBA partition in the rubbery phase gradually decreases in the order of DGEBA/DDM/VR, DGEBA/DDS/ESR, and DGEBA/DDM/ESR. The phase inversion in these formulations therefore requires a gradually higher rubber content.

All formulations with a rubber content below that required for phase inversion form only small rubbery particles. The initial homogeneous formulations reach their saturation point at a later (advanced) curing stage, when most of the DGEBA monomer has been already consumed. The phase-separation process, therefore, proceeds practically without DGEBA partition in the rubbery phase. At a higher rubber content (above phase inversion), the phase-separation process follows a similar mechanism, but only small DGEBA particles are formed.

In summary, the results show that a small variation in the nature (i.e., polarity and functionality) of the diamine or the epoxidized triglyceride oil leads to a big difference in thermoset morphology in terms of particle-size distribution (Table V) and phase inversion (Table VI). In addition to the nature of the diamines and the oils, other factors, such as their reactivity, are expected to influence the phase-separation process. Although we do not have data, a small difference in the cure rates of the DDM formulations at 75 °C and the DDS formulations at 150 °C might affect both the particle-size distribution and phase inversion.

Epoxidized Linseed Oil (ELO).　ELO has a molecular structure and molecular weight similar to those of ESO, but it has a much higher average epoxy functionality, in the range of 7.5 to 8.5.

As expected, the initial ELO does not undergo phase separation from DGEBA/DDM and DGEBA/DDS formulations. Surprisingly, the epoxidized linseed liquid rubber (ELR), based on ELO/DDM soluble oligomers, does not phase-separate either. This result is due to the higher polarity and lower molecular weight of ELR as compared with VR and ESR. ELR therefore has better solubility and compatibility with the polar DGEBA phase. A higher-molecular-weight ELR cannot be obtained because cross-linking occurs at a low conversion as a result of the high ELO epoxy functionality.

Comparison of the three liquid rubbers shows that the less polar VR, which has a lower epoxy functionality and perhaps a higher molecular weight, has a lower solubility in the polar DGEBA/DDM and DGEBA/DDS matrices and undergoes easier phase separation than the ESR and ELR rubbers. The more polar ELR, which has a higher epoxy functionality and a lower molecular weight, does not undergo phase separation at all under the same conditions. ESR has miscibility characteristics that are intermediate between those of VR and ELR, and ESR formulations exhibit intermediate types of morphology.

Epoxidized Crambe Oil (ECO).　The preliminary preparation of liquid rubbers is a major shortcoming for using epoxidized triglyceride oils (such as soybean oil) to toughen commercial epoxy thermosets. Unfortunately, only epoxy formulations that contain ESR or VR undergo phase separation and form two-phase epoxy thermosets. As mentioned above, the direct introduction of ESO, ELO, and VO into DGEBA/DDM or DGEBA/DDS formulations results in one-phase (homogeneous) epoxy thermosets. These three oils consist of triglycerides that contain predominantly C18 fatty acids residues. They contain only a small amount of C16 and do not contain C20 or C22 fatty acid residues.

On the basis of the phase-separation process described in the foregoing discussion, one could expect that epoxidized triglyceride oils with a higher molecular weight, composed of fatty acid residues longer than C16/C18, would not require preliminary preparation of their liquid rubbers and might undergo phase separation when introduced directly into the epoxy formulations. For this purpose, we carried out an initial study (8) on the phase separation of DGEBA/DDM formulations that contained epoxidized crambe oil.

The initial (nonepoxidized) crambe oil has a higher molecular weight than soybean, linseed, and vernonia oils and a unique fatty acid composition (7). Its triglycerides consist predominantly (about 62%) of C22 fatty acid residues (erucic acid):

$$CH_3(CH_2)_7CH=CH(CH_2)_{11}COOH$$

Efforts to develop crambe oil as an industrial crop are carried out at present by the U.S. Department of Agriculture. Fully epoxidized ECO was obtained as described by Chang (7).

Five DGEBA/DDM/ECO(x) thermosets, with x = 0, 5, 10, 20, and 30 wt%, were prepared. In contrast to the DGEBA/DDM/ECO(0), which forms homogeneous (one-phase) thermosets, phase separation is observed for the other four formulations that contain ECO. The two-phase thermosets with an ECO content below 30%, DGEBA/DDM/ECO(<30), consist of a rigid DGEBA/DDM matrix and randomly distributed ECO/DDM particles. Large craters left by debound crambe particles are clearly distinguished (Figure 4). The number of craters increases with increasing ECO content. The apparent volume fraction of ECO particles in these DGEBA/DDM/ECO(<30) thermosets, however, is significantly lower than the actual ECO content. Obviously, the phase separation of ECO is incomplete, and further research is required for optimization of ECO formulations and curing conditions.

DGEBA/DDM/ECO(30) thermosets exhibit a morphology that is characteristic of phase-inversion phenomena. Smaller (2–10 μm) and much larger (30–1000 μm) DGEBA particles in a continuous crambe rubbery phase were distinguished in SEM (Figure 5).

These initial results indicate that ECO formulations have an advantage over ESO, ELO, and VO formulations. ECO has a higher molecular weight

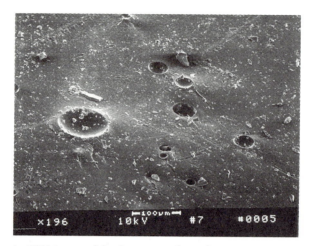

Figure 4. SEM image of the fracture surface of a DGEBA/DDM/ECO(20) thermoset.

than the other three oils, and its phase separation does not require preliminary preparation of crambe liquid rubber. Another potential advantage of ECO is its low epoxy functionality. This allows a higher conversion of ECO with DDM before gelation occurs and the preparation of higher-molecular-weight "crambe" liquid rubbers (ECR). ESR should undergo better and perhaps complete phase separation from the DGEBA matrix.

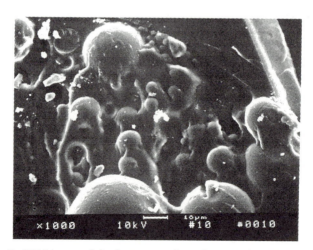

Figure 5. SEM image of the fracture surface of a DGEBA/DDM/ECO(30) thermoset.

Conclusions

Particle-size distribution and phase inversion of the two-phase DGEBA thermosets depend on the miscibility of their two phases (DGEBA and triglyceride) and can be regulated in a broad range by the nature of the epoxidized triglyceride oil (or liquid rubber) and the diamine used for their preparation. Further research shows that the physicomechanical, chemical, electrical, and other properties of these two-phase epoxies depend on their morphology (*1*). Commercial epoxy thermosets and coatings toughened with epoxidized triglyceride oils, especially with ESO, which is commercially available at a low price ($0.50/lb), have outstanding physicomechanical and other properties and appear to be attractive for industrial applications (*2*).

Acknowledgments

We thank the U.S. Agency for International Development, the South Coast Air Quality Management District, and the U.S. Department of Agriculture (USDA) High Erucic Acid Development Effort (HEADE) Fund for financial support, and K. Carlson (USDA) for fruitful discussions and supplying samples of vernonia and crambe oils.

References

1. Frischinger, I.; Dirlikov, S. In *Interpenetrating Polymer Networks;* Klempner, D.; Sperling, L. H.; Utracki, L. A., Eds.; Advances in Chemistry Series 239; American Chemical Society: Washington, DC, 1994; pp 517–538.
2. Frischinger, I.; Dirlikov, S. In *Interpenetrating Polymer Networks;* Klempner, D.; Sperling, L. H.; Utracki, L. A., Eds.; Advances in Chemistry Series 239; American Chemical Society: Washington, DC, 1994; pp 539–556.
3. Frischinger, I.; Dirlikov, S. *Polym. Mater. Sci. Eng.* **1993,** 69, 32.
4. Frischinger, I.; Dirlikov, S. *Polym. Mater. Sci. Eng.* **1993,** 69, 390.
5. Frischinger, I.; Dirlikov, S. *Polym. Mater. Sci. Eng.* **1993,** 69, 392.
6. Frischinger, I.; Dirlikov, S. *Polym. Mater. Sci. Eng.* **1993,** 69, 394.
7. Chang, S.-P. *J. Am. Oil Chem. Soc.* 1979, 56, 855.
8. Letasi, R.; Chen, Z.; Dirlikov, S. *Polym. Mater. Sci. Eng.* **1994,** 70, 332.

Effect of Rubber on Stress-Whitening in Epoxies Cured with 4,4′-Diaminodiphenyl Sulfone

Bum Suk Oh[1], Ho Sung Kim[2]*, and Pyo Ma[2]

[1]Department of Welding Technology, Cheon-An Junior College, Cheon-An City, Chung-Nam 330-240, South Korea
[2]Department of Mechanical Engineering, The University of Newcastle, Callaghan, Newcastle, NSW 2308, Australia

Stress-whitening of U-notched, four-point-bending specimens made from rubber-modified epoxies was studied. The epoxies were cured using 4,4′-diaminodiphenyl sulfone. Stress-whitening was observed to be a major response to deformation. The size of the stress-whitened zone at the root of a notch decreased with increasing rubber content. The stress-whitening is shown to be caused by hydrostatic stress. Also, two different species were found in the stress-whitened zone corresponding to two different shapes at the root of the U-notch. A circle-shaped species was reversible by heating below the glass-transition temperature and was deduced to be due to matrix cavitation, whereas a dendrite-shaped species was partly irreversible by heating and was due mainly to highly cavitated rubber particles and shear bands.

THE FRACTURE TOUGHNESS OF BRITTLE EPOXIES can be improved by the addition of rubber particles. The modes of deformation, in the vicinity of a crack, responsible for the improvement in toughness have been identified. They include shear-band formation between rubber particles, cavitation of rubber particles, and stretching of rubber particles (*1*). The epoxy systems that can be effectively toughened, however, are confined to those with a relatively low cross-link density. The molecular mobility of these epoxy systems is higher than that of those with a high cross-link density, which results in increased ductility, efficiently improving the toughness (*2*). The molecular mobility of the highly cross-linked brittle epoxies is restricted, so that plastic deformation such as shear yielding is limited. Accordingly, toughening such epoxies by a

*Corresponding author.

rubber modification is ineffective (3). Although there has been little progress in the toughening of highly cross-linked brittle epoxies, epoxies with a high cross-link density have advantages such as improved resistance to chemical attack and high glass-transition temperatures (T_gs) (1).

The deformation responsible for the improvement of toughness in polymers is generally accompanied by stress-whitening. The term stress-whitening referred to here should be distinguished from that used in describing fracture surfaces (4). Here, it describes subsurface discoloration of the material in response to deformation. It is due to the scattering of light from free surfaces created during deformation. Although stress-whitening is a well-known phenomenon in polymers (5–7), its study in rubber-modified epoxies with high T_gs has not been well advanced.

This chapter examines the nature of stress-whitening in such epoxies.

Experimental Details

The material used was a diglycidyl ether of bisphenol A (DGEBA) based epoxy resin (Ciba-Geigy, GY250) cured using stoichiometric amounts of 4,4'-diaminodiphenyl sulfone (DDS). The rubber used for the modifications was Hycar carboxy-teminated butadiene–acrylonitrile (CTBN) rubber (1300 × 13). The curing schedule for all the rubber-modified epoxy–DDS systems was as follows: first the rubber and then DDS were mixed with the epoxy resin and stirred at 135 °C until the DDS was dissolved; the systems were cured for 24 h at 120 °C and then postured for 4 h at 180 °C. The control epoxies were cured according to the same schedule.

Molded sheets of the epoxies were cut into double-notch four-point bend (DN-4PB) specimens in accordance with the dimensions given in Figure 1. Because more plastic deformation can be promoted in a U-notch than in a crack, U-notched specimens were used. The U-notched DN-4PB specimens were loaded at

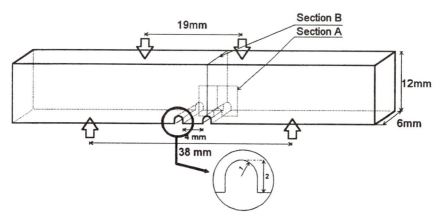

Figure 1. Shape and dimensions of test specimen.

Table I. Notation for Rubber-Modified Epoxies

Designation	Rubber (phr[a])	DDS (phr)	T_g (°C)
RF0	0	32	222[b]
RF5	5	32	217[b]
RF10	10	32	195[c]
RF15	15	32	216[c]
RH0	0	16	76[c]
RH15	15	16	106[c]

[a]phr denotes parts per hundred of resin by weight.
[b]Determined by differential thermal analysis at 20 °C/min using powder samples.
[c]Determined by differential thermal analysis at 20 °C/min using block samples.

a crosshead speed of 0.5 mm/min until one of the notches fractured. The designations of the specimens and their compositions are given in Table I.

For preparation of thin sections, the specimens were sectioned, and then the sectioned surfaces were progressively polished using different grades of abrasive. The polishing was finished with 0.05-μm alumina powder abrasive. The next step involved mounting the polished specimens onto glass slides. The glass slides were cleaned and subsequently coated with a thin layer of an epoxy adhesive to bond with the polished surfaces. The mounted specimens were cut to 1.5 mm thick, and then a thinning procedure similar to the polishing was used. Thinning was carried out until the sections were ≈35 μm thick.

Results and Discussion

Two series of specimens, designated RF and RH, were used. The difference between the two series was in the amount of curing agent (*see* Table I), which resulted in high T_gs for the RF series and relatively low T_gs for the RH series. As shown in Figure 2, the bending force at the fracture, P_f, of the RF series specimens decreased with increasing rubber content, whereas the P_f of RH series specimens with a rubber content of 15 phr increased compared with that of control specimens. These results indicate that in the RF series specimens, the fracture stress at the U-notch root decreases with increasing rubber content, whereas in specimen RH15 it increases.

Throughout this work, the surviving U-notches of the DN-4PB rubber-modified specimens were used for microscopic investigations of stress-whitening. The configurations of the sections, designated A and B, are shown in Figure 1. The control specimens did not display any detectable stress-whitening. In rubber-modified epoxies, stress-whitening at the root of the U-notch occurred at Section A, as shown in Figures 3a and 3b, but not at the outer surface. The stress-whitened zones of RF series specimens were circular (Figure 3a) whereas that of specimen RH15 was rather dendritic (Figure 3b). Furthermore, contrary to expectations, the extension of stress-whitened zones in RF series decreased with increasing rubber content. Figure 4 shows measure-

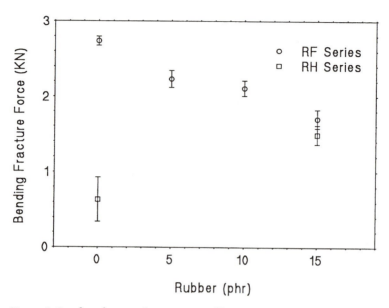

Figure 2. Bending fracture force versus rubber content for U-notched specimens.

ments of the vertical extent of the stress-whitened zone from the root of the U-notch for each rubber content.

To obtain more information, stress-whitened zones were prepared at Section B. For all the rubber-modified specimens, the size of the stress-whitened zone increased from zero at the outer surface to a maximum at the midsection (Section A). Figure 5 shows the whitened zone at Section B of specimen RF5. The whitened zone of RF series specimens can be three-dimensionally visualized, as shown in Figure 6. The shape of the whitened zone is in opposition to that of the plastic zone ahead of a crack tip in dense materials, in which the size of the plastic zone decreases to a minimum at the midsection because of the state of plane strain. At the outer surface there will always be plane stress, and hence the stress in the thickness direction, σ_z, is zero at the surface. Concurrently, plane strain prevails in the interior, thus increasing the σ_z in the interior. It can accordingly be seen that the maximum hydrostatic stress is found at the midsection (Section A). Thus, stress-whitening appears to be due to the hydrostatic stress components rather than the deviatoric stress components.

To identify the types of damage occurring in the stress-whitened zones, thin sections were examined under a bright field using a transmission optical microscope. Figure 7a reveals details of the whitened zone in specimen RH15 shown in Figure 3b. This type of damage is well described elsewhere (4). The cavitated particles appear dark because of light diffraction, and shear bands, which are birefringent, have formed between these particles. In RF series

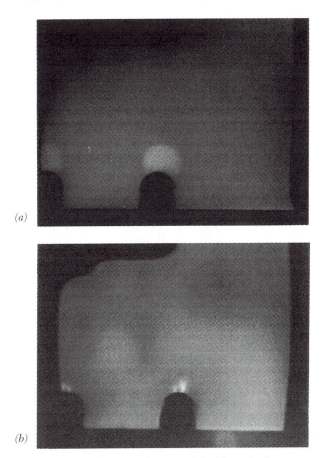

(a)

(b)

Figure 3. Stress-whitened zone at the root of the U-notch of specimens after polishing. The specimens are viewed under a stereomicroscope. (a) Specimen RF5, Section A. (b) Specimen RH15, Section A.

specimens, the rubber content did not seem to affect the microscopic features of the whitened zone, although it affects its size. Also, in contrast to specimen RH15, no apparent shear bands were found in the stress-whitened zones, and some cavitated rubber particles were present (*see* Figure 7b). These cavitated rubber particles were also found outside the stress-whitened zone, and thus no border, in terms of the density of rubber cavitations, was observed between the stress-whitened zone and the rest. The whitened zones of the RF series were not detectable under microscopic dark-field, bright-field, and cross-polarized light. This stress-whitening does not therefore appear to be due to cavitated rubber particles. It can be deduced that the stress-whitening is due to matrix cavitation.

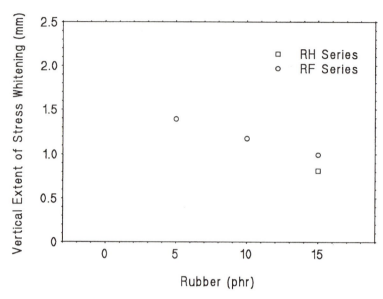

Figure 4. Vertical extent of stress-whitening from the root of the U-notch, Section A.

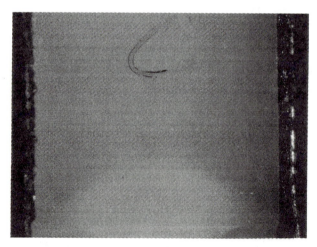

Figure 5. Stress-whitened zone at the root of the U-notch of specimen RF5, Section B, after being polished. The specimen is viewed under a stereomicroscope.

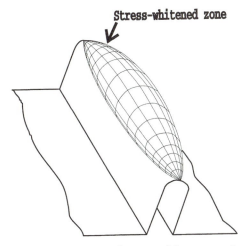

Figure 6. *Three-dimensional visualization of the stress-whitened zone of RF series specimens.*

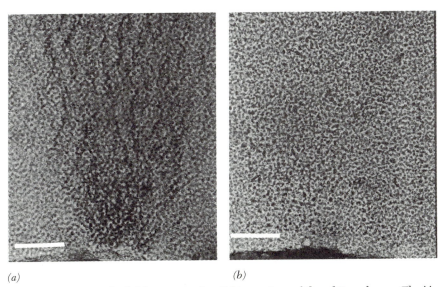

(a)   *(b)*

Figure 7. *Bright-field micrographs of thin sections of the whitened zone. The U-notch root is at the bottom. The scale bar represents 50 µm. (a) Specimen RH15, Section A. (b) Specimen RF15, Section A.*

Finally, to examine the reversibility of stress-whitening, reactions of samples to slow heating in an oven were observed. The thin sections mounted on glass slides were used as samples for fast reaction to heating. Stress-whitening vanished in all RF series specimens when the temperature rose from 130 to 150 °C (the duration was about 10 min), which is below the T_gs of the samples. For specimen RH15, however, although the size of the stress-whitened zone was reduced noticeably when the temperature rose from 100 to 130 °C, the rest did not change until the temperature rose further to 150 °C, which is above the T_g of the sample. This observation indicates that the stress-whitening in specimen RH15 has two components: one, which is similar to the whitening seen in specimen RF5, is reversable by heating, and the other is irreversible by heating.

Conclusions

This study of stress-whitening in rubber-modified epoxies showed that the size of the whitened zone at the root of a notch decreases with increasing rubber content. Stress-whitening has been shown to be caused by hydrostatic stress. Two different species of stress whitening were found. One is reversible by heating and is deduced to be due to matrix cavitation; the other is irreversible by heating and is due to highly cavitated rubber particles and shear bands.

Acknowledgment

We thank K. J. Doolan of the research laboratory at the Broken Hill Proprietary Company, Ltd., for providing the glass-transition temperatures of the samples.

References

1. Garg, A. C.; Mai, Y. W. *Compos. Sci. Technol.* **1988,** *31,* 179.
2. Pearson, R. A.; Yee, A. F. *J. Mater. Sci.* **1989,** *24,* 2571.
3. Sue, H. J. *Polym. Eng. Sci.* **1991,** *31,* 275.
4. Pearson, R. A.; Yee, A. F. *J. Mater. Sci.* **1986,** *21,* 2475.
5. Breuer, H.; Haaf, F.; Stabenow, J. *J. Macromol. Sci. Phys.* **1977,** *B14(3),* 387.
6. Smith, J. W.; Kaiser, T. *J. Mater. Sci.* **1988,** *23,* 3833.
7. Lee, Y. W.; Kung, S. H. *J. Appl. Polym. Sci.* **1992,** *46,* 9.

9

Thermal Shock Behavior and Evaluation of Epoxy Resins Toughened with Hard Particulates

Masatoshi Kubouchi[1], Ken Tsuda[1], and Hidemitsu Hojo[2]

[1]Department of Chemical Engineering, Tokyo Institute of Technology, 2-12-1, O-okayama, Meguro-ku, Tokyo 152, Japan
[2]Engineering Management Department, College of Industrial Technology, Nihon University, 1-2-1, Izumi-cho, Narashino-shi, Chiba-ken 275, Japan

The effects of ceramic particles and filler content on the thermal shock behavior of toughened epoxy resins were studied. Thermal shock tests were performed with notched-disk specimens to evaluate thermal shock resistance. Ceramic particles that were used as filler were silicon nitride, silicon carbide, silica, and alumina. Test results were evaluated by using fracture mechanics. The mechanism of thermal shock fracture is discussed on the basis of analysis of scanning electron microscopic observations. In resins filled with stiff and strong particles, filler content had a remarkable effect on thermal shock resistance. With strong particles, the crack propagated through a space between particles in a zigzag way, whereas with easily broken fillers, the crack propagated in a straight path by breaking particles. With weakly bonded particles, the crack propagated through the debonded interface caused by the difference of thermal expansion that results from thermal shocking. The thermal shock resistance was calculated on the basis of fracture mechanics and compared with experimental values.

EPOXY RESIN IS CURRENTLY USED as a high-performance insulating material in the electric (1) and electronic (2) fields. The resins generally show low toughness, but their fracture toughness is improved by adding second-phase materials. Toughened materials are often composed of either a soft rubbery material (3, 4) or a hard inorganic particulate (4–6) combined with the matrix resin. The second phase also improves other properties, including electrical properties, thermal properties, flame resistance, etc. In many applications of epoxies, high resistance to thermal shock is required. Failure by thermal shock is a complex phenomenon because it is greatly affected by the geometry of the test sample and the thermal and mechanical properties of the material.

We proposed a new test method to evaluate the thermal shock resistance of epoxy resin (7). This test method uses a notched-disk specimen, and the thermal shock resistance can be evaluated analytically on the basis of linear fracture mechanics (8). In our previous studies, we reported on the use of our proposed thermal shock test and evaluation methods (8, 11) to determine the thermal shock resistance of toughened epoxy with a soft second phase (9, 10), and also with hard particulates (11).

This chapter discusses the behavior, under thermal shock conditions, of epoxy resins toughened with ceramic particulates. Alumina Al_2O_3 and silica SiO_2, which are usually used as filler for insulation materials, and the new ceramic materials silicon carbide SiC and silicon nitride Si_3N_4 are employed. For these toughened epoxy resins, the thermal shock resistance is evaluated by using fracture mechanics. The difference between experimental and calculated values of the thermal shock resistance is discussed from a fractographic point of view.

Experimental Details

Materials. Epoxy resins filled with ceramic particulates were used. A bisphenol A epoxy resin (epoxy equivalent weight 370–435) and an alicyclic epoxy resin (epoxy equivalent weight 157) were mixed as the matrix resin. A phthalic anhydride was added as a hardener. The weight ratio of bisphenol resin:alicyclic resin:hardener was chosen to be 9:1:3. Alumina, silica, silicon carbide, and silicon nitride were used as fillers. These ceramic fillers, shown in Figure 1, are angular-shaped particulates without surface treatment. The average diameter (weight mean particulate diameter at 50%) of each of these fillers was about 10 μm, and each had almost the same particulate size distribution. In order to investigate the effects of filler content on thermal shock resistance, volume fractions (V_fs) were varied from 0 to about 40%. The physical and mechanical properties of these test materials, obtained experimentally, are shown in Table I. The fracture toughness was determined with notched-beam specimens loaded in three-point bending (6).

Test Method. Thermal shock tests were conducted by using the same method described in the preceding papers (7–10), that is, employing a sharp notched-disk specimen 60 mm in diameter and 10 mm thick (Figure 2a). At first, the specimen with balsa heat insulators attached on both flat sides (see Figure 2b), which causes only radial heat transfer, was heated in a high-temperature air bath. Then the specimen was quickly put into the cooling bath. As the low-temperature bath, a dry ice–pentane system, whose temperature was approximately 200 K, was used. The temperature difference of the thermal shock test was obtained by changing the initial air-bath temperature. The difference in thermal expansion between the outer part and inner part makes a Mode I problem of fracture mechanics at the tip of the notch. After 5 min of cooling, the specimen was checked to determine whether the crack initiated at the notch.

We used visual observation to detect the initiation of the crack in the neat resin due to thermal shock. But because the filled resin was opaque, all specimens were observed by color checking. Actually, the crack initiated from the notch and propagated along the radial direction, as in neat resin or rubber-modified resin

Figure 1. SEM photographs of filler ceramic particulates: (a) silicon nitride (Si₃N₄), (b) silicon carbide (SiC), (c) silica (SiO₂), (d) alumina (Al₂O₃).

(7, 10). This fact supports the availability of the analysis of a Mode I fracture mechanics problem.

   Fracture-specimen surfaces were examined after thermal shock testing and fracture toughness testing by use of scanning electron microscopy (SEM).

Results and Discussion

   Thermal Shock Resistance. Thermal shock test results were similar to those obtained in neat epoxy resin or toughened epoxy with a soft second-phase material *(7–11)*. Typical test results of epoxy resin filled with a ceramic particulate are shown in Figure 3. The figure shows the relation between the temperature difference ΔT and the normalized notch length c/R of 178 phr (parts per hundred of resin by weight) resin filled with silicon carbide particulate ($V_f = 34.2\%$), where c is the initial notch length and R is the radius of the disk specimen. The critical temperature difference ΔT_c, which corresponds to the smallest temperature difference to initiate the crack from the notch, is clearly observed. This critical temperature difference, shown in the curve between the open and solid symbols, decreases with increasing notch length and

Table I. Thermal and Mechanical Properties of Particulate-Filled Resins

Filler	Si_3N_4					SiC					SiO_2				Al_2O_3				
Filler content (phr[a])	25	90	177	200	222	50	87	178	200	232	50	100	150	200	50	100	150	200	250
Volume fraction (%)	6.9	21.0	34.2	36.7	39.5	17.5	20.0	34.2	36.4	40.0	—	29.7	38.8	45.8	10.5	19.1	26.1	32.0	37.1
Young's modulus (GPa)	9.84	—	9.59	10.2	13.4	7.04	9.79	15.40	17.76	20.04	—	6.67	9.58	10.89	5.19	6.42	8.51	9.89	11.57
Poisson's ratio	0.37	—	0.32	0.31	0.33	0.34	0.34	0.38	0.33	0.28	0.34	—	0.35	0.31	0.35	0.35	0.31	0.32	0.30
Fracture toughness ($MPa\sqrt{m}$)	1.28	—	2.07	2.19	2.61	1.31	2.02	2.34	2.52	2.71	—	0.92	1.82	1.91	1.04	1.00	1.12	1.22	1.25
Thermal conductivity (W/m·K)	0.26	—	0.78	0.80	0.81	0.39	0.64	1.24	1.27	1.36	0.38	0.49	0.56	0.66	—	0.52	0.66	0.90	1.06
Coefficient of thermal expansion ($10^6/K$)	63.0	—	36.0	33.6	30.5	52.9	47.7	36.4	34.0	29.8	—	40.6	34.3	29.0	40.4	34.9	29.3	24.0	18.5

[a] phr is parts per hundred of resin by weight.

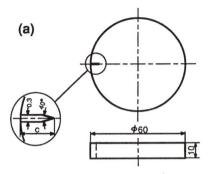

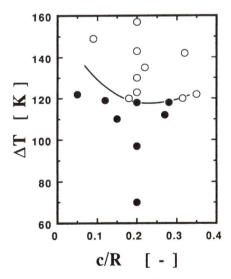

Figure 2. Schematic of the notched-disk specimen and its holder.

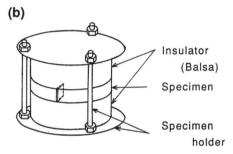

Figure 3. Thermal shock test result for 178 phr epoxy resin filled with SiC. Key: ○, crack initiation and propagation from the notch; ●, no crack propagation.

has a minimum value at about $c/R = 0.2$. This minimum critical temperature difference can be used for evaluating the thermal shock resistance.

A comparison of critical temperature differences of resins filled with several ceramic particulates is shown in Figure 4. The volume fraction of all these composites is 34.2%. The critical temperature difference of epoxy filled with hard particulates was classified into three groups on the basis of thermal shock resistance. Composites filled with a strong particulate, such as silicon nitride or silicon carbide, showed high thermal shock resistance. Some improvement in thermal shock resistance was recognized for silica-filled composites. Composites filled with alumina or aluminum nitride showed almost comparable or lower resistance compared with the neat resin.

Figure 5 shows the effects of filler content on thermal shock resistance at $c/R = 0.2$ for composites of silicon nitride, silicon carbide, silica, and alumina. The thermal shock resistance of resin filled with silicon nitride increases linearly with the volume fraction. The value of the thermal shock resistance is high, especially at higher volume fraction ($V_f > 40\%$), that is, thermal shock resistance reaches 140 K (Figure 5a). The thermal shock resistance of composite filled with silicon carbide increases rapidly with the increase of filler content, and it reaches 135 K at V_f of 40%, which is similar to the case of silicon nitride (Figure 5b). In the case of silica-filled composites there is also an increase, but above a 30% volume fraction a plateau is reached (Figure 5c). Alumina-filled composites show a decrease in thermal shock resistance with filler content, then an almost constant value starting at $V_f = 20\%$ (Figure 5d).

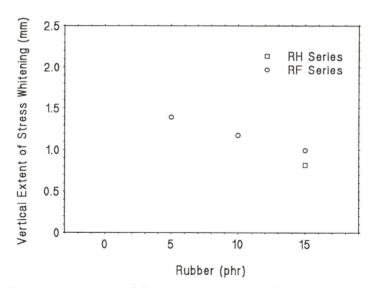

Figure 4. Comparison of the critical temperature difference, ΔT_c, of ceramic-particle-filled epoxy resins at $V_f = 34.2\%$.

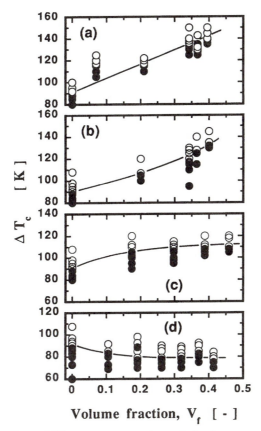

Figure 5. Effect of filler contents on thermal shock resistance at c/R = 0.2: (a) Si_3N_4, (b) SiC, (c) SiO_2, and (d) Al_2O_3.

Fractography. In order to discuss the effect of filler content on the thermal shock resistance of each composite, fracture surfaces were observed by SEM after thermal shock test and compared with fracture toughness test specimens. At lower particulate content (V_f < 20%), specimens that had undergone the two tests were similar in appearance for the four composites, that is, they had a smooth surface without the appearance of particulates. At a higher particulate content (V_f > 30%), however, more complicated fractographies were recognized. The fracture surfaces of the resins filled with ceramic particulates are shown in Figures 6 and 7. In resins filled with silicon nitride and silicon carbide, rough fracture surfaces were observed in both tests at the higher particulate contents (Figure 6, a–d). Detailed observation confirmed that thin resin layers remained on the particulate surface. The presence of the layers means that the particulate and matrix are bonded

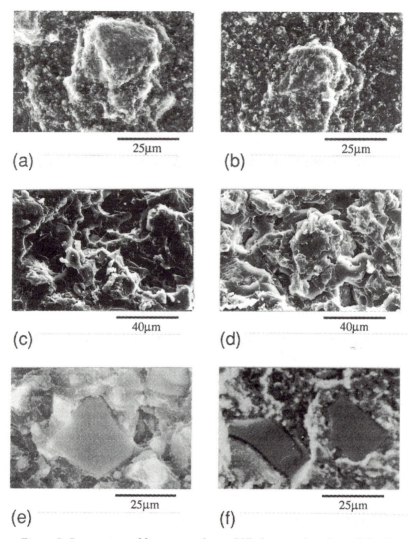

Figure 6. Comparison of fracture surfaces of filled resins after thermal shock testing (left) and fracture toughness testing (right): (a, b) Si_3N_4, (c, d) SiC, (e, f) SiO_2.

strongly, so the crack propagates through a space between the particulates in a zigzag way.

In silica-filled resins, flat surfaces with a particulate appearance were observed in both tests at a higher filler content. Also, flat fracture surfaces of particulates themselves were observed instead of rough ones, as shown in Figure 6, e and f. This observation means that the crack propagated straight forward by breaking silica particulates.

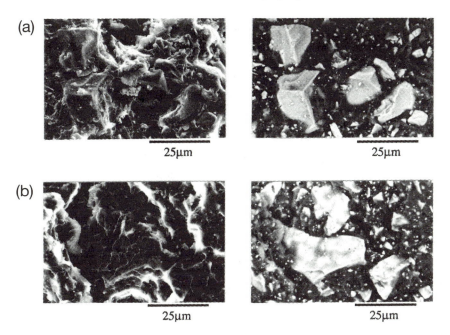

Figure 7. Comparison of fracture surfaces after thermal shock testing (a) and fracture toughness testing (b) of Al$_2$O$_3$-filled epoxy resin. Left: SEI. Right: BEI.

In alumina-filled resins, on the other hand, different surfaces were observed in thermally shocked and fracture-toughness-tested specimens at a higher filler content. The SEM photographs on the left side of Figure 7 are secondly electron micrographic images (SEIs) which are ordinary SEM images, and those on the right side are backscattering electron micrographic images (BEIs), which depend on the atomic mass and show the morphology just under the surface. The fracture surface in Figure 7 shows debonding of the particulate–matrix interface after thermal shock testing (a), because the particulate surface can be observed in SEI, whereas the surface is rough after fracture toughness testing. We cannot find particulates in the SEI, whereas they are obvious in the BEI at the corresponding position (Figure 7b). This fact suggests that the crack propagates through the matrix resin in the fracture toughness test of alumina-particulate-filled resin, and thus particle interfacial debonding is not recognized, as is the case with silicon carbide composites. Thus the crack in alumina seems to be easy to propagate by interfacial debonding as a result of thermal shocking.

These fracture mechanisms are shown schematically in Figure 8. In the particulate-filled resins that have a strong particulate–matrix interface, both thermal shock resistance and fracture toughness are improved. When the particulate itself is strong enough to withstand crack propagation, as is the case

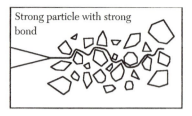

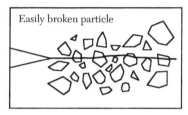

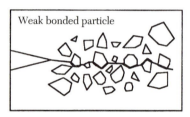

Figure 8. Schematic of the three types of crack-propagation processes that occur in ceramic-particulate-filled resins under thermal shock conditions.

with silicon nitride or silicon carbide, thermal shock resistance increases rapidly with filler content. But if the particulate is broken by crack propagation, as in silica, the tendency of thermal shock resistance to increase is not so remarkable at higher filler content.

On the other hand, in weakly bonded particulates such as alumina, thermal shock resistance decreases with increasing filler content. In this case, interfacial debonding occurs because of the difference in thermal expansion, and then the crack propagates through the interface preferentially.

Evaluation of Thermal Shock Resistance. The results of the thermal shock tests are evaluated by the method based on fracture mechanics. Thermal shock resistance $(\Delta T_c)_{cal}$, can be calculated as follows (8, 10):

$$(\Delta T_c)_{cal} = \frac{(1-\nu)K_{Ic}}{E\alpha}\frac{C_1}{\sqrt{R}} + \frac{k(1-\nu)K_{Ic}}{E\alpha}\frac{C_2}{h\sqrt{R^3}} \tag{1}$$

where K_{Ic} is Mode I fracture toughness, n is Poisson's ratio, a is the coefficient of thermal expansion, E is the elastic modulus, k is thermal conductivity, h is the heat transfer coefficient, and C_1 and C_2 are constants determined by the conditions of the thermal shock test.

The relationship between the ratio $(\Delta T_c)_{exp}/(\Delta T_c)_{cal}$ and the volume fraction V_f is shown in Figure 9, where $(\Delta T_c)_{exp}$ is thermal shock resistance obtained by experiment. Where $(\Delta T_c)_{exp}/(\Delta T_c)_{cal} = 1$, it means that the thermal shock resistance can be evaluated well with equation 1. The thermal shock resistance of composites filled with silicon nitride, silica, and alumina can be especially evaluated at lower volume fractions. However, for silicon carbide, the ratio is not unity but greater than 1, and so the prediction of thermal shock resistance made with equation 1 is a conservative evaluation. In the case of alumina, on the other hand, the ratio $(\Delta T_c)_{exp}/(\Delta T_c)_{cal}$ decreases remarkably with increasing V_f. These values of $(\Delta T_c)_{exp}/(\Delta T_c)_{cal}$ are almost constant at lower volume fractions in every case.

Because the alumina composites show $(\Delta T_c)_{exp}/(\Delta T_c)_{cal} < 1$, which indicates that the thermal shock resistance is overestimated by equation 1, the fracture behavior of alumina-filled composites is examined in further detail. As shown in Figure 7, debonding of the interface is observed in the thermal-shock test specimen but not in the fracture-toughness test specimen Therefore, for the evaluation of thermal shock resistance by equation 1 without overestimation, K_{Ic} should be measured under the condition in which the same fracture pattern as that seen in the thermal shock test is obtained.

A fracture toughness test was performed using a pre-thermal-shocked specimen in order to obtain the fracture toughness measured under same conditions as the thermal shock test. That is, using the specimen for the fracture toughness test (a rectangular bar with a notch for three-point bending), moderate thermal shock was conducted before the fracture toughness test. The

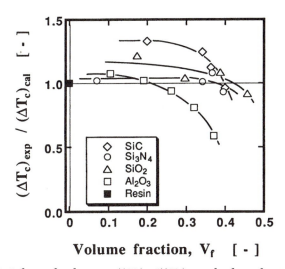

Figure 9. Relationship between $(\Delta T_c)_{exp}/(\Delta T_c)_{cal}$ and volume fraction.

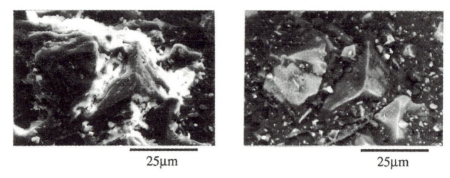

25μm   25μm

Figure 10. SEM photographs of fracture surfaces obtained by fracture toughness testing with a pre-thermal-shocked specimen of Al_2O_3-filled resin. Left: SEI. Right: BEI.

thermal shock of going from room temperature to a dry ice–pentane system (200 K) was used to prepare the pre-thermal-shocked specimen.

Figure 10 shows SEIs and BEIs from the fracture toughness test that was performed with the pre-thermal-shocked specimen. A comparison of fracture toughness measured in the normal and pre-thermal-shocked specimens is summarized in Table II. As shown in Figure 10, particulate–matrix interfacial debonding can be observed in alumina composites, but it is not detected in other particulates. The values of the fracture toughness are also changed only in alumina-filled resins. Values of $(\Delta T_c)_{exp}/(\Delta T_c)_{cal}$ modified using $K_{Ic}(T)$, obtained from the fracture toughness test conducted with a pre-thermal-shocked specimen, are also shown in Table II. The modified values are close to unity.

Table II. Comparison of K_{Ic} Obtained in Fracture Toughness Testing of Pre-Thermal-Shocked and Normal Specimens, and Modified Values of $(\Delta T_c)_{exp}/(\Delta T_c)_{cal}$.

Filler	$K_{Ic}(T)/K_{Ic}$	$(\Delta T_c)_{exp}/(\Delta T_c)_{cal}$ at $V_f = 34.2\%$	
		Normal Value	Modified Value
$Al_2O_3{}^a$	0.85	0.70	0.78
$Al_2O_3{}^b$	0.89	0.65	0.88
SiO_2	0.99	1.08	—
Si_3N_4	1.01	1.01	—
SiC	1.01	1.25	—

[a]Mean diameter of the Al_2O_3 particulate is 27.3 μm.
[b]Mean diameter of the Al_2O_3 particulate is 3.4 μm.

Conclusions

The effects of ceramic particles and filler content on the thermal shock behavior of toughened epoxy resins have been studied. Resins filled with stiff and strong particles, such as silicon nitride and silicon carbide, show high thermal shock resistance, and the effect of filler content is remarkable. At higher volume fractions ($V_f > 40\%$), the thermal shock resistance of these composites reaches 140 K, whereas that of neat resin is about 90 K. The highest thermal shock resistance is obtained with silicon nitride. The thermal shock resistance of silica-filled composites also increases with increasing filler content, but above 30% of volume fraction it comes close to a certain value. On the contrary, in alumina-filled resin, the thermal shock resistance shows a decrease with increasing filler content.

In the case of strong particulates, such as silicon nitride or silicon carbide, cracks propagate through the space between particulates in a zigzag way. In the case of easily broken filler such as silica, however, cracks propagate straight forward by breaking particulates. In the case of weakly bonded particulates such as alumina, cracks propagate through the debonded interface caused by the difference of thermal expansion at thermal shocking.

The thermal shock resistance was calculated on the basis of fracture mechanics and compared with experimental values. Except for alumina-filled epoxy resin, the values were in good agreement. For the alumina-filled resin, the calculated thermal shock resistance was an overestimation. The disagreement between the calculated and experimental values for the alumina-filled composites is based on the difference in fracture behavior seen in the thermal shock and fracture toughness tests. By performing the fracture toughness test on a pre-thermal-shocked specimen, relatively good results were obtained. These results imply that our proposed thermal shock test and evaluation method can be applied to any epoxy resin systems toughened with a hard particulate.

References

1. Kaiser, T. *Chimia* **1990,** *44,* 354–359.
2. Nagai, A.; Eguchi, S.; Ishii, T.; Numata, S.; Ogata, M.; Nishi, K. *Proc. ACS, Div. PMSE,* **1994,** *70,* 55–56.
3. Hojo, H. *J. Mater. Sci. Soc. Jpn.,* **1982,** *18,* 281–286.
4. Moloney, A. C.; Kausch, H. H.; Kaiser, T.; Beer, H. R. *J. Mater. Sci.* **1987,** *22,* 381–393.
5. Hojo, H.; Toyoshima, W.; Tamura, M.; Kawamura, N. *Polym. Eng. Sci.* **1974,** *14,* 604–609.
6. Nakamura, Y.; Yamaguchi, M.; Kitayama, A.; Okubo, M.; Matsumoto, T. *Polymer,* **1991,** *32,* 2221–2229.
7. Hojo, H.; Kubouchi, M.; Tamura, M.; Ichikawa, I. *J. Thermoset. Plast. Jpn* **1988,** *9,* 133–140.

8. Kubouchi, M.; Hojo, H. *J. Soc. Mater. Sci. Jpn* **1982,** *39,* 202–207.

9. Kubouchi, M.; Hojo, H. *Proc. ACS Div. PMSE* **1990,** *63,* 200–204.

10. Kubouchi, M.; Hojo, H. *Toughened Plastics I,* ACS Advances in Chemistry Series 233, 1993, 365–379.

11. Kubouchi, M.; Tsuda, K.; Tamura, M.; Ichikawa I.; Hojo, H. *J. Thermoset. Plast. Jpn* **1992,** *13,* 215–225.

10

The Molecular Network System: A Toughened, Cross-linked Polyester System

F. J. McGarry and R. Subramaniam

Department of Materials Science and Engineering, Massachusetts Institute of Technology, Cambridge, MA 02139

A solution of unsaturated polyester–styrene monomer, an epoxy resin (diglycidyl ether of bisphenol A), and a reactive liquid rubber (amino-terminated butadiene–acrylonitrile), each with its own curative, reacts at 100 °C to form a transparent, cross-linked solid. A two-phase structure results, in which one phase contains no rubber, and in the second phase, the rubber is uniformly distributed in domains a few hundred angstroms in size. Below 9% rubber content by weight, the rubber-modified phase is discontinuous but comprises as much as 70% of the total volume present; the rubber-free phase is continuous. Above 9%, a phase inversion occurs, and now the continuous phase contains rubber while the discontinuous one does not. Apparently, the amino end groups of the ATBN can react slowly with the unsaturated bonds in the polyester, whereas the terminal carboxylic acid groups of the polyester react more rapidly with the epoxy groups. The network that is formed results from chain extension of the polyester by the epoxy and cross-linking of the polyester by the styrene. The epoxy–ATBN reactions provide both chain extension and cross-linking. The properties of the network vary with rubber content in a systematic way; the critical-energy release rate increases from 0.35 to 3.60 lb/in., and tensile strain to failure increases from 1.6 to 12% as the content goes to 20%. Both modulus and strength decline.

POLYESTER RESINS FORM HIGHLY CROSS-LINKED NETWORKS, which result in brittle composites prone to microcracking and damage under impact. A better system is one that retains the desirable properties of polyester—stiffness and chemical resistance—and at the same time has higher impact resistance and increased strain to failure. Synthesis of a simultaneous interpenetrating polymer network (IPN) with one flexible and one rigid component was considered an attractive way of achieving this goal. The system explored here is a combination of rubber-modified epoxy and a styrene cross-linked polyester.

A simultaneous IPN of epoxy with unsaturated polyester, cross-linked using microwave radiation, has been reported (*1*). The radiation was intended to yield a more isotropic structure because of uniform heating; also, it could reduce grafting reactions between the two networks and produce a more classical IPN structure. Unfortunately, the mechanical properties and morphology of the network were not reported.

IPNs of unsaturated polyester and polyurethanes have been used to improve impact strength (*2*). These are single-phase materials because of grafting reactions between the polyurethane and the polyester. They display optical clarity and a single glass-transition temperature, I_g, the location and breadth of which depend on the composition of the polymer (*3*). In fact, all of the materials were transparent without any apparent heterogeneity, and the grafting reactions were shown to be important for improvements in impact strength and fracture toughness, K_{Ic}, of the polyester (*4*).

Experimental Details

The polyester resin used in this study, MR 13006 (Aristech Corporation), was supplied as a 60-wt% solution in styrene monomer. The epoxy resin, a diglycidyl ether of bisphenol A (Epon 828), was obtained from Shell Chemical Company. The reactive liquid rubber, an amino-terminated butadiene–acrylonitrile copolymer (ATBN 1300 × 16), was provided by the BFGoodrich Company. The resin was mixed with additional styrene monomer to maintain the ratio of reactive unsaturation in the polyester-to-styrene monomer at 1 to 3. We added 1.5 wt% of *tert*-butylperbenzoate initiator to the solution, which we then degassed under vacuum. The mixture was poured between vertical, Teflon-coated, aluminum plates and cured under atmospheric pressure at 100 °C. In the modified compositions, the rubber was first dissolved in the styrene monomer, and then all the other components were added and the solution cured as described. In all the compositions, the ratio of the amine functions with respect to the epoxy functions was kept at 1 to ensure complete cure of the epoxy.

The reaction between the epoxy and the end groups of the polyester was modeled by using 1,2-epoxy-3-phenoxypropane and *n*-octanoic acid. An equimolar solution of the two was prepared in tetrahydrofuran (THF) with a catalytic amount of pyridine; a few drops of this solution were placed on a NaCl window and the THF allowed to evaporate. The reaction was carried out at 100 °C, and the IR spectrum of the mixture was recorded every 15 min. Similar experiments were performed without a catalyst. The reaction between the ATBN and the polyester unsaturation was studied by using piperazine and diethylfumarate as the model compounds.

The complex modulus of samples of the neat resins was determined at different temperatures and frequencies using dynamic mechanical analyzer (Seiko DMS 200). The sample was heated at 2 °C/min while holding the amplitude of the oscillation at 30 μm for most samples. All tests were carried out in an atmosphere of nitrogen gas.

Transmission electron microscopy (TEM) was used to study the morphology of the cross-linked matrix and to determine the size of the rubber domains. Specimens were microtomed and exposed to osmium tetroxide vapor to stain the rubber-rich portions of the network. The fracture surfaces of specimens were coated with gold and examined with a scanning electron microscope (SEM).

Results and Discussion

IR data from the model compounds indicate that the amino end groups of the ATBN can react with the double bond in the polyester. In fact, the 1645 cm^{-1} absorption due to the double bond almost disappeared after 40 min at 100 °C, and the absorption due to the amino groups (N–H) also diminished considerably (Figure 1). A Michael-type addition reaction between the secondary amino hydrogen of piperazine derivatives and the activated double bonds of diethylfumarate in the polyester is known (5, 6).

The reaction between the epoxy groups and carboxyl groups in the polyester proceeds rapidly in the presence of pyridine as a catalyst, as shown by the disappearance, after 50 min, of the epoxy absorption at ~915 cm^{-1} and the carbonyl absorption (due to COOH) at ~1710 cm^{-1}, and the appearance of a carbonyl absorption at 1738 cm^{-1} (due to the ester carbonyl) (Figure 2). In the absence of a catalyst the reaction takes much longer. The epoxy and the unsaturated polyester can also interact through hydrogen bonding between the carbonyl groups on the polyester and the epoxy groups. The carbonyl absorption of the polyester shifted from 1725 cm^{-1} to 1728 cm^{-1} upon addition of the epoxy resin (7). IR studies on oligomers of polyester and their blends with epoxy resin show that the carbonyl groups on the polyester interact with the hydroxyl groups on the epoxy (8). These model-compound studies suggest that the cross-linked matrix that results is not a pure interpenetrating network and that a considerable amount of intersystem grafting also can occur. In the actual system, the reaction conditions are more complex; the viscosity of the reaction medium increases sharply with the onset of gelation, and reactions that are otherwise possible become limited by diffusion. A definitive understanding of the structure remains to be established.

The dynamic mechanical spectra show only one tan δ or loss peak in each of the compositions studied (Figure 3). This result suggests that the components

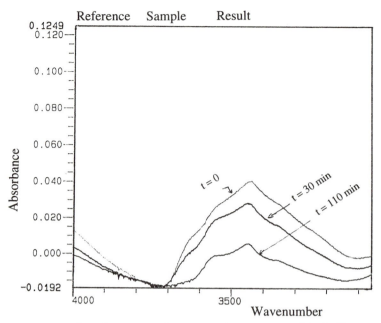

Figure 1. Infrared spectrum of a mixture of piperazine and diethylfumarate after heating at 100 °C, showing changes in absorption due to the amino groups in piperazine.

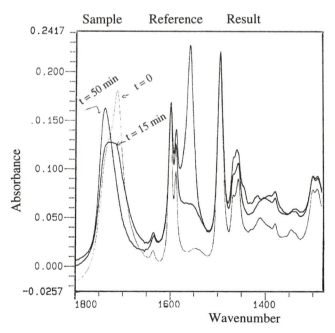

Figure 2. Infrared spectrum of a mixture of 1,2-epoxy-3-phenoxypropane and n-octanoic acid at 100 °C as a function of time, showing the rapid disappearance of acid carbonyl absorption in the presence of pyridine as a catalyst.

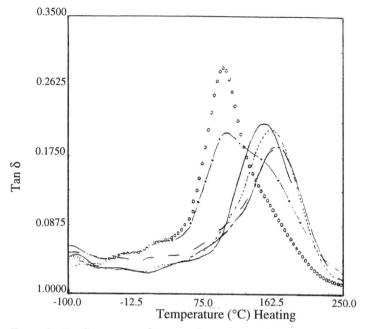

Figure 3. Tan δ curves as a function of temperature for specimens containing different percentages of rubber. Key: ——, *0%;* — · —, *12.5%;* ○, *22%;* ·······, *5%;* — —, *8%; and* — — —, *10%.*

are intimately mixed on a molecular scale and that the mobility of the ATBN segments is restricted by the glassy matrix around them. The loss peaks are very broad, which is a characteristic of interpenetrating networks (9). The width at half-maximum is at least 75 °C and often more than 100 °C. This observation arises from the microheterogeneous morphology caused by the cross-linking and phase separation, which results when the change in free energy of mixing is near zero and much of the material locates at a phase boundary (10), producing diffuse domains on the order of 200 to 400 Å in size. The composition of these domains varies, and the glass-transition temperatures therefore vary within the same material. If the two components have widely separated T_gs, the tan δ peak for the material is broad.

As shown in Figure 4, an abrupt change in T_g occurs when the rubber content increases from 10% to 12.5%. This change is due to a phase inversion: the rubber-modified material switches from discontinuous to continuous and the glassy phase does the opposite. Figure 5 shows the loss passing through a minimum in the same region. Figure 6 shows that the rubbery modulus, measured above T_g, also changes rapidly. These data suggest that the mobility of the short rubber segments is increasingly restricted until phase inversion takes place, increasing thereafter. The cause of the scatter in the rubbery-modulus

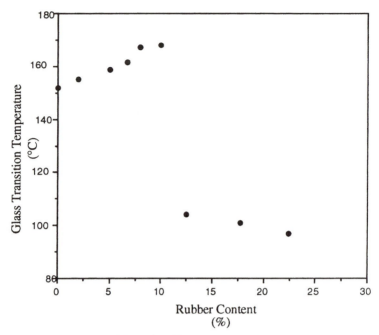

Figure 4. Variation of T_g *with rubber content, as determined from dynamic mechanical analysis. The temperature at which tan δ reached a maximum value was taken as* T_g. *The experiment was conducted in the tensile mode at a heating rate of 2 °C/min at 5 Hz.*

data in the inversion region (Figure 6) is not clear. It could be due to an increase in the molecular weight between cross-links from the reaction of the ATBN with the polyester, which would decrease the rubber modulus, E_r. On the other hand, an increase in cross-link density due to interpenetration would increase E_r (*11*). It is further complicated by the error associated with the measurement of E_r, which can be as high as 30% in some cases. Above T_g, the strains produced in the sample are beyond the limit of linear response of the instrument, and this fact leads to an error in the measured value of E_r (*12*).

Transmission electron micrographs of specimens stained with OsO_4 show the heterogeneous phase structure of the material (Figures 7 through 9). When rubber is present, it is in domains on the order of a few hundred angstroms in size. In the rubber-modified phase, rubber-free inclusions can also be found whose size varies widely. (The diffuse dark regions in the glassy phase are due to differences in the elevation of the surface.) A similar morphology has been observed in interpenetrating networks of poly(methyl methacrylate) and nitrile rubber (*13*). The transmission micrograph of the sample containing styrene, epoxy, and polyester (Figure 9), shows no effect of

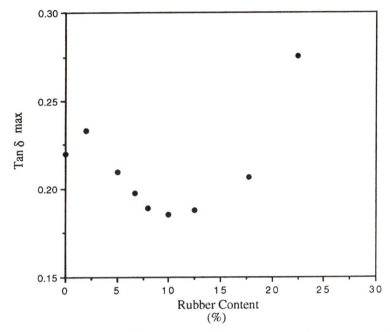

Figure 5. Variation of the maximum value of tan δ with rubber content. Values of tan δ were obtained using a Seiko DMS 200 dynamic mechanical analyzer in the tensile mode. The heating rate was 2 °C/min at 5 Hz.

staining because no double bonds are attacked; the system reacts fully. Scanning electron micrographs of OsO_4-stained fracture surfaces show phase inversion (Figures 10 and 11). The glassy phase is continuous at rubber contents below 10%, whereas the rubber-rich phase is continuous at higher rubber contents.

Morphology.   Phase inversion in polymer mixtures occurs when the volume fraction of the dispersed phase becomes equal to or exceeds 0.5 (*14*). The driving force is to minimize the interfacial energy of the system. This is not the case here because the volume fraction of the rubber-rich phase at phase inversion is about 0.85. After inversion, the fraction of the continuous rubber-rich phase is only 0.28, and it increases to 0.63 at 12.5% rubber content. Initially, the components are fully soluble and compatible, but as the reactions proceed, the molecular weight of the products increases and phase separation results. The ability to separate and invert is dependent on the viscosity of the medium. The unsaturated polyester forms a gel at conversions as low as 2 to 5%, and both the ability to separate and to invert is impeded. Thus the morphology depends on the two competing effects of phase inversion and

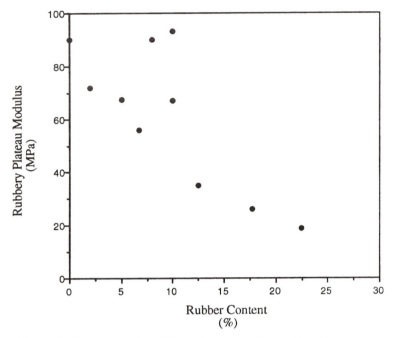

Figure 6. Variation of the rubbery plateau modulus with rubber content, measured using dynamic mechanical analysis in the tension mode at a heating rate of 2 °C/min. The two values at 10% rubber content were obtained from two different runs under similar conditions.

polymerization/chain extension. The situation is further complicated by the fact that the temperature increases, finally reaching 100 °C, which also affects compatibility, phase separation, and reaction rates. Apparently, the ATBN, epoxy, and polyester react first, forming a rubber-rich matrix, at temperatures below which the polyester–styrene reaction takes place. This reaction continues until the temperature reaches the initiation point for the free-radical curing of the polyester, after which the latter dominates. At 8% rubber content, the epoxy–ATBN and ATBN–polyester reactions do not produce enough rubber-modified matrix for phase inversion to occur before gelation of the polyester. At 10% and higher, the phase inversion does occur, apparently before initiation of the polyester cure, while the mixture is still fluid.

In the TEM samples, the rubber domains are uniformly distributed and on the order of a few hundred angstroms in size. The micrographs also show the presence of domains that have no rubber in them. The rubber-free domains probably contain polyester and epoxy that have reacted, but it is not possible to confirm this possibility. Styrene-cross-linked polyester has a T_g of 185 °C. When the epoxy is introduced in a 1:2 ratio (epoxy:polyester), T_g de-

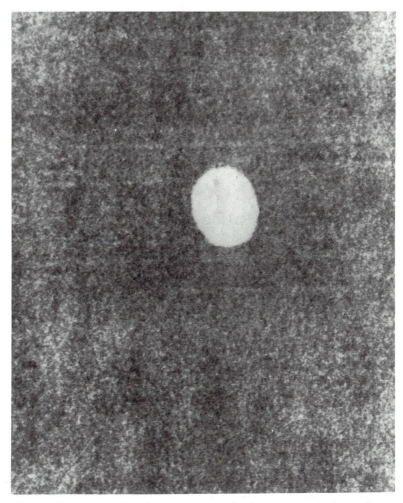

Figure 7. Transmission electron micrograph of an OsO_4-stained sample containing 22% rubber, showing the widely distributed rubber phase. The size of the rubber domains is on the order of a few hundred angstroms. The light region in the micrograph is the brittle polyester–epoxy phase domains. Magnification 60,000×.

creases to 152 °C. Figure 4 shows it increasing again as rubber is added, until at 10% it is nearly 170 °C. Then it drops abruptly to about 105 °C as the content is further increased. As seen in Figure 5, the tan δ variation is more gradual, and the fact that it remains a single damping peak is consistent with the absence of any large agglomerations of rubber throughout the entire composition range. Agglomeration in epoxy–acrylate rubber systems can be prevented

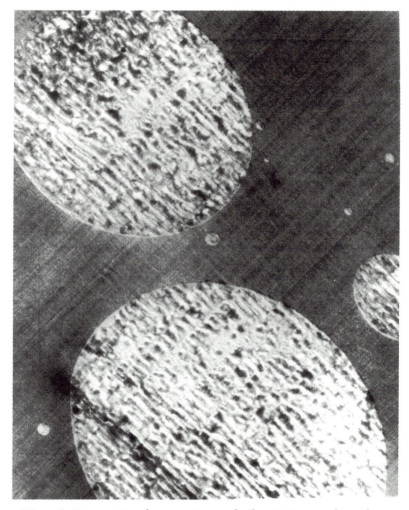

Figure 8. Transmission electron micrograph of an OsO₄-stained sample contain-ing 22% rubber, showing the brittle-phase domains as circular bright regions. The diffuse dark regions in the brittle phase are due to differences in the elevation of the surface. Magnification 7500×.

by as little as 3% grafting (*15*), and in the epoxy–polyester–rubber mixture, this possibility was demonstrated by the IR studies on the model compounds. This could explain the small rubber domains.

The network that is formed is highly grafted. A schematic of the final structure is shown in Figure 12, which should be recognized as just one specu-lation of many. We do not know the precise molecular structure of the molecu-lar network system, beyond the fact that the rubber in it is always uniformly

Figure 9. Transmission electron micrograph of a sample containing no rubber. Magnification 12,000×.

distributed regardless of its concentration. The morphology is more explicit, as shown in Figure 13; the rubber-containing phase has few if any small, glassy occlusions in it. Conversely, the larger regions of the glassy phase rarely contain rubber-modified occlusions. It appears that the two phase types are distinct and consistent; the interfaces are well-defined and the morphology is reproducible.

The castings of the molecular network system are transparent to translucent. As the rubber content increased, a brownish hue became apparent. This change often indicates the presence of a degraded amine. The transparency is

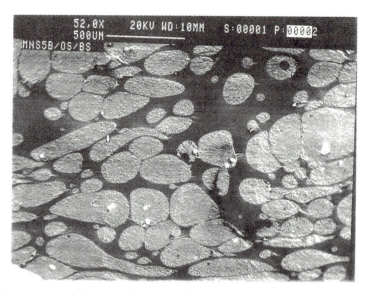

Figure 10. Back-scattered image of the OsO₄-stained fracture surface of a sample containing 8% rubber. The rubber-rich domains appear bright because of the osmium in these regions. Note the presence of small inclusions of the brittle phase within the rubber-rich domains.

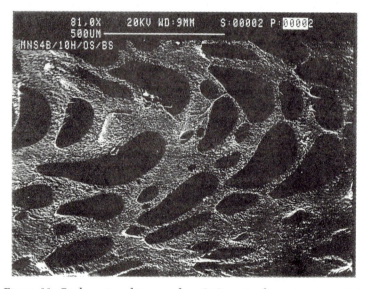

Figure 11. Back-scattered image of an OsO₄-stained specimen containing 10% rubber, showing the rubber-rich continuous phase as bright regions.

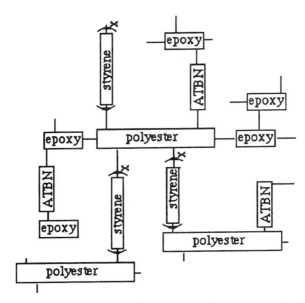

Figure 12. Schematic of the grafting reactions that are possible during cross-linking of the matrix. The value of x is normally between 1 and 5. Some polystyrene also probably forms but is not shown here.

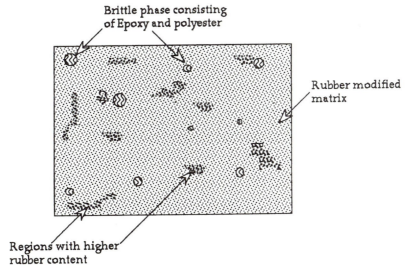

Figure 13. Proposed morphology of the molecular network system after curing. The domains are shown clearly separated merely for purposes of illustration. In the actual matrix the domains may not be so clearly demarcated.

Table I. Mechanical Properties of a Matrix with a Polyester-to-Epoxy Ratio of 2:1.

Rubber Content (%)	Ultimate Tensile Strength[a] (psi)	Modulus[a] (Msi)	Strain[a] (%)	K_{Ic} ksi·$\sqrt{in.}$	G_{Ic} (lb/in.)
0	8100	0.59	1.6	0.45 (0.01)	0.34
2	8500 (1300)	0.59 (0.02)	1.72 (0.43)	0.46 (0.01)	0.35
5	6700 (1000)	0.54 (0.02)	1.53 (0.17)	0.58 (0.04)	0.62
6.7	8100 (1000)	0.48 (0.03)	2.35 (0.86)	0.85 (0.02)	1.50
8	7300 (200)	0.55 (0.03)	1.50 (0.07)	0.56 (0.04)	0.57
10	7200 (300)	0.41 (0.02)	2.79 (0.35)	0.73 (0.06)	1.30
12.5	6100 (60)	0.35 (0.01)	3.02 (0.11)	0.79 (0.02)	1.80
15	4300 (300)	0.33 (0.01)	1.77 (0.23)	0.86 (0.05)	1.46
17.7	4950 (85)	0.27 (0.01)	5.32 (0.75)	0.84 (0.03)	2.63
20	4700 (65)	0.26 (0.01)	4.69 (0.93)	0.93 (0.04)	3.33
22.4	3200 (60)	0.19 (0.00)	11.2 (2.08)	0.82 (0.04)	3.57

NOTE: Numbers in parentheses are standard deviations.
[a]Tensile properties were measured using ASTM D–638. Msi, million pounds per square inch; ksi, thousand pounds per square inch.
[b]Plane strain fracture toughness was measured using ASTM E–399.

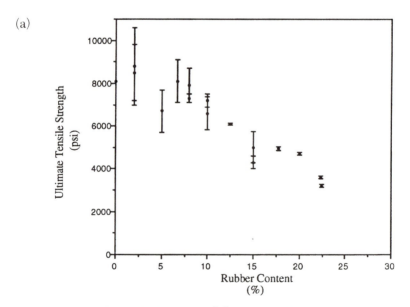

(a)

Figure 14. Variation of ultimate tensile strength (a).

(b)

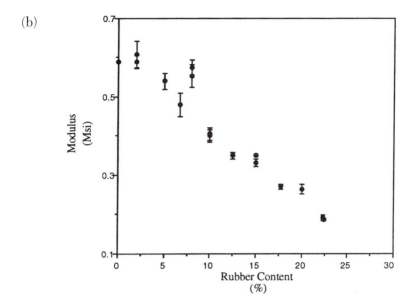

(c)

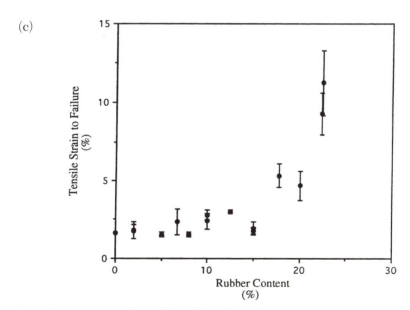

Figure 14. Tensile modulus, (b), and tensile strain to failure (c) with rubber content, for samples with an epoxy-to-polyester ratio of 1:2. Tensile properties were measured using ASTM D–638 on dog-bone-shaped samples at a speed of 0.2 in./min.

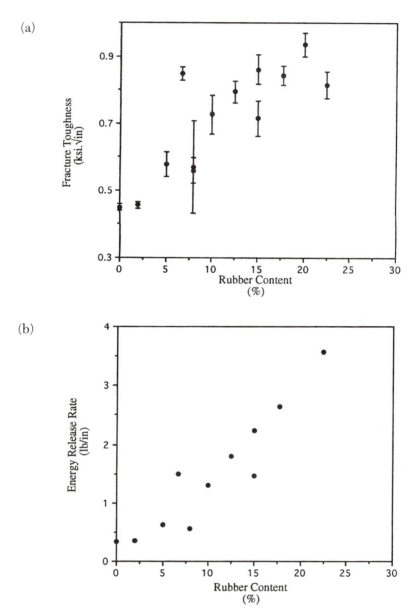

Figure 15. (a) Variation of plane strain fracture toughness with rubber content for samples with an epoxy-to-polyester ratio of 1:2. Compact tension specimens were tested at 0.2 in./min. using ASTM E–399. (b) Variation of strain-energy release rate, G_{Ic}, with rubber content for samples with an epoxy-to-polyester ratio of 1:2.

consistent with the small rubber domains and with the absence of optical discontinuities.

Mechanical Properties.   The tensile properties of the cast specimens were determined using dog-bone-shaped specimens following ASTM D-638, at a crosshead speed of 0.2 in./min on a tensile testing machine (Instron 4505). The strain to failure of the specimens was measured using an extensometer. The tensile strength and tensile modulus were also determined.

The plane strain fracture toughness was determined using ASTM E-399 with compact tension specimens. Specimens were precracked using a razor blade. The ratio of the precrack length to the total length of the sample was kept within the specified limits. Only samples with straight precracks were used for testing, which was done at a rate of 0.2 in./min.

The mechanical properties of the cured castings with a polyester-to-epoxy ratio of 2:1 are presented in Table I and Figures 14 and 15. Both the ultimate tensile strength and the modulus decrease with increasing rubber, whereas the tensile strain to failure, the fracture toughness (K_{Ic}) and the critical-strain-energy release rate increase. Some properties appear to fluctuate at the phase-inversion region, but this possibility needs further study. The critical-energy release rate, G_{Ic}, was calculated from the fracture toughness; it increases more than tenfold for the samples with the highest rubber content. This result is due to the enhanced plastic flow and blunting of the crack tip. The variations of K_{Ic} and G_{Ic} with rubber content are shown in Figure 15.

Summary

A new combination of unsaturated polyester, styrene monomer, DGEBA epoxy, and ATBN completely reacts to form a solid, transparent product containing small, dispersed, rubber domains. On a larger scale, at least two different phases can be identified. Macroscopically, as the rubber content increases, toughness and ductility increase, whereas modulus and strength decline. When reinforced with glass fibers, the toughened matrix should offer attractive damage resistance.

References

1. Gourdenne, A.; Heintz, P. *Proc. IUPAC, I.U.P.A.C., Macromol. Symp., 28th* **1982,** 486.
2. Frisch, H. L.; Frisch, K. C.; Klempener, D. *Polym. Eng. Sci.* **1974,** *14,* 646.
3. Meyer, G. C.; Mehrenberger, P. Y. *Eur. Polym. J.* **1977,** *13,* 383.
4. Hsieh, K. H.; Tsai, J. S.; Chang, K. W. *J. Mater. Sci.* **1991,** *26,* 5877.
5. Tanzi, M. C.; Pornaro, F.; Miuccio, A.; Grassi, L.; Danusso, F. *J. Appl. Polym. Sci.* **1986,** *31,* 1083.
6. Tanzi, M. C.; Levi, M.; Danusso, F. *J. Appl. Polym. Sci.* **1991,** *42,* 1371.
7. Lin, M.-S.; Chang, R. *J. Appl. Polym. Sci.* **1992,** *46,* 815.

8. Suspene, L.; Pascault, J. P.; Lam, T. M. *J. Appl. Polym. Sci.* **1990,** *39,* 1347.
9. *Interpenetrating Polymers and Related Materials;* Sperling, L. H., Ed.; Plenum: New York, 1981.
10. Fox, R. B.; Fay, J. J.; Sorathia, U.; Sperling, L. H. In *Sound and Vibration Damping with Polymers;* Corsaro, R. D.; Sperling, L. H., Eds.; ACS Symposium Series 424, American Chemical Society: Washington, DC, 1990; pp 359–365.
11. Tsenoglou, C. In *Crosslinked Polymers: Chemistry, Properties, and Applications;* Dickie, R. A.; Labana, S. S.; Bauer, R. S., Eds.; ACS Symposium Series 367; American Chemical Society: Washington, DC, 1988; pp 59–65.
12. Fox, R. B.; Bitner, J. L.; Hinkley, J. A.; Carter, W. *Polym. Eng. Sci.* **1985,** *25,* 157.
13. Butta, E.; Levita, G.; Marchetti, A.; Lazzeri, A. *Polym. Eng. Sci.* **1986,** *26,* 63.
14. Bucknall, C. B. In *Toughened Plastics;* Applied Science: London, 1977; Chapter 4, p 71.
15. Scarito, P. R.; Sperling, L. H. *Polym. Eng. Sci.* **1979,** *19,* 297.

Elastomer-Modified Vinyl Ester Resins: Impact Fracture and Fatigue Resistance

A. R. Siebert[1], C. D. Guiley[1], A. J. Kinloch[2], M. Fernando[2], and E. P. L. Heijnsbrock[2]

[1]Industrial Specialties Division, BFGoodrich Company, Brecksville, OH 44141
[2]Department of Mechanical Engineering, Imperial College, London SW7 2BX, United Kingdom

Vinyl esters are preeminent chemical- and corrosion-resistant materials. Various products are available that are based primarily on epoxy resins that have been addition-esterified with methacrylic acid and diluted with styrene monomer. These products are based on diepoxide structure, molecular weight, and modifier type (e.g., nitrile rubber, urethane, and glycol); and blends with styrenated resins are also available. Elastomer modification gives a fourfold improvement in the fracture energy of vinyl esters over that of unmodified vinyl esters. Impact fracture energy and fatigue properties of elastomer-modified vinyl ester resins that are further modified with epoxy-terminated butadiene–acrylonitrile rubber are greatly enhanced and do not appear to be rate-sensitive. Additive tougheners greatly alter the morphology of cured specimens, providing clues to toughness enhancement.

V INYL ESTER RESINS (VERs, epoxy methacrylates) are a major class of styre-nated, free radically curable, corrosion- and chemical-resistant thermoset resins. They are largely used in fiber-reinforced structural applications, and they have a substantial history of long-term service in numerous environments at elevated temperatures and pressures, usually under load.

VERs are available as both rigid and flexible epoxy resins. The flexible epoxy resins generally have a depressed glass-transition temperature, T_g, and inferior chemical resistance. Nitrile-rubber-modified VER (*1*) appeared on the market in the mid- to late 1970s. These elastomer-modified VERs show improved fatigue resistance over unmodified VER (*2*).

Recent work has shown that the addition of epoxy-terminated butadi-ene–acrylonitrile (ETBN) liquid rubbers to elastomer-modified VER gives a fivefold increase in fracture energy over that of elastomer-modified VER (*3*).

It is noteworthy that this increase is in addition to a fourfold increase already found for the elastomer-modified VER compared with an unmodified VER.

This chapter describes our work in determining the impact fracture energies of elastomer-modified VER that has been further blended with various amounts of ETBN, and measuring the fatigue resistance of these materials.

Experimental Details

Materials.   The elastomer-modified VER used in this work was Derakane 8084 (Dow Chemical Co.), which has about 7.5 phr (parts per hundred resin by weight) of a carboxyl-terminated butadiene–acrylonitrile (CTBN) liquid polymer reacted into the resin base (1). The reactive liquid rubber used as the additive was ETBN 1300×40. The ETBN was made by reacting CTBN 1300×8, 17% bound acrylonitrile, with liquid diglycidyl ether of bisphenol A epoxy resin (Epon 828), at a molar ratio of 1:2. Because of the high viscosity of ETBN, it was dissolved in styrene to give a 50% solids solution before addition to the vinyl ester resin. A combination of cobalt naphthenate and MEK peroxide was used to cure the samples. Sheets were made in Teflon-coated aluminum molds having a sample thickness of about 6 mm for fatigue measurements and 13 mm for impact testing. Recipes were cured for 1 h at 80 °C and then for 2 h at 120 °C.

Preparation of Test Specimens.   For slow measurements of fracture energy, G_{Ic}, the compact tension (CT) specimens, 25.4 × 31.75 mm, were machined from the 6-mm plaques. A natural crack was made by tapping gently with a fresh razor blade. Identical CT specimens were prepared for studies of fatigue crack growth. For impact G_{Ic} measurements, single-edged, notched-bar (SENB) specimens were prepared by machining the 13-mm sheets into bars of dimensions 80 × 13 × 10 mm. Cracks were then inserted into the bars, in the center of the 80 × 10 mm face, by machining an initial notch using a narrow milling cutter and then gently tapping a fresh razor blade into the notch so as to propagate a sharp, natural crack ahead of the razor blade. Cracks of various lengths, a, were inserted using these techniques.

Slow- and Fast-Speed Testing.   Slow-speed testing was performed at 21 μm/s. Values of fracture toughness, K_{Ic}, and of G_{Ic} were calculated by the standard procedures (4). The impact tests were conducted using a commercial instrumented machine (Ceast, Turin, Italy) at 0.5 km/s, which corresponds to a typical front-end of a car folding up after a crash at 30 mph (5). The fracture energy, G_{Ic}, is related to the test parameters by the following equation (6):

$$G_{Ic} = U_c/BW\Phi \tag{1}$$

where U_c is the stored elastic strain energy at the onset of crack growth, B is the thickness of the specimen, W is the width of the specimen, and Φ is calculated from published tables (6) and is a function of a/W and L/W, where a is the crack length and L is the length or span of the test specimen between support points. The value of G_{Ic} may be determined by using several SENB specimens containing different crack lengths and then plotting U_c versus $BW\Phi$, which should be linear and pass through the origin. For each test condition, triplicate tests were performed, and the typical variation in the value of G_{Ic} was ±8%.

Fatigue-Crack Propagation Test.   CT specimens were used in Mode I, tension–tension fatigue tests. The CT specimens were $60 \times 62.5 \times 6$ mm, and cracks were inserted into the specimens as just described. Testing was undertaken at a frequency of 20 Hz and under displacement control using an R-ratio (minimum to maximum load) of 0.6. This ratio was chosen so that the results obtained in the present work could be compared with those reported by Blankenship et al. (2). A constant-amplitude, sinusoidal waveform was applied using a Mayes servohydraulic machine. The advancing crack was followed during fatigue testing using a calibrated traveling microscope. The number of cycles was recorded at observed increments in growth. The range of the applied stress-intensity factor, K, was calculated using the standard equations (7). Duplicate tests were undertaken, and all the test results obtained are shown in Figure 6. The reproducibility of the duplicate tests was excellent.

Glass-Transition Measurements.   The glass-transition temperature, T_g, was measured using a Mettler TA3000 differential scanning calorimetry instrument with a scan rate of 20 °C/min; the second scan was measured to determine T_g.

Fractography.   Micrographs from the surfaces of the CT specimens were examined using a JEOL scanning electron microscope (SEM). Transmission electron micrographs were obtained from microtomed sections stained with OsO_4 using a JEOL 100SX transmission electron microscope (TEM).

Results and Discussion

Fracture-Energy Testing.   Table I gives the recipes and the fracture energies measured under slow and fast rates of test, for the elastomer-modified VER and for ETBN additions to the elastomer-modified VER. Also given are the total amounts of reactive liquid rubber, from both ETBN addition and CTBN reacted directly into the VER, for each recipe. For reference, the unmodified VER has a slow G_{Ic} of 0.11 kJ/m². Finally, the T_g obtained from differential scanning calorimetric measurements is given for each recipe.

Table I.   Recipes and Fracture Energies of Four Elastomer-Modified Vinyl Ester Resins Blended with Various Amounts of ETBN

Sample No.	1	2	3	4
Elastomer-modified VER	100	100	100	100
ETBN	0	6	10.7	15.6
Cobalt naphthenate	0.5	0.5	0.5	0.5
MEK peroxide	2	2	2	2
Rubber level	7.5	10	12	14
G_{Ic} (kJ/m²), slow	0.43	1.63	2.08	2.31
G_{Ic} (kJ/m²), fast	0.57	1.22	2.0	2.3
T_g (°C), differential scanning calorimetry	110	109	110	109

NOTE: Units for the recipes are phr.

The G_{Ic} values from the slow and fast fracture tests are generally in good agreement, which suggests that there is no great effect of test rate on the measured toughness of these different recipes. Thus, the same trend is observed irrespective of the test rate, in that the value of G_{Ic} steadily increases as the level of CTBN additive increases, as seen in Figure 1. There are no significant differences in the values of T_g for the various recipes. Hence, increasing the addition of CTBN has no deleterious effect on the short-term thermal properties of the materials.

Microscopic Studies.   Figure 2 shows a TEM image of an OsO_4-stained sample of recipe 1, the elastomer-modified VER. Particles on the order of 20 nm are seen. Thus, it is not surprising that the cured sample of recipe 1 is optically clear.

Figures 3 and 4 show SEM images of recipes 2 and 3 respectively, taken from the CT samples that were fractured under fatigue loading (see the next section). The SEMs for recipe 2 reveal that a distinct particulate phase now exists. There are particles about 1 μm in diameter and a few larger particles. The

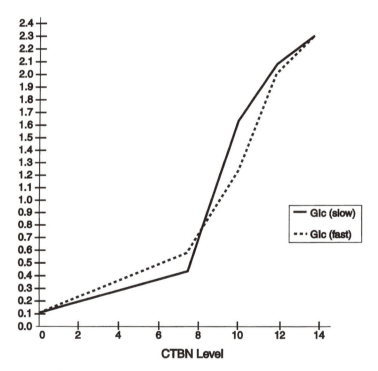

Figure 1. Fracture energy, G_{Ic}, measured at slow and fast testing rates versus total CTBN content.

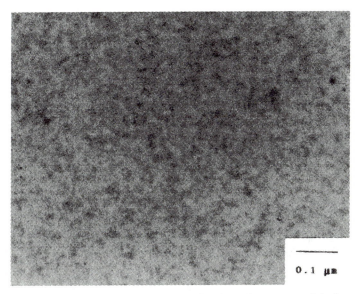

Figure 2. TEM image of a cured sample of an elastomer-modified vinyl ester resin (Table I, recipe 1). The sample shown is an ultrathin, OsO₄-stained section. Magnification 60,000×.

small holes of about 1 μm are presumably formed by dilatation of the particles. This sample showed a moderate amount of stress-whitening on the surface during testing. Recipe 3 (Figure 4) shows two distinct ranges of particle sizes. The small holes of about 1 μm are presumably formed by dilatation of the smaller particles. The larger particles range in size from about 10 to 15 μm and appear to be made up of multiple small particles. This sample showed a considerable amount of stress-whitening on the surface of the CT specimen.

Figure 5 is an SEM image of recipe 4, with a total CTBN content of 14 phr. This SEM image also shows two distinct particle-size ranges, with the small holes, of about 1 to 2 μm, again presumably being formed by dilatation of the small particles. The larger particles range in size from about 15 to 40 μm and appear to be made up of multiple small particles. This sample also showed a considerable amount of stress-whitening on the surface of the CT specimen.

Fatigue-Crack Propagation Testing.   The fatigue-crack propagation results, log da/dN, where a is the crack length and N is the number of cycles, are plotted against log ΔK in Figure 6 for recipes 1, 2, and 3. Fatigue-crack propagation tests were not performed on recipe 4. Recipes 1 and 3 have similar fatigue behavior except that the fatigue curve for recipe 3, with a total CTBN content of 12 phr, extends to much higher levels of log da/dN because

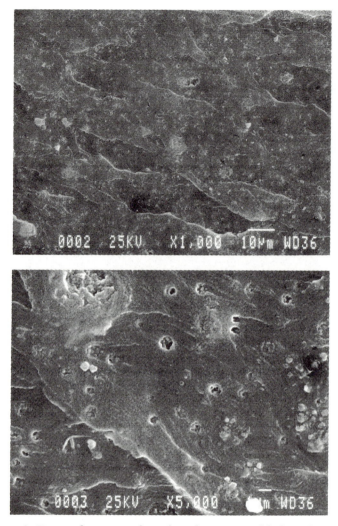

Figure 3. Fatigue-fracture surface of an elastomer-modified vinyl ester resin to which ETBN has been added (Table I, recipe 2). The magnifications of these SEM images of cured samples are 1000× (top) and 5000× (bottom).

of its higher G_{Ic} value. Recipe 2, with a total CTBN of 10 phr, shows a significant increase in fatigue performance, with log da/dN versus log ΔK shifted to the right and showing a somewhat lower slope.

The reason for the improved fatigue performance of recipe 2 (with 10 phr of total CTBN content) compared with that of recipe 3 (with 12 phr of total CTBN content) is not fully understood. However, it may be due to the tough-

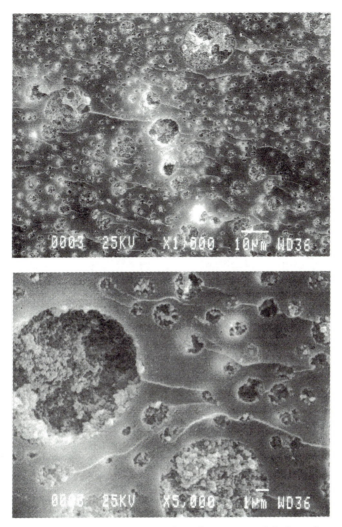

Figure 4. Fatigue-fracture surface of an elastomer-modified vinyl ester resin to which ETBN has been added (Table I, recipe 3). The magnifications of these SEM images of cured samples are 1000× (top) and 5000× (bottom).

ening mechanism involving cavitation and debonding of the rubbery particles. While such mechanisms may contribute to the overall toughness of the material, by enabling subsequent plastic-hole formation in the matrix (8, 9), the presence of voids or holes associated with these mechanisms may readily promote fatigue-crack growth through the voided material. This proposed explanation is supported by the micrographs shown in Figures 3 and 4. These fig-

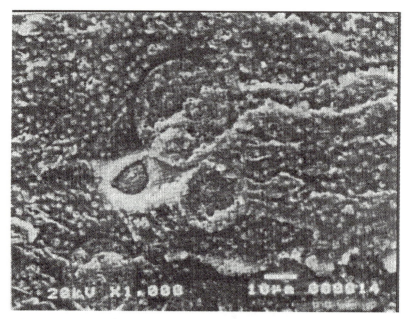

Figure 5. Fatigue-fracture surface of an elastomer-modified vinyl ester resin to which ETBN has been added (Table I, recipe 4). The magnification of this SEM image of a cured sample is 1000×.

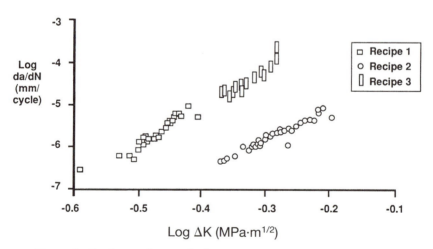

Figure 6. Crack growth rate, da/dN, versus stress intensity range, ΔK. Testing was performed at a frequency of 20 Hz and an R-ratio of 0.6.

ures clearly illustrate that the tougher material (recipe 3) has more voids, and it is this material that exhibits poorer fatigue resistance.

Finally, it should be noted that stress-whitening arises because these voids, created by cavitation and debonded particles, scatter light. Hence, it is observed that the most intense stress-whitening is present in the recipe that undergoes the greatest extent of particle cavitation and debonding.

Whatever the fundamental mechanisms involved, the practical implication is that the balance between high initial toughness and good fatigue-crack resistance must be considered when formulating these CTBN-modified materials. In the study of crack-propagation behavior in toughened plastics under fatigue loadings, the introduction of a second rubbery phase has been reported (*10*) to increase fatigue resistance as well as impact toughness. In some polymers, however, increasing the amount of rubbery phase may continue to increase the impact toughness without producing a further increase in fatigue resistance. However, nearly all of these studies have been concerned with thermoplastic polymers. Recently, Kinloch and Osiyemi (*11*) studied the toughness and fatigue behavior of two rubber-toughened, thermosetting epoxy adhesives. They found that the tougher adhesive possessed inferior fatigue behavior, which is in broad agreement with the observations just mentioned on the thermosetting vinyl ester resin materials.

Summary

The rate of test has little effect on the fracture energy, G_{Ic}, when CTBN is added as ETBN (see the section "Materials") to an elastomer-modified vinyl ester resin. Thus, these systems do not appear to be rate-sensitive. The added CTBN has a significant effect on G_{Ic} of the cured system, but such an addition does not lead to a loss of glass-transition temperature, T_g, with increasing amounts of added CTBN. The systems with added CTBN show both large and small particles, with the small particles having undergone cavitation and so promoting plastic deformation in the highly cross-linked matrix resin.

The sample having 10 phr of total CTBN (2.5 phr from ETBN addition and 7.5 phr from the elastomer-modified VER) shows a significant improvement in fatigue behavior over the elastomer-modified vinyl ester resin itself. The higher rubber level of 12.0 phr of total CTBN, 4.5 phr from ETBN and 7.5 phr from the elastomer-modified vinyl ester resin, gives a fatigue resistance that is somewhat similar to that of the elastomer-modified vinyl ester resin itself.

References

1. Dow Chemical Co., U.S. Patent 3 892 819, 1975; Fibercast Corp., U.S. Patent 3, 928 491, 1975.

2. Blankenship, L. T.; Barron, D. L.; Kelley, D. H. Presented at the 44th Annual Conference of the Composites Institute, Society of Plastics Industries, Dallas, February 1989; Session 14C.
3. Siebert, A. R.; Guiley, C. D.; Egan, D. R. Presented at the 47th Annual Conference on the Composites Institute, Society of Plastics Industries, Dallas, February 1992; Session 17C.
4. ASTM D-5045-91 A, Plane Strain Fracture Toughness.
5. Kinloch, A. J.; Kodokian, G. A.; Jamarani, M. B. *J. Mater. Sci.* **1987,** *22,* 4111.
6. Williams, J. G. *Fracture Mechanics of Polymers;* Ellis Horwood: Chichester, England, 1984; p 62.
7. Kinloch, A. J.; Young, R. J. *Fracture Behavior of Polymers;* Applied Science: London, 1983; p 204.
8. Kinloch, A. J.; Shaw, S. J.; Hunston, D. L. *Polymer* **1983,** *24,* 1355.
9. Huang, Y.; Kinloch, A. J. *J. Mater. Sci.* **1992,** *27,* 2763.
10. Hertberg, R. W.; Manson, J. A. *Fatigue of Engineering Plastics;* Academic: Orlando, FL, 1980; p 209.
11. Kinloch, A. J.; Osiyemi, S. O. Presented at the Structural Adhesives in Engineering III Conference, Plastics and Rubber Institute, London, 1992; paper 31.

Crazing and Dilatation-Band Formation in Engineering Thermosets

H.-J. Sue[1], P. C. Yang[2], P. M. Puckett[2], J. L. Bertram[2], and E. I. Garcia-Meitin[2]

[1]Department of Mechanical Engineering, Texas A&M University, College Station, TX 77843–3123
[2]B-1603, Dow Chemical USA, Freeport, TX 77541

The crazing phenomenon was observed in a high-performance, thermosetting, 1,2-dihydrobenzocyclobutene and maleimide (BCB–MI) resin system. The craze-fibril diameter, band thickness, and size of the damage zone observed in the BCB–MI matrix were all larger than those of the polystyrene craze. As a result, the BCB–MI had a much higher fracture toughness than polystyrene, bisphenol A polycarbonate, and many other rubber-toughened thermosets. Unusual craze-like dilatation bands were also detected in several engineering thermosets. These dilatation bands were, however, less effective in fracture-energy dissipation than those due to craze bands. Only when the formation of dilatation bands was extensive did the toughening effect become significant. The possible cause(s) and conditions resulting in crazing and the formation of dilatation bands in thermosets are discussed. The importance of the present findings for the toughening of high-performance thermosets is also addressed.

ONE OF THE PREDOMINANT MECHANISMS of toughening in thermoplastics is crazing. The stretching, disentanglement, and fibril formation of high molecular weight, linear, thermoplastic polymers are well understood and characterized. In the literature, many researchers have claimed that crazing can occur in thermosets and is a major toughening mechanism for thermosets (*1–5*). For instance, Sultan and McGarry (*1*) indicated that crazing could be a predominant flow mechanism in rubber-modified epoxies when the rubber particle is large and the stress field is tensile. Bucknall and Yoshii (*4, 5*) showed signs of crazing in a carboxy-terminated butadiene–acrylonitrile (CTBN), rubber-modified epoxy matrix. However, they did not convincingly demonstrate that crazing had taken place and was a major source of toughening. Consequently, in the area of thermoset toughening, it is still believed that crazing is not likely to occur in engineering thermosets (*6–9*) because these highly cross-linked molecules cannot undergo significant molecular stretching and disentanglement (5, 10). Therefore, no known efforts have focused on toughening thermosets via crazing or craze-like damage.

Recent progress and technological breakthroughs in polymer synthesis, modification, and characterization have led to new findings of previously unreported toughening mechanisms in thermosets as well as a better understanding of why and how these mechanisms operate (11–16). For instance, a phenomenon we designate "croiding" is observed in a core–shell rubber (CSR) modified diglycidyl ether of bisphenol A (DGEBA) epoxy resin cured with piperidine (Figure 1) (11). At the optical microscopic scale (a practical resolution of about 1 μm), these croids resemble the well-known crazes observed in polystyrene. However, when transmission electron microscopy (TEM) is used,

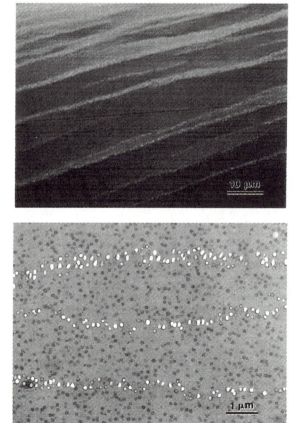

Figure 1. DN–4PB plane-strain damage zone of epoxy that has been modified with core–shell rubber. Top: ROM image obtained under Nomarski interference contrast. Bottom: TEM image of the croids in the plane-strain region. In both images the crack propagates from the upper right to the upper left.

the croids are found to be composed of linear arrays of dilatation bands containing numerous, highly cavitated, CSR particles (Figure 1, bottom). These findings suggest the possibility of toughening engineering thermosets via massive crazing or craze-like toughening mechanisms.

The present work focuses on characterizing the fracture mechanism(s) of several moderately cross-linked engineering thermosets whose molecular weights (M_c) range from 560 to 920 g/mol. Characterization was performed using the double-notch, four-point-bend (DN–4PB) technique (*17*) together with various microscopic and spectroscopic tools. The goal of this work is to gain an understanding of how and why crazing and dilatation bands occur in engineering thermosets, and to use that knowledge to effectively toughen thermosets via promotion of crazes or craze-like dilatation bands.

Experimental Details

Materials.   The thermosetting resins investigated include (1) 1,2-dihydrobenzocyclobutene and maleimide (BCB–MI) resin, (2) a modified version of the BCB–MI resin (BCB–MI–M), (3) cross-linkable epoxy thermoplastic (CET, which is chemically similar to TACTIX 695 epoxy resin from Dow Chemical Co.), and (4) DGEBA epoxy (D.E.R. 332, Dow Chemical Co.) cured with piperidine. The chemical structures of the resins are shown in Chart I.

The detailed syntheses, curing schedules, and physical testing conditions of these resins can be found elsewhere (*11, 12, 14, 15, 18*). For convenience, the basic physical properties of these resins are given in Table I.

Mechanical Testing.   The thermosetting resins were cured and pour-cast into rectangular plaques, which were then machined into (1) 12.7 × 1.27 × 0.635 cm bars for the DN–4PB experiments (*17*), and (2) 6.35 × 1.27 × 0.635 cm bars for the single-edge-notch, three-point-bend (SEN–3PB), fracture-toughness measurements. These bars were notched with a notching cutter (250 μm radius), and then tapped with a liquid-nitrogen-chilled razor blade to wedge open a sharp crack. A Sintech-2 screw-driven mechanical testing machine was used to conduct both the SEN–3PB and DN–4PB experiments at a crosshead speed of 0.0508 cm/min. When the DN–4PB experiment was conducted, care was taken to ensure that the upper contact-loading points were touching the specimen simultaneously.

Microscopy.   The DN–4PB damage zone of the subcritically propagated crack was cut along the crack-propagation direction but perpendicular to the fracture surface using a diamond saw. The plane-strain core region and plane-stress surface region of the crack-tip damage zone were prepared for transmitted optical microscopy (TOM), reflected optical microscopy (ROM), and TEM following the procedures described by Sue and co-workers (*20, 21*). TOM and ROM were performed using an Olympus Vanox-S microscope. TEM was conducted using a JEOL 2000FX ATEM operated at an accelerating voltage of 100 kV.

The density inside the dilatation bands was measured using electron energy loss spectroscopy (EELS) with a Gatan 666 parallel EELS (PEELS), which was attached to the TEM. An accelerating voltage of 100 kV was used. The density of the dilatation band was calculated using the following equation (*22*):

$$\rho_{\text{craze}} = \rho_{\text{matrix}} \frac{\ln\left(\dfrac{I_{T,c}}{I_{0,c}}\right)}{\ln\left(\dfrac{I_{T,m}}{I_{0,m}}\right)} \tag{1}$$

where ρ is the density, I_T is the total spectral intensity, I_0 is the zero loss intensity, the subscript c indicates either the craze band or the dilatation band, and the subscript m indicates the bulk matrix. To ensure that our approach gives reasonable results, the same method was also used to measure the density of a craze band in a tensile-loaded polystyrene sample.

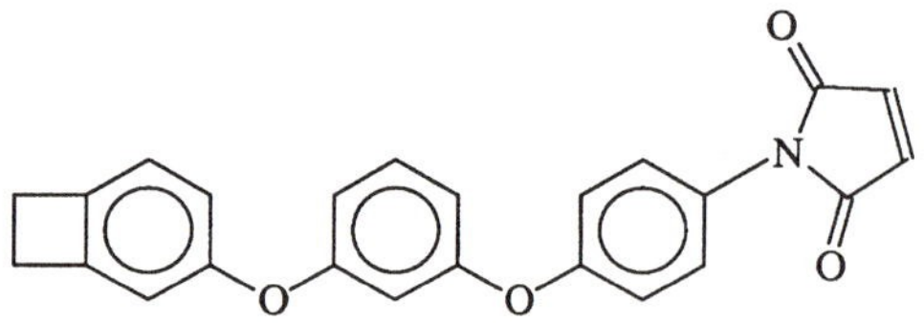

1,2-Dihydrobenzocyclobutene and maleimide (BCB–MI)

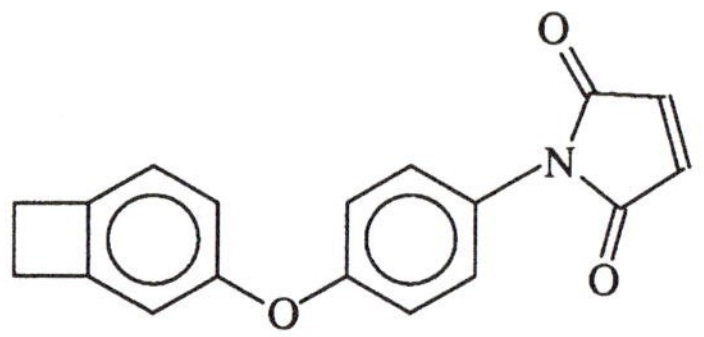

Modified BCB–MI (BCB–MI–M)

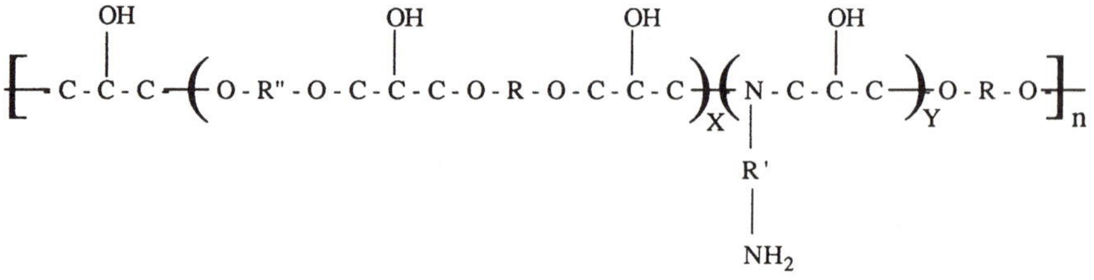

Cross-linkable epoxy thermoplastic (CET)

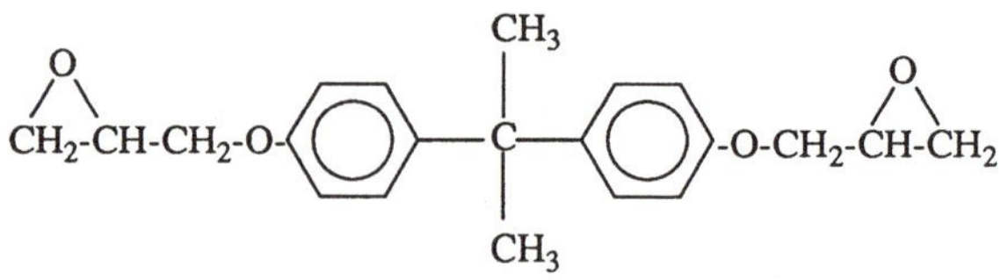

Diglycidyl ether of bisphenol A (DGEBA)

Chart **I**

Table I. Selected Properties of the Thermosets Investigated

Sample	T_g^a (°C)	G'^b (MPa)	M_c^c (g/mol)	Modulus (MPa)	K_{1c} (MPa·m$^{1/2}$)
BCB–MI	201	2.56	920	3520[d]	3.95[f]
BCB–MI–M	271	4.65	560	3380[d]	1.12
CET	145	3.50	730	3250[e]	1.00
DGEBA–piperidine	95	4.00	560	3250[e]	0.77

[a]The temperature at which the tan δ curve at a frequency of 1 Hz is a maximum; tan δ is the ratio of loss modulus to storage modulus.
[b]Rubbery plateau shear-storage modulus.
[c]Nielsen's equation is used (*19*).
[d]Flexural modulus.
[e]Tensile modulus.
[f]The plane-strain criterion is not satisfied.

Results and Discussion

The present work focuses on investigating the possible formation of crazing or dilatation bands in several moderately cross-linked, neat, engineering thermoset resins. Even though the thermoset resins investigated possess relatively low cross-link densities (M_c ranges from 560 to 920 g/mol), they have all the physical characteristics of ordinary thermosets (Table I). Except for the DGEBA–piperidine epoxy resin, all the resins are currently under evaluation for aerospace structural applications. This section discusses the conditions under which crazing and dilatation bands form, and describes the fracture behaviors of these high-performance resins.

BCB–MI Resin.   It has been previously indicated that the high-T_g and high-modulus BCB–MI resin has an extremely high fracture toughness (*14*). The source of this toughness remains an enigma. In order to find out how the BCB–MI resin fractures, DN–4PB experiments were conducted. As shown in Figure 2, the TOM image (taken in the plane-strain region) indicates that numerous dark dilatation bands, which scatter light effectively, have formed around the crack-tip damage zone. These bands may be caused by the formation of multiple cracks or crazes. To identify the features inside these dark bands, a TEM investigation was performed.

The TEM images obtained from the plane-strain crack-tip damage zone clearly demonstrate widespread and fully grown crazes (Figure 3). At the crack tip, the matrix is highly stretched, which leads to crack-tip blunting (see the arrow in Figure 3, top). A few micrometers away from the crack tip, fully grown crazes are observed (Figure 3, middle). These crazes exhibit sharp boundaries with the matrix and are bridged by fully grown craze fibrils. Some of the crazes also contain partially grown cracks (see arrows). When foreign inclusions, such as silicone rubber, are present, the crazes can be effectively triggered by these

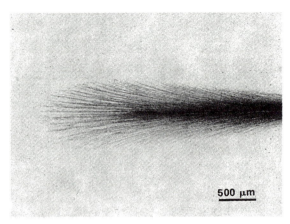

500 μm

Figure 2. Bright-field TOM image of BCB–MI resin, showing the plane-strain core region of the DN–4PB crack-tip damage zone. The crack propagates from right to left.

inclusions (Figure 3, bottom). This resembles the crazing phenomenon in rubber-toughened polystyrene (5). The crazes formed in BCB–MI have band thicknesses of up to about 1 μm and appear to have much coarser craze fibrils and larger damage zones than those observed in the polystyrene fracture. The reason these crazes can outgrow those in polystyrene is probably due to the moderately low cross-link density of the BCB–MI network (23). Consequently, BCB–MI has a much higher K_{Ic} than that of polystyrene (~3.95 MPa · m$^{0.5}$ vs. 0.85 MPa · m$^{0.5}$).

PEELS was utilized to measure the variation in density inside the craze band. Because the thin sections are prepared from the DN–4PB damage zone of a bulk sample, the thickness of the thin sections is believed to be approximately uniform throughout. Therefore, the variations in the measured PEELS spectra intensities must be due to local density differences instead of deviations in thickness.

When the PEELS measurement was conducted, an abrupt drop in density was observed at the interface between the matrix and the craze bands (Figure 4). In addition, a drop of approximately 50% in density was found at the base of the already unloaded craze band. This observation implies that an extension ratio of at least 2 exists for the craze fibrils. This phenomenon is not uncommon for thermoplastic crazes (5, 10). To ensure that the PEELS method gives reasonable results, the density of the craze band inside a polystyrene tensile specimen was measured (Figure 5) using the same sample-preparation procedures described in the section "Experimental Details." The measured density of the craze band in the unloaded polystyrene was found to be about 0.62 g/cm^3, which is in good agreement with the number reported in the literature (5, 10, 24).

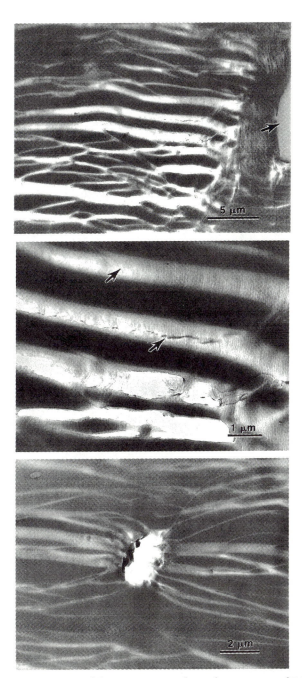

Figure 3. TEM images of the DN–4PB crack-tip damage zone of BCB–MI. The crack propagates from right to left. Top: In addition to extensive crazing in the damage zone, significant crack-tip blunting is observed (arrow). Middle: In a region a few micrometers ahead of the crack tip, craze fibrillation (arrows) is clearly observed. Bottom: Farther away from the crack tip but still inside the damage zone, crazes are nucleated and propagate from a silicone rubber inclusion, which was pulled out during the TEM thin-sectioning process.

PEELS Line Scan of BCB-MI

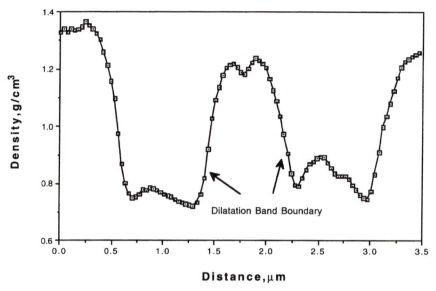

Figure 4. PEELS line scan of the craze bands observed in BCB–MI. A sharp transition in density is detected at the boundaries between the dilatation bands and the matrix. The slight increase in density at the center region of the bands may result from the stretched molecules inside the dilatation bands partially snapping back after unloading.

When the stress state was mainly biaxial (i.e., the plane-stress state), the crazing phenomenon in BCB–MI was suppressed. No signs of crazing were observed on the surface region of the crack-tip damage zone (Figure 6). This finding strongly suggests that the crazing phenomenon found in BCB–MI is not exactly the same as that observed in polystyrene, where crazes can form on the sample surface as well as in thin films (23). Therefore, we surmise that the BCB–MI is less prone to crazing than polystyrene. In other words, the critical crazing stress for BCB–MI may be slightly higher than its shear yielding stress, whereas for polystyrene, the crazing stress is known to be much lower than its shear yielding stress (25). When the material is under constrained conditions (e.g., a thick specimen with a sharp crack in it), the shear yielding stress component is suppressed (21). Instead, the crazing phenomenon is promoted. As a result, crazing prevails in the plane-strain region of BCB–MI.

BCB–MI–M Resin. The structure of BCB–MI–M is similar to that of BCB–MI resin except that the former has one less oxy-1,3-phenylene spacer. The alteration of this chemical structure results in a polymer with much

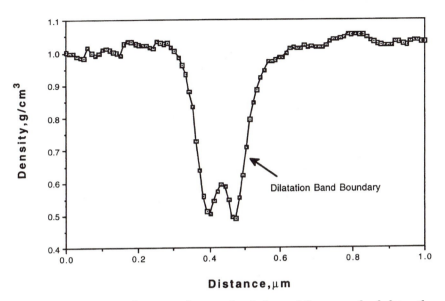

Figure 5. PEELS line scan of a craze band obtained from an unloaded, tensile-pulled polystyrene.

lower fracture toughness (K_{Ic} = 1.12 MPa · m$^{0.5}$) and a diminishing crack-tip damage zone (Figure 7). When this small damage zone was investigated using TEM, only a few dilatation bands were observed around the crack-tip region (Figure 8, top). No signs of crazing are found in this system. In this system, M_c ≈ 560 g/mol. At this relatively high level of cross-linking, it is surprising to observe the formation of dilatation bands, in particular for this high-T_g (271 °C), high-performance, thermoset resin (Table I). The K_{Ic} value of BCB–MI–M is more than double that of a comparable high-performance polycyanate system with similar thermal performance (26). Thus, the promotion of dilatation bands seems to be an attractive alternative route for toughening high-performance thermosets and should not be ignored.

The PEELS experiment was conducted to measure the density inside the unloaded dilatation band of BCB–MI–M. A density of about 1.17 g/cm^3, as opposed to 1.28 g/cm^3 for the bulk, was found at the base of the unloaded dilatation band. This value translates into an 8.5% drop in density. Since the drop in density in the dilatation bands is not as significant as that in BCB–MI or polystyrene, these dilatation bands cannot scatter light as effectively as the crazes. Consequently, the dilatation bands exhibit low contrast with respect to the surrounding matrix (Figure 7). This type of dilatation band could be a precursor

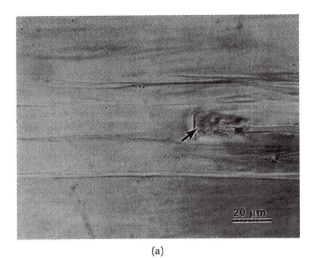

(a)

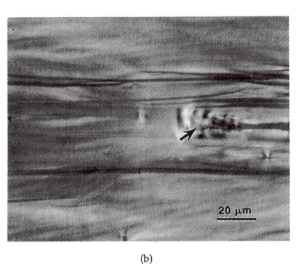

(b)

Figure 6. TOM images of the DN–4PB plane-stress damage zone of BCB–MI taken under bright field (a), and crossed polars (b). The arrow indicates the crack tip. The crack propagates from right to left.

of a fully grown craze. That is, if the BCB–MI–M were not so highly cross-linked with respect to BCB–MI, these dilatation bands would have grown into crazes.

CET Resin.   The neat CET resin has an M_c value of about 730 g/mol. This value indicates that the CET resin may be capable of forming either craz-

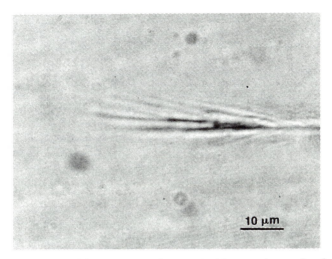

Figure 7. Bright-field TOM image of BCB–MI–M resin taken in the plane-strain core region of the DN–4PB crack-tip damage zone. The crack propagates from right to left.

ing or dilatation bands. As expected, when the neat CET was fractured using a DN–4PB specimen, a limited number of dilatation bands were found around the crack tip (Figure 8, middle). These dilatation bands resemble those of BCB–MI–M (Figure 8, top). The K_{Ic} of this system is also close to that of BCB–MI–M (Table I).

It is interesting that, upon rubber modification, the CET resin matrix can no longer form dilatation bands (*18*). Only rubber-particle cavitation and matrix shear yielding are detected. This observation implies that a dilatational stress component is required to trigger the formation of dilatation bands. In other words, upon rubber-particle cavitation, the dilatational stress component in the matrix is reduced. This suppresses the formation of dilatation bands. This conjecture finds support in the work of Glad (*27*), who investigated thin-film deformation of epoxy resins with various cross-link densities and could not find any signs of dilatation bands in his study.

DGEBA–Piperidine. The DGEBA–piperidine system has an M_c value of about 560 g/mol. Therefore, this system is likely to form dilatation bands. Indeed, when the DN–4PB and PEELS experiments were conducted, the dilatation bands were found around the crack tip (Figure 8, bottom). The density inside the dilatation bands was about 1.08 g/cm³, which is about 7% lower than that of the surrounding matrix. This dilatation band feature resembles those of the BCB–MI–M and CET resins.

The DGEBA–piperidine investigated here is a commonly studied epoxy

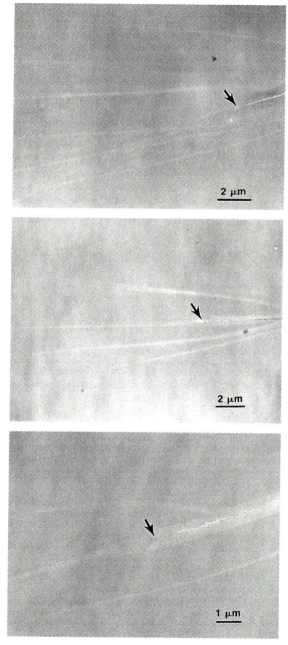

Figure 8. TEM images of the crack tip (arrow) of the DN–4PB damage zone of BCB–MI–M (top), CET (middle), and DGEBA–piperidine (bottom). The crack propagates from right to left in all three images.

system (*6, 8, 11, 28*). It has been repeatedly shown that this system is ductile in tension but becomes brittle when under mode I fracture. Only the so-called "furrow" or "hackle" region due to crack arrest is found on the fracture surface (*29*). This furrow region is thought to be caused by localized plastic deformation prior to the crack jump. When the DGEBA–piperidine system is toughened by rubber particles, the K_{Ic} can be improved severalfold (*6*). In this case, the toughening is created by rubber-particle cavitation followed by shear yielding of the epoxy matrix. When glass spheres, glass fibers, and other rigid fillers are utilized to toughen the epoxy, only crack bridging, crack pinning, and crack deflection are reported as the major toughening mechanisms (*30, 31*). There has not been any reported data indicating the formation of dilatation bands in this epoxy system, either with or without toughening agent present. This fact clearly demonstrates the usefulness of the present experimental approach— that is, the DN–4PB experiment and careful microscopic observations—for the definitive study of fracture mechanisms in polymers.

The present study makes it clear that crazing and dilatation bands, although surprising, can take place in a variety of engineering thermosets. However, not all thermosets, under all conditions, can undergo crazing and dilatation-band formation (*21, 27*). It appears that crazing or dilatation bands can form when M_c is 560 g/mol or greater. The stress state of the material also needs to be highly triaxial (*27*). Other important factors not addressed in this study that may cause crazing and dilatation-band formation include heterogeneity of the thermoset network, flexibility and mobility of the molecular backbone chain, and the toughening particles utilized (*32*). They will be discussed elsewhere (*33*).

Importance of the Findings. The present work indicates that the formation of crazing and dilatation bands in moderately cross-linked engineering thermosets can be quite common. Therefore, it is important to determine whether crazing and dilatation bands are effective in toughening thermosets. It is equally important to find out how to promote these types of fracture mechanisms, either via modification of network architecture or incorporation of toughening agents.

The present study shows that if massive crazing takes place in engineering thermosets, the resulting materials are tougher than rubber-modified thermosets (*6, 14, 18, 26*). If dilatation bands can extensively form around the crack-tip region, the toughening effect may also become significant (*32*). Moreover, the most recent work by Sue et al. (*34*) indicates that it is possible to further promote dilatation-band formation by adding appropriate rigid toughener particles. About a twofold improvement in K_{Ic} can be achieved. Toughening thermosets via crazing and dilatation bands should therefore not be overlooked. These mechanisms provide effective alternative routes to the toughening of high-performance thermosets.

In this regard, because most high-performance thermosets are used in applications such as composites, adhesives, and thick structural parts, the matrix is inevitably under high triaxial stress (35). If the modulus, yield stress, and creep resistance are not to be compromised, the most likely route for effective toughening of high-performance thermosets would be via either crazing or dilatation-band formation. This assertion is based on the fact that the most effective energy-absorbing fracture mechanism is plastic deformation of the matrix, which can occur by shear yielding or by plastic dilatation. For shear yielding to take place in a highly constrained environment, for example, in composite form, in adhesive form, or at the crack tip, the hydrostatic-tension component must be relieved in the matrix (6, 11, 21). This will generally lead to the compromise of using rubber modification. A better alternative is to use plastic dilatation, such as crazing or dilatation bands, to toughen the matrix. It is therefore important to further understand and promote the formation of craze and dilatation bands in engineering thermosets. It is hoped that new engineering thermosets can evolve to meet the demands of the aerospace and automotive industries.

Conclusions

The fracture behavior of a variety of moderately cross-linked engineering thermosets shows that either crazing or dilatation bands are formed in the damage zone upon mode I fracture. The toughening effect due to these rather unusual mechanisms, if properly promoted, can reach that due to shear banding. The present study suggests that when M_c is about 560 g/mol or higher and the stress state is highly triaxial, it is likely that the thermoset will undergo either crazing or dilatation-band formation. Although it is still uncertain how to best optimize crazing and dilatation-band formation in thermosets, it is clear that this method of toughening thermoset polymers will be best suited for applications such as composites, adhesives, and other structural parts.

Acknowledgments

We thank M. T. Bishop, C. C. Garrison, L. L. Walker, E. T. Vreeland, K. J. Bruza, C. E. Allen, L. J. Laursen, R. D. Peffley, W. R. Howell, and L. M. Kroposki for their discussion and support of this work. We also thank W. Bonfield for permitting the use of work previously published in *J. Mater. Sci. Letter.*

References

1. Sultan, J.; McGarry, F. *Polym. Eng. Sci.* **1973,** *13,* 19.
2. Morgan, R. J.; Mones, E. T.; Steele, W. J. *Polymer* **1982,** *23,* 295.
3. Lilley, J.; Holloway, D. G. *Philos. Mag.* **1973,** *28,* 215.
4. Bucknall, C. B.; Yoshii, T. *Br. Polym. J.* **1978,** *10,* 53.

5. Bucknall, C. B. *Toughened Plastics;* Applied Science: London, 1977.

6. Yee, A. F.; Pearson, R. A. *J. Mater. Sci.* **1986,** *21,* 2462, 2475.

7. Garg, A. C.; Mai, Y. W. *Composites Sci. Technol.* **1988,** *31,* 179.

8. Kinloch, A. J. In *Advances in Polymer Science;* Dusek, K., Ed.; Springer-Verlag: Berlin, 1986; Vol. 72, p. 45.

9. Narisawa, I.; Murayama T.; Ogawa, H. *Polymer* **1982,** *23,* 291.

10. Kambour, R. P. *J. Polym. Sci.: Macromol. Rev.* **1973,** *7,* 1.

11. Sue, H.-J. *J. Mater. Sci.* **1992,** *27,* 3098.

12. Bertram, J.L.; et al. U.S. Patent 4,594,291, 1986.

13. Dewhirst, K. C. U.S. Patent 4,786,668, 1988.

14. Kirchhoff, R. A.; Bruza, K. J. In *Progress in Polymer Science;* Corley, L. S., Ed.; Pergamon: London, 1993; Vol. 18, p. 85.

15. Henton, D. E.; Pickelman, D. M.; et al. U.S. Patent 4,778,851, 1988.

16. Yang, P. C.; Pickelman, D. M. U.S. Patent 4,894,414, 1990.

17. Sue, H.-J.; Yee, A. F. *J. Mater. Sci.* **1993,** *28,* 2975.

18. Sue, H.-J.; Bertram, J. L.; Garcia-Meitin, E. I.; Walker, L. L. *Colloid Polym. Sci.* **1994,** *272,* 456.

19. Nielsen, L. E. *J. J. Macromol. Sci.* **1969,** *C3,* 69.

20. Sue, H.-J.; Burton, B. L.; Garcia-Meitin, E. I.; Garrison, C. C. *J. Polym. Sci., Polym. Phys. Ed.* **1991,** *29,* 1623.

21. Sue, H.-J.; Garcia-Meitin, E. I.; Pickelman, D. M. In *Elastomer Technology Handbook;* Cheremisinoff, N. P., Ed.; CRC: Boca Raton, FL, 1993; Chapter 18.

22. Egerton, R. F. *Electron Energy Loss Spectroscopy in the Electron Microscope;* Plenum: New York, 1986, p. 291.

23. Henkee, C. S.; Kramer, E. J. *J. Polym. Sci., Polym. Phys. Ed.* **1984,** *22,* 721.

24. Kambour, R. P. *J. Polym. Sci. A* **1964,** *2,* 4159.

25. Whitney, W.; Andrews, R. D. *J. Polym. Sci. C* **1967,** *16,* 2981.

26. Yang, P. C.; Pickelman, D. M.; Woo, E. P.; Sue, H.-J. *36th International SAMPE Conference Proceedings;* San Diego, CA, April 1991.

27. Glad, M. D. Ph.D. Thesis, Cornell University, Ithaca, NY, 1986.

28. Levita, G. In *Rubber-Toughened Plastics;* Riew, C. K., Ed.; Advances in Chemistry 222; American Chemical Society: Washington, DC, 1989; Chapter 4, pp. 93–118.

29. Vakil, U. M.; Martin, G. C. *J. Mater. Sci.* **1993,** *28,* 4442.

30. Kinloch, A. J.; Maxwell, D. L.; Young, R. J. *J. Mater. Sci.* **1985,** *20,* 4169.

31. Low, I.-M.; Mai, Y.-W.; Bandyopadhayay, S.; Silva, V. M. *Mater. Forum* **1987,** *10,* 241.

32. Sue, H.-J.; Yang, P. C.; Bishop, M. T.; Garcia-Meitin, E. I. *J. Mater. Sci. Lett.* **1993,** *12,* 1463.

33. Sue, H.-J.; Bertram, J. L. *Polym. Eng. Sci.* to be submitted.

34. Sue, H.-J.; Yang, P. C.; Bishop, M. T.; Garcia-Meitin, E. I. *Polym. Mater. Sci. Eng.; Proc. Am. Chem. Soc. Div. Polym. Mater. Sci. Eng.* **1994,** *70,* 256.

35. Shi, Y.-B.; Chen, L. P.; Yee, A. F.; Sue, H.-J. *Proceedings of the International Symposium on Polymer Alloys and Composites;* 1992; p. 145.

Determination and Use of Phase Diagrams for Rubber- or Thermoplastic-Modified Poly(cyanurate) Networks

ZhiQiang Cao, Françoise Mechin, and Jean-Pierre Pascault*

Laboratoire des Matériaux Macromoléculaires, UMR CNRS 5627,
Institut National des Sciences Appliquées de Lyon, Bât. 403
20, Avenue A. Einstein, 69621 Villeurbanne Cedex, France

Blends of two cyanate ester monomers [4,4'-dicyanato-1,1-diphenylethane (DPEDC) or 4,4'-dicyanato-2,2-diphenylpropane (DCBA)] with several initially miscible reinforcing additives were studied as a function of cyanate conversion. The phase diagrams (temperature vs. conversion for the different transitions: phase separation, gelation, and vitrification) provide a good overview of the systems and allow easy comparisons. DCBA is a better solvent than DPEDC. Rubber systems based on butadiene–acrylonitrile have an upper critical solubility temperature (UCST) behavior, whereas poly(ether sulfone)s (PESs) induce a lower critical solubility temperature (LCST) behavior. The acrylonitrile content of the rubbers and the molar mass of the PES additives also have a great influence on their miscibility. During an isothermal cure, phase separation always occurs before vitrification; in rubber it generally occurs before gelation, and in PES it occurs together with gelation. The temperature and viscosity at which phase separation occurs are critical for the final morphology. Reactive additives accelerate the curing process and modify this morphology by inducing a complex matrix–particle interface, and a substructure inside the dispersed particles; these modifications produce the best toughening effects.

B RITTLE THERMOSETS ARE BEST TOUGHENED by the introduction of a rubbery or thermoplastic dispersed phase (*1–4*). The dispersed phase can be produced by two methods: in situ reaction and adding preformed particles. The first method is used much more because it is easy and it can create specif-

---

*Corresponding author.

0-8412-3151-6

ic morphologies. In the first method, the initial mixture of monomer(s) and additive is homogeneous at the isothermal curing temperature, T_i, and the dispersed phase is obtained by the phase separation induced by polycondensation of the monomer(s).

The main factors that determine the degree of toughening are the final polymer blend morphology and the adhesion between the two phases. For initially miscible reactive systems, the first factor (including diameters, number, and volume fraction of the dispersed spherical particles) is determined by the competition between the rates of phase separation and polycondensation. Adhesion between the two phases depends on the chemical and physical properties of both the additive and the monomer (5–8). The phase-separation process induced by the reaction and the formation of the morphology are intricate phenomena; for thermosetting systems, their study is focused mainly on rubber-toughened polyepoxy matrices (1, 3). An analysis of the experimental results leads to the following conclusions:

1. The concentration of dispersed particles decreases with T_i.
2. The volume fraction of dispersed phase, V_D, remains practically constant, goes through a maximum, or decreases with T_i.
3. The number-average diameter, $\bar{D}$, of the spherical particles increases with T_i.
4. $\bar{D}$ and V_D increase with the initial volume fraction of the additive, $\phi_{a,0}$.
5. But V_D is always greater than $\phi_{a,0}$. This fact means that the dispersed phase is not formed only of pure rubber additive, R, but also contains some epoxy–amine copolymer. It can be about 50 wt%. A second phase separation has been demonstrated to occur inside the dispersed particles. (9)

On the other hand, the initial miscibility of the monomer(s) with the additive has a great influence on both the phase separation and the final morphologies. Many experimental results (5–10) show that phase separation occurs well before gelation time, t_{gel}, or conversion, x_{gel}, and before vitrification (t_{vit}, x_{vit}). The better the initial miscibility, the higher the conversion at the cloud point, x_{cp}, where the phase separation induced by polycondensation occurs at a given value of T_i. Initial miscibility and T_i control the "chemical quench". The viscosity at the cloud point, η_{cp}, obviously increases with the conversion, x_{cp}. The viscosity, η_{cp}, can affect the nucleation and growth of the dispersed-phase particles and consequently their average diameter (6, 11), and sometimes also the shape of the particles, which is not necessarily spherical.

Cyanate ester (CE) resins are the key monomers for a new type of high-performance polymer. They were developed by Hi-Tek Polymers during the 1980s, then Rhône-Poulenc, and now Ciba-Geigy. The polycyclotrimerization of cyanates can take place by simple heating, or it can be catalyzed by transition metal cations together with labile hydrogen compounds, phenols, alcohols, or amines (*12*).

Like epoxy systems, CE systems can be toughened with (*1*) engineering thermoplastics (*13–18*), including polysulfones or poly(ether sulfone)s, polyimides, and polyesters or polyarylates, and (*2*) rubbers, including butadiene–acrylonitrile copolymers (*13*), polysiloxanes (*14*), and preformed particles (*15, 16*). We can conclude that the different concepts, rubber- or thermoplastic-modified thermosets, work as well with CE as with epoxy. But unlike epoxy, only a few results in the literature give relations between the initial miscibility of monomer and additive, the curing conditions, and the morphologies and properties. This chapter examines the influence of rubber and thermoplastic additives (nature and end groups) on polymerization kinetics and, mainly, on the evolution of phase diagrams. Work on morphologies and properties is examined elsewhere (*19*).

Experimental Methods

Monomers. The CE monomers, 4,4′-dicyanato-1,1-diphenylethane (DPEDC) and dicyanate of bisphenol A (DCBA), were provided by Ciba-Geigy (Louisville, KY) in reference Arocy L10 and Arocy B10, and were used as received. Liquid DPEDC contains 2–3% impurities (trimer, monophenol–monocyanate, and ortho–para-substituted isomers). DCBA is a high-purity (>99.5%) crystalline powder (melting temperature 79 °C). The chemical structures of these monomers are shown in Chart I.

Additives. Acrylonitrile–butadiene rubbers were provided by BFGoodrich (Brecksville, OH). The amino-terminated butadiene–acrylonitrile (ATBN) rubber was obtained by reacting carboxyl-terminated butadiene–acrylonitrile (CTBN) with an excess diamine, Unilink 4200 (from UOP, El Dorado Hills, CA); consequently, free diamine molecules always remained in the rubber. The rubbers have almost the same molar mass but different end groups, which have been characterized in a previous work (*20*). Their structures are given in Chart I, and they are described in Table I. The two poly(ether sulfone)s (PESs) (Victrex, from ICI, United Kingdom) used in this study are described in Table II.

The two PES additives have different molar masses and different concentrations of phenolic end groups. By ^{1}H-NMR spectroscopy we have estimated that PES4100P is an unreactive oligomer, in contrast to PES5003P, which has approximately one OH per molecule.

Preparation of the Blends. The CE blends, which contain 15 wt% additive, were prepared by manual stirring at room temperature for rubber additives, and with dichloromethane as a solvent for PES additives.

Polymerization Reaction. The differential scanning calorimetry (DSC) pan, containing 10–15 mg of sample, was sealed under air. It was then

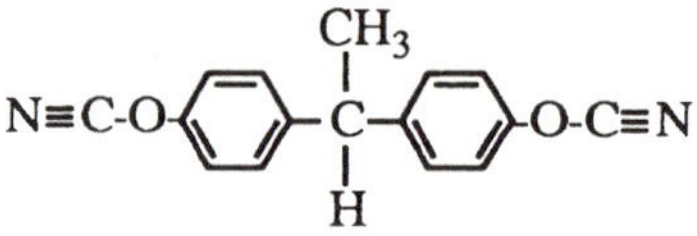

DPEDC, 264 g/mol

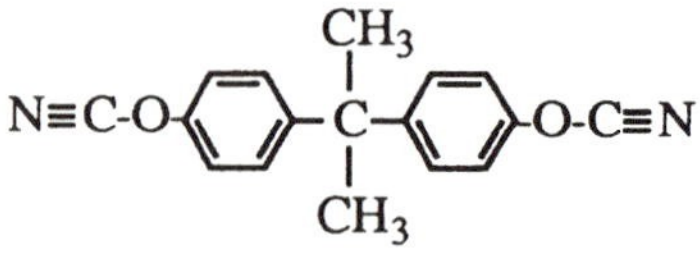

DCBA, 278 g/mol

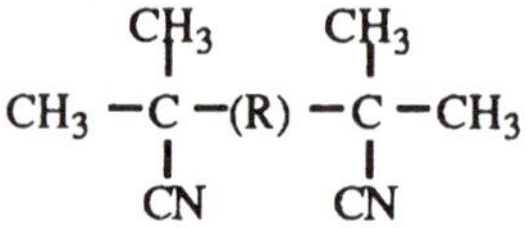

NFBN

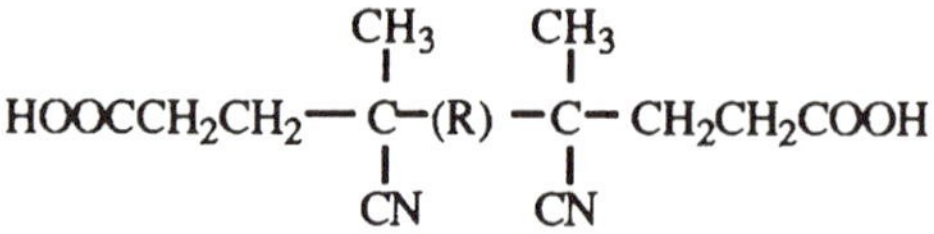

CTBN

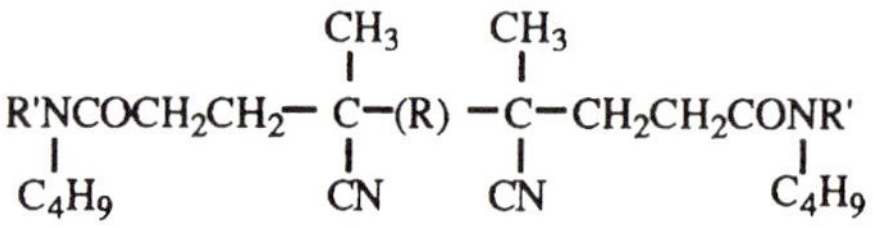

ATBN

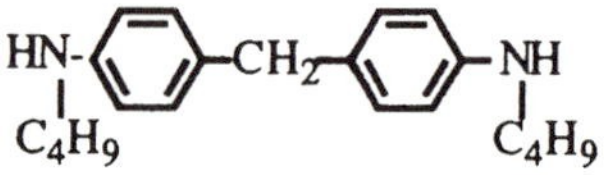

Unilink 4200

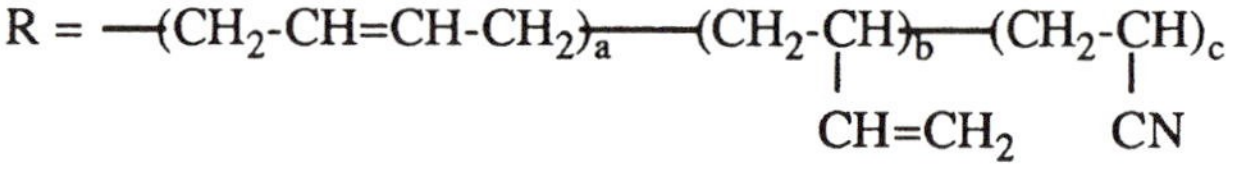

R = —(CH$_2$-CH=CH-CH$_2$)$_a$——(CH$_2$-CH)$_b$—(CH$_2$-CH)$_c$

CH=CH$_2$ CN

R

R' = —⟨ ⟩— CH$_2$ —⟨ ⟩— NHC$_4$H$_9$

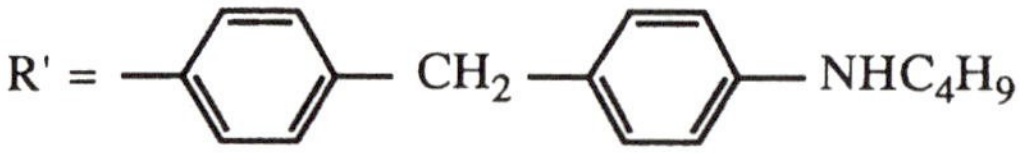

R'

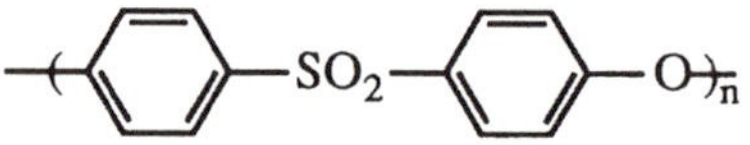

PES additive

Chart I. Chemical structures and characteristics of dicyanate monomers, elastomer additives, and PES additives.

Table I. Characteristics of the Elastomer Additives

Characteristic	NFBN	CTBN	ATBN	Unilink 4200
Functionality	0	1.8 (COOH)	1.9 (NHR')	2.0 (NHC$_4$H$_9$)
% AN	17.2	18.0	18.0	0
$\overline{M}_n$ (g/mol)	3600	3600	3600 + 10% excess Unilink 4200	310
$\overline{M}_w/\overline{M}_n$	2.1[a]	1.9[a]	2.8 (polymer peak)[a]	1.0
T_g (°C)	−67	−60	−47	—[b]

NOTE: Chemical structures are shown in Chart **I**.
[a]From size-exclusion chromatography measurements (polystyrene standards).
[b]Not measured.

placed in the oven at 180 °C, and the reaction allowed to proceed. In order to measure the temperature at the cloud point, T_{cp}, during polymerization, about 2 g of sample in a test tube was isothermally polymerized *in situ* inside the light-transmission device (*21*).

Cure of Plates. In an oven, ~240 g of blend in a 6-mm-thick mold was precured at 180, 150, or 120 °C for a time necessary to arrest the phase separation, and then it was postcured with the same cycle at higher temperatures until completion of the polycondensation of CE monomer without any degradation. The evolution of the real temperature of the plate was followed by use of a thermocouple.

Analytical Techniques. The cloud point of the blends was determined with a light-transmission device (*21*). Once the blend was cloudy, the test tube was taken out and chilled in ice, so that the time and conversion at the cloud point, t_{cp} and x_{cp}, could be obtained. The T_g value and conversion were measured by DSC (Mettler TA3000 microcalorimeter) (*22*). The gel time, t_{gel}, of rubber–cyanate blends was determined as the time at which insolubles appeared in tetrahydrofuran (THF). That of PES–cyanate was determined by dynamic mechanical analysis (Rheometrics RDA700).

We observed the fracture surfaces using a scanning electron microscope (SEM, JEOL 840A) after coating them with a gold sputterer. Transmission electron micrographs (TEMs) were obtained by using a JEOL 1200 device for ultrami-

Table II. Characteristics of the PES Additives

Reference	$\overline{M}_n{}^a$	$\overline{M}_w/\overline{M}_n$	T_g (°C)[b]
PES4100P	15,200	2.7	216
PES5003P	21,900	2.1	222

NOTE: The chemical structure of PES is shown in Chart **I**.
[a]From size-exclusion chromatography measurements in dimethylformamide at 80 °C.
[b]From DSC measurements.

crotomed samples. Using a procedure previously described (23), we calculated the following parameters:

$$\text{number-average diameter:} \bar{D} = \frac{\Sigma n D}{\Sigma n}$$

$$\text{volume fraction of the dispersed phase: } V_D = (\pi/4)\frac{\Sigma n D^2}{A_T}$$

where A_T is the total area of the micrograph region under analysis.

Tensile and compressive tests were performed at room temperature with a tensile testing machine (DY25, Adamel-Lhomargy). For the tensile tests, strain measurements were performed with an extensometer (EX-10) at a strain rate of 3.3×10^{-4} s^{-1}, using ISO-60 standard specimens. Samples of dimension $20 \times 12 \times 6$ mm were deformed in a compression cage between polished steel plates. The nominal strain was determined by averaging the results from two linear variable differential transformer (LVDT) transducers. The strain rate used was 8.3×10^{-4} s^{-1}.

Fracture toughness was studied with opening mode I tests and in a plane strain rate. The samples were loaded to failure using a crosshead speed of 1 mm/min (5.2×10^{-4} s^{-1}), at room temperature. Values for the critical stress intensity factor, K_{Ic}, were calculated by using the following formula:

$$K_{Ic} = \sigma_c \sqrt{\pi a}\ Y\left(\frac{a}{w}\right) \tag{1}$$

where σ_c is the critical fracture stress, a is the crack length, and w is the specimen width. The correction factor is given by:

$$Y\left(\frac{a}{w}\right) = 1.09 - 1.735\frac{a}{w} + 8.2\left(\frac{a}{w}\right)^2 - 14.18\left(\frac{a}{w}\right)^3 + 14.57\left(\frac{a}{w}\right)^4 \tag{2}$$

Once K_{Ic} was determined, the fracture energy, G_{Ic}, was calculated with Young's modulus and the Poisson coefficient, which were determined experimentally for every formulation following usual procedures.

Results and Discussion

Initial Miscibility of the Blends.   In Figure 1, we have plotted the cloud-point curves obtained with the following substances:

1. two rubbers with the same percentage of acrylonitrile (AN):
   - a nonfunctional rubber, nonfunctional butadiene–acrylonitrile (NFBN)
   - a carboxy-terminated rubber, CTBN
2. two different reactive monomers:
   - a CE, DPEDC
   - bisphenol A diglycidyl ether, DGEBA (DER 332 from Dow)

All the blends have an upper critical solubility temperature (UCST) behavior. It is evident that the CE monomer is a better solvent of rubbers than the epoxy prepolymer.

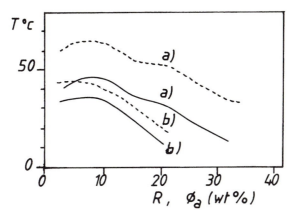

Figure 1. Initial cloud-point curves for rubber–DGEBA (DER 332) (- - - -) and rubber–DPEDC (Arocy L10) (———), with NFBN rubber (a) and CTBN rubber (b). R is rubber, and ϕ_a is mass fraction.

The NFBN rubber is less miscible than the CTBN rubber. An amino-terminated rubber, ATBN8, with the same amount of AN, 18 wt%, is quite miscible with DPEDC or DGEBA, showing that chain ends of reactive liquid oligomers can have a strong effect on their miscibility (20). Finally, by decreasing the percentage of AN, the miscibility window decreases drastically (21). For example, ATBN31, with only 10 wt% AN, is quite immiscible with DPEDC and DGEBA at temperatures lower than 100 °C.

PES is miscible with DGEBA or DPEDC at room temperature, and PES–DGEBA blends have a lower critical solubility temperature (LCST) behavior (Figure 2). But because of CE reactions at temperatures greater than 100 °C, the comparison between CE and DGEBA is not possible.

Polymerization. To plot the evolution of the phase diagram, it is necessary to have a good knowledge of the reaction rates and conversions at different isothermal curing temperatures. These kinetic parameters were determined in a previous study (22), using T_i = 180 °C for systems based on DPEDC, and T_i = 200 °C for those based on DCBA (which is less reactive). Our experiments revealed that in the first part of the curing process, that is, as long as it remained kinetically controlled, the unreactive additives (with respect to the cyanate functions, i.e., NFBN, CTBN, and PES4100P) did not modify the overall kinetics compared with the neat monomer. Systems based on DCBA could be modeled (24) by the following equation:

$$\frac{dx}{dt} = k_1(1-x)^2 + k_2x(1-x)^2 \qquad (3)$$
$$\underset{\text{second order}}{\hphantom{xxxxx}} \quad \underset{\text{second order autocatalyzed}}{\hphantom{xxxxxxxxxxx}}$$

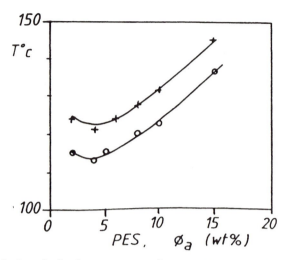

Figure 2. Initial cloud-point curves for PES–DGEBA (DER 332). Key: +, PES4100P; and ○, PES5003P.

Equation 3 is often used in the literature to account for the kinetic behavior of several common dicyanate monomers (*25, 26*). The autocatalytic character of cyanate polycyclotrimerization is now quite well established; we noticed that it was especially enhanced in samples with a large surface directly exposed to air (*24, 27, 28*). The kinetics of DPEDC blends were more difficult to describe because they were perturbed by several residual impurities in the monomer. Together with adventitious air moisture, the latter increased gel conversion with respect to the theoretical value (0.5).

In contrast to these unreactive modifiers, the additives that were likely to react with cyanates (ATBN via its secondary amine function, and PES5003P via its phenol function) were shown to appreciably accelerate their cure, even in its very first steps. Phenols (*12*) usually react with cyanates to yield iminocarbonates; the latter are generally proposed as intermediates in the polycyclotrimerization mechanism:

$$Ar\text{–}OH + Ar'\text{–}O\text{–}C\equiv N \rightarrow Ar\text{–}O\text{–}\underset{\substack{\| \\ NH}}{C}\text{–}O\text{–}Ar'$$

$$Ar\text{–}O\text{–}\underset{\substack{\| \\ NH}}{C}\text{–}O\text{–}Ar' + 2Ar'\text{–}O\text{–}C\equiv N \rightarrow Trimers + xAr\text{–}OH + yAr'\text{–}OH$$

In our group (*29*), experiments run with model monofunctional phenols confirmed these possible first steps. The same kind of studies (*29*) were carried out with a monofunctional aromatic amine and suggested that in ATBN, the following reactions could presumably be involved:

$$R-\langle\bigcirc\rangle-\underset{\underset{C_4H_9}{|}}{N}-H + AR-O-C\equiv N \xrightarrow{\text{low } T} R-\langle\bigcirc\rangle-\underset{\underset{C_4H_9}{|}}{N}-\overset{\overset{\displaystyle NH}{\parallel}}{C}-O-AR \xrightarrow{\text{high } T} Ar-OH + R-\langle\bigcirc\rangle-\underset{\underset{C_4H_9}{|}}{N}-C\equiv N$$

$$Ar-OH + Ar-O-C\equiv N \rightarrow \cdots\cdots$$

These reactions could explain the reaction activation observed for the blends of DPEDC with PES5003P and ATBN, especially as the latter also contains excess free diamine. Considering the 10% residual Unilink 4200 in ATBN, we determined the reaction rate associated with a 1.74% Unilink 4200–DPEDC blend, which simulates the concentration of the excess free Unilink 4200 in a 15-wt% ATBN8–DPEDC blend. The results are not presented here, but in the presence of Unilink 4200, DPEDC reacted faster, in the same range of reactivity as ATBN–Unilink (22).

All the rubber or thermoplastic-modified cyanates studied were initially miscible at the curing temperature, T_i. At a certain conversion, phase separation occurred as the molar mass of the cyanate oligomers increased and more-polar cyanate functional groups were transformed into less-polar cyanurate rings. The phase-separation phenomena could be observed simply by the fact that the solution became cloudy (cloud point). We measured the conversions at which the isothermally cured blends became cloudy (x_{cp}) and the gel conversions (x_{gel}); the x_{gel} values were practically the same for neat systems and blends, considering experimental error. For all the blends, phase separation occurred well before gelation. Therefore, we conclude that in these cases phase separation did not disturb the polymerization kinetics or the network buildup.

For PES4100P and PES5003P, the conversions at the cloud point were respectively 0.44 and 0.40 for T_i = 180 °C. Once again, we conclude that phase separation is unlikely to influence the polycyclotrimerization kinetics.

Finally, we noticed that in the last stages of the reaction, in contrast to the neat systems for which vitrification limited the conversion to about 0.9 at T_i = 180 °C, the conversion of all the rubber-modified systems easily reached 0.95 or more, a result indicating that the vitrification effect was delayed and suggesting that part of the rubber dissolved in the matrix and decreased its $T_g(t)$. The same conclusion could be drawn from the glass-transition temperatures of the final networks ($T_{g\infty}$), which are displayed in Table III. Such an effect was not observed for the Unilink 4200–DPEDC mixture (without rubber).

Moreover, the maximum conversion of the 15-wt% PES5003P–DPEDC blend was lower than that of the 15-wt% PES4100P–DPEDC blend. This difference was attributed to the possible reactions between additive chain ends

Table III. Phase-Separation Process, Morphologies, and Properties for Neat or Modified Poly(cyanurate) Networks

Reactive System	x_{cp}	x_{gel}	η_{cp}, Pa·s	$\bar{D}$, μm	$T_{g\infty}$, °C	K_{Ic}, $MPa\sqrt{m}$	G_{Ic}, J/m^2	E,[a] GPa	σ_y,[b] MPa
Neat DPEDC	—[c]	0.63	—	—	268	0.71	148	3.00	138
1.74 wt% Unilink–DPEDC	—	—	—	—	263	0.82	250	2.40	136
15 wt% NFBN–DPEDC	0.32	0.61	0.015	7.2±0.5	260	0.87	470	1.43	68
15 wt% ATBN8–DPEDC	0.47	0.61	0.2	1.8±0.2	248	1.40	840	2.00	100
15 wt% PES5003P–DPEDC	0.40	0.59	1.0	1.1±0.2	246	1.06	390	2.55	128

NOTE: Blends were precured at 180 °C for 2 h and postcured at 260 °C.
[a] E, Young's modulus.
[b] σ_y, yield stress.
[c] Not measured.

and cyanate monomer: the mass fraction of PES5003P dissolved in the matrix after phase separation may be higher than that of PES4100P; thus the vitrification of the former occurs earlier.

Evolution of Phase Diagrams. The increase in the molar mass of the cyanate ester with conversion obviously modifies the phase diagram of blends. The evolution of phase diagrams of nonfunctional rubber was first studied.

At $T_i = 180$ °C for NFBN–DPEDC and at $T_i = 200$ °C for NFBN–DCBA (isothermal polymerization), the conversions, x_{cp}, where blends with different fractions of additive became cloudy, were measured. Figure 3 shows that the evolution of x_{cp} with the additive fraction is weak, and x_{cp} is always lower than the gel conversion of the cyanates, about 0.6. For a polymer blend with two components, the free energy of mixing per unit volume of blend can be expressed by the Flory–Huggins equation. Elsewhere *(30)*, we estimate the Flory–Huggins interaction parameter χ by fitting the experimental points.

For a given fraction of additive, the T versus x phase diagrams would be more useful for practical applications and for better control of the phase-separation process. In these phase diagrams, it is possible to plot the three transformations together: phase separation, T_{cp} versus x_{cp}; gelation, $x_{gel} = 0.6$; and vitrification, T_g versus x.

A 15-wt% NFBN–DPEDC blend at $T_i = 180$ °C, and a 15-wt% NFBN–DCBA blend at 200 °C were isothermally polymerized until a certain conversion, then cooled until the blends became cloudy. (Cooling was achieved by simply stopping the heating of the oven in the light-transmission device. The

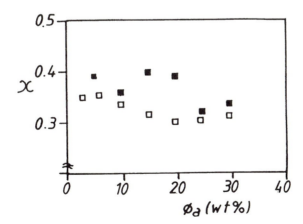

Figure 3. Experimental phase diagram showing conversion (x) versus the mass fraction of rubber (ϕ_a). The rubber is NFBN. Key: □, DPEDC, $T_i = 180$ °C; and ■, DCBA, $T_i = 200$°C.

cooling rate depends on the temperature of the oven.) In this way, T_{cp} and the corresponding x_{cp} were obtained. Figure 4 shows that the increase in T_{cp} seems to be linear with respect to x (in the range of T_i studied, because in fact it is an ascending concave curve), which agrees with the UCST-type initial phase diagrams of the blends. On the other hand, at the same T_{cp}, the x_{cp} of NFBN–DCBA is always higher than that of NFBN–DPEDC, which means that DCBA is a better solvent for NFBN than DPEDC.

Furthermore, the evolution of T_g for the continuous phase of the blends can be plotted in the same figure for both the 15-wt% NFBN–DPEDC and 15-wt% NFBN–DCBA blends (Figure 4). It appears that at a given x, T_{cp} is always higher than T_g. In other words, during an isothermal cure, phase separation occurs before vitrification of the blends. Figure 4 also shows that x_{cp} is always lower than x_{gel} for $T < 220$ °C.

To study the influence of rubber reactivity on phase separation, we also measured the phase diagrams of the blends with rubbers ATBN8 and CTBN. As in the 15-wt% NFBN blend, for the blends with ATBN8 and CTBN, the overall T versus conversion phase diagrams, including phase separation (T_{cp}, x_{cp}), gelation ($x_{gel} \approx 0.6$), and vitrification (T_g, x) can be plotted (Figure 5). At one isothermal curing temperature, T_i, it appears that x_{cp} is in the order ATBN8 > CTBN > NFBN. This result is consistent with the order of the initial miscibilities of the rubbers with DPEDC.

To isolate these different effects, we did the same experiments with 15-wt% ATBN rubber by replacing small amounts of ATBN8 with ATBN31, which contains only 10% AN (compared with 18% AN in the former) and seems to be immiscible with DPEDC. Figure 6 shows that the AN content of ATBN had a drastic effect on the miscibility of ATBN with DPEDC. Similarly,

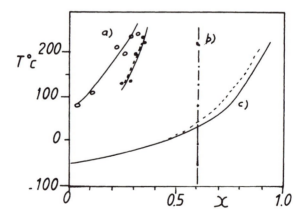

Figure 4. Experimental phase diagram showing temperature versus conversion. Key: o, *DPEDC–NFBN (15 wt%);* ●, *DCBA–NFBN (15 wt%); a, phase separation; b, gelation (x ≈ 0.6); and c, vitrification (T_g) versus* x.

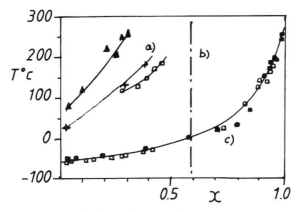

Figure 5. Experimental phase diagram showing temperature versus conversion for rubber (15 wt%)–DPEDC blends. Key: ▲, NFBN; +, CTBN; □, ATBN8. Curves (a), (b), and (c) are defined as in Figure 4.

Chen et al. (20) found that the miscibility of CTBN with DGEBA decreased with the decrease in the AN content of the rubber.

Figure 7 shows that in contrast to rubber–DPEDC blends, T_{cp} decreases almost linearly as x increases (once again in the range of T_i studied, because in fact it is a descending concave curve). This result confirms that the PES–CE blends show LCST behaviors. Moreover, the T_{cp} versus x curve of the 15% PES4100P–DPEDC blend is to the left of that of the 15% PES5003P–

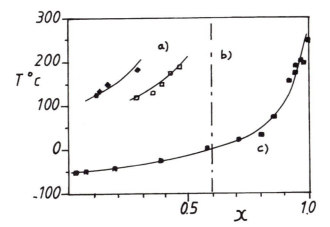

Figure 6. Experimental phase diagram showing temperature versus conversion for rubber (15 wt%)–DPEDC blends. Key: □, ATBN8; and ◆, ATBN8–ATBN31 (5 wt% in the rubber blend). Regions (a), (b), and (c) are defined as in Figure 4.

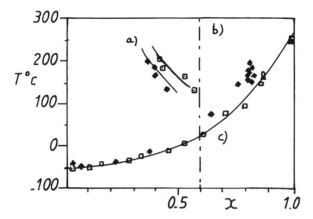

Figure 7. Experimental phase diagram showing temperature versus conversion for PES (15 wt%)–DPEDC blends. Key: ◆, PES5003P; and □, PES4100P.

DPEDC blend. These results indicate that PES4100P is more miscible with DPEDC than PES5003P, which is consistent with the fact that PES4100P has a smaller molar mass.

In addition, the evolution of T_g of the continuous phase of the blends can be plotted in Figure 7 with the T_{cp} versus x phase diagram. Once again, phase separation occurs before the vitrification of the blends. For a sample cured at 90 °C, phase separation is expected in the same range as gelation ($x_{cp} \approx x_{gel}$), and in fact no phase separation is observed. Pellan and Bloch (*31*) also reported that this blend was always clear when precured at 90 °C, even when this precuring up to gelation was followed by a postcuring at 200 °C. We think that this observation results from high viscosity during polymerization and gelation of the system.

Morphologies and Properties. The phase diagrams of temperature versus conversion (or vs. time) are useful for controlling the curing of a blend. We said in the introduction that the morphologies are mainly controlled by the temperature, T_{cp}, or the viscosity, η_{cp}, at which phase separation occurs. Generally, when T_{cp} decreases or η_{cp} (or x_{cp}) increases, the dispersed particles become smaller. In Figure 8 we show a series of micrographs obtained by scanning electron microscopy of the fracture surfaces of different samples. Samples are blends of DPEDC and different additives. They were precured at $T_i = 180$ °C to control phase separation and gelation, then postcured at high temperature (260 °C) in order to reach full conversion ($x \approx 1$). As we saw in the phase diagrams, the conversion (x_{cp}) or viscosity (η_{cp}) at the cloud point depends on the additive used (Table III).

Table III shows the following: With NFBN, $x_{cp} \approx 0.30$ and the particles

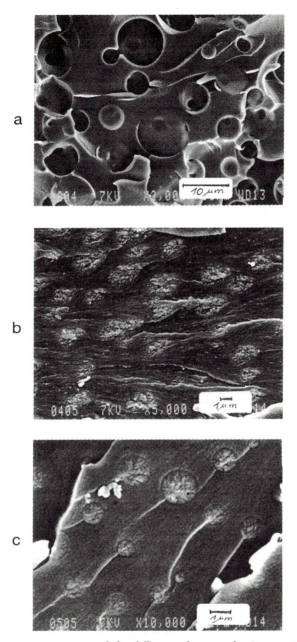

Figure 8. SEM images of the fully cured materials. Part a: 15 wt% NFBN–DPEDC. Part b: 15 wt% ATBN8–DPEDC. Part c: 15 wt% PES5003P–DPEDC.

are spherical and large ($\bar{D} \approx 7$ μm). With ATBN, on the other hand, $x_{cp} \approx 0.47$ and the particles are smaller ($\bar{D} \approx 1.7$ μm), but they also have like a branch substructure. In this case, we expect to see not only an effect of the viscosity but certainly also of the reaction between the rubber and cyanate functions before network formation.

Furthermore, with PES5003P, $x_{cp} \approx 0.40$ and the particles are also smaller than with NFBN. We also see an effect of the viscosity and of the reactive chain ends of the thermoplastic. But because T_i is 180 °C, which is lower than the T_g of the thermoplastic, one important effect is certainly also the vitrification of the dispersed particles when phase separation occurs. The consequence is that only with thermoplastic additives is an evolution of the morphology observed between precuring and postcuring processes.

With the help of these phase diagrams, it is now possible to define different precuring and postcuring schedules. The effects of the curing process on the final morphologies are presented elsewhere (*19*).

The introduction of any additive improves the toughness of the material; in particular, K_{Ic} of ATBN8–DPEDC is more than twice that of neat DPEDC. In contrast, different additives result in different losses in Young's modulus, from 17% to 50%, and different losses in the final material glass-transition temperature compared with neat DPEDC (Table III). The fact that T_g is lower for ATBN8- or PES5003P-based materials than for those obtained with NFBN obviously indicates that with the latter, the continuous phase contains smaller amounts of additive because of poorer miscibility. However, the residual fraction of PES in the final matrix [$(\phi_a)_m \infty$] calculated from Fox's equation for the PES5003P–DPEDC blend is above the maximum theoretical value. Thus, we believe that the decrease in T_g for the PES5003P–DPEDC blend results not only from the presence of residual additive but also from the modification of the poly(cyanurate) network because of a possible reaction between PES5003P and DPEDC.

In fact, the system with 15-wt% ATBN8 has the best-balanced properties of all the modified systems in this study: This sample shows an excellent improvement of the toughness without too large of a loss in T_g and in the modulus, while the system based on NFBN has worse mechanical properties.

Conclusions

Rubbers and PESs are initially miscible with cyanate ester monomers. Phase separation occurs during the reaction. By plotting the phase diagrams (temperature vs. conversion), it is possible to compare the effects of chain ends and AN content in butadiene–acrylonitrile random copolymers and the effect of molar mass in PES. The cyanate ester monomer based on bisphenol A is a better solvent than DPEDC, and both dicyanates are better solvents than DGEBA.

With the use of phase diagrams, we are also able to control the temperature and the viscosity at which phase separation occurs. The final morphologies of the three systems based on the same dicyanate monomer and modified with NFBN, ATBN, and PES are quite different and have different interfaces. When the additive can react with the monomer before network formation, a two-level structure is observed: a primary structure (dispersed particles), and a substructure inside the dispersed particles. The complex morphology obtained in this case gives the best toughening effect.

References

1. *Rubber Modified Thermoset Resins*; Riew, C. K.; Gillham, J. K., Eds.; Advances in Chemistry **208**; American Chemical Society: Washington DC, 1984; and references therein.
2. *Rubber-Toughened Thermosetting Polymers in Structural Adhesives—Developments in Resins and Primers;* Kinloch, A. J., Ed.; Applied Science Publishers: London, 1980; Chapter 5, pp. 127–162.
3. *Rubber-Toughened Plastics;* Riew, C. K., Ed.; Advances in Chemistry **222,** American Chemical Society: Washington DC, 1989; and references therein.
4. *Toughened plastics I: Science and Engineering;* Riew, C. K.; Kinloch, A. J., Eds.; Advances in Chemistry **233,** American Chemical Society: Washington DC, 1993.
5. Williams, R. J. J.; Borrajo, J.; Ababbo, H. E.; Rojas, A. J. In *Rubber Modified Thermoset Resins*; Riew, C. K.; Gillham, J. K., Eds.; Advances in Chemistry **208**, American Chemical Society: Washington DC, 1994; pp. 195–213.
6. Montarnal, S.; Pascault, J. P.; Sautereau, H. In *Rubber-Toughened Plastics;* Riew, C. K., Ed.; Advances in Chemistry **222,** American Chemical Society: Washington DC, 1989; pp. 193–224.
7. Verchère, D.; Sautereau, H.; Pascault, J. P.; Moschiar, S. M.; Riccardi, C. C.; Williams, R. J. J. In *Toughened plastics I: Science and Engineering;* Riew, C. K.; Kinloch, A. J., Eds.; Advances in Chemistry **233,** American Chemical Society: Washington DC, 1993; pp. 335–363.
8. Rozenberg, B. A. *Makromol. Chem. Macromol. Symp.* **1991,** *41*, 165–177.
9. Chen, D.; Pascault, J. P.; Sautereau, H.; Vigier, G. *Polym. Int.*, **1993,** *32*, 369–379.
10. Manzione, L. T.; Gillham, J. K.; McPherson, C. C. *J. Appl. Polym. Sci.* **1981,** *26*, 889–907.
11. Ruseckaite, R. A.; Hu, L. J.; Riccardi, C. C.; Williams, R. J. J. *Polym. Int.* **1993,** *30*, 287.
12. Grigat, E.; Pütter, R. *Angew. Chem. Int. Ed.* **1967,** *6(3)*, 206–218.
13. McConnell, V. P. *Adv. Compos.* **1992,** *May–June*.
14. Arnold, C.; McKenzie, P.; Malhotra, V.; Pearson, D.; Chow, N.; Hearn, M.; Robinson, G. *Proceedings of the 37th International SAMPE Symposium*; Society for the Advancement of Material and Process Engineering: Covina, CA 1992; pp. 128–136.
15. Yang, P. C.; Pickelman, D. M.;, Woo, E. P. *Proceedings of the 35th International SAMPE Symposium*; Society for the Advancement of Material and Process Engineering: Covina, CA 1990; p. 408.
16. Yang, P. C.; Woo, E. P; Laman, S. A.; Jakylowski, J. J.; Pickelman, D. M.; Sue, H. J. *Proceedings of the 36th International SAMPE Symposium*; Society for the Advancement of Material and Process Engineering: Covina, CA 1991, p. 437.

17. Shimp, D. A.; Christenson, J. R. *Plastic–Metal–Ceramics*; Hornfeld, H. L., Ed.; Society for the Advancement of Material and Process Engineering: Switzerland, 1990; pp. 81–93.
18. Srinivasan, S. A.; McGrath, J. E. *High Perform. Polym.* **1993**, *5*, 259–274.
19. Cao, Z. Q.; Mechin, F.; Pascault, J. P. *ACS Polym. Mater. Sci. Eng. Div. Prepr.* **1994**, *71*, 752–753.
20. Chen, D.; Pascault, J. P.; Bertsch, R. J.; Drake, R. S.; Siebert, A. R. *J. Appl. Polym. Sci.* **1994**, *51*, 1959.
21. Verchère, D.; Sautereau, H.; Pascault, J. P.; Moschiar, S. M.; Riccardi, C. C.; Williams, R. J. J., *Polymer* **1989**, *30*, 107.
22. Cao, Z. Q.; Mechin, F.; Pascault, J. P. *Polym. Int.* **1994**, *34*, 41–48.
23. Verchère, D.; Sautereau, H.; Pascault, J. P.; Moschiar, S. M.; Riccardi, C. C.; Williams, R. J. J. *J. Appl. Polym. Sci.* **1990**, *41*, 467.
24. Georjon, O.; Galy, J.; Pascault, J. P. *J. Appl. Polym. Sci.* **1993**, *44*, 1441.
25. Gupta, A. M. *Macromolecules* **1991**, *24*, 3459–3461.
26. Simon, S. L.; Gillham, J. K. *J. Appl. Polym. Sci.* **1993**, *47*, 461.
27. Mirco, V.; Cao, Z. Q.; Mechin, F.; Pascault, J. P. *ACS Polym. Mater. Sci. Eng. Div. Prepr.* **1992**, *66*, 451.
28. Mirco, V.; Mechin, F.; Pascault, J. P. *ACS Polym. Mater. Sci. Eng. Div. Prepr.* **1994**, *71*, 688–689.
29. Mirco, V. PhD Thesis, INSA de Lyon, 1995.
30. Borrajo, J.; Riccardi, C. C.; Williams, R. J. J.; Cao, Z. Q.; Pascault, J. P.
31. Pellan, L.; Bloch, B. In *C.-R. Journ. Natl. Comp.* **1992**, *November*, 161–172.

14

Mechanical Behavior of a Solid Composite Propellant During Motor Ignition

Y. Traissac[1], J. Ninous[1], R. Neviere[2], and J. Pouyet[3]

[1]SNPE, BP 57, 33166 St. Médard en Jalles Cedex, France
[2]SNPE, 91710 Vert le Petit, France
[3]Laboratoire de Mécanique Physique, Unité de Recherch Associée No. 867, Université de Bordeaux I, 33400 Talence, France

In order to understand the behavior of composite propellants during motor ignition, we conducted a study of the mechanical and ultimate properties of a propellant filled with hydroxy-terminated polybutadiene under imposed hydrostatic pressure. The mechanical response of the propellant was examined by uniaxial tensile and simple shear tests at various temperatures, strain rates, and superimposed pressures from atmospheric pressure to 15 MPa. The experimentally observed ultimate properties were strongly pressure-sensitive. The data were formalized in a specific stress-failure criterion.

THE CONCEPT OF A SOLID PROPELLANT GRAIN used for rocket propulsion (Figure 1) implies a mechanical design that satisfies specific needs (long-term storage, thermal cycles, internal pressure, and acceleration). The mechanical integrity of the grain has to be assessed for thermomechanical loading because of the casting process and for pressure loading during the ignition phase. Thermomechanical loading arises from the difference between the thermal-expansion coefficients of the propellant and the case into which the grain is cast.

The failure behavior of composite propellants, which are filled elastomers, is complicated by the presence of filler particles. Under loading, phenomena such as cavitation and debonding can arise at or near the filler–matrix interface. (*1, 2*) Identification of a practical failure criterion for such

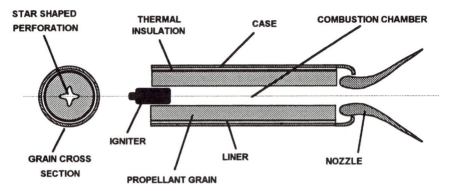

Figure 1. Schematic of a solid-propellant rocket motor with a star-shaped geometry.

materials has been the objective of extensive studies for many years (3–5). During the ignition phase, an internal pressure gradient produces a rapid (a few milliseconds) tensile strain field in the combustion chamber. If this strain field is excessive, it may lead to propellant degradation, which would be dangerous for the integrity of the structure. Thus, an optimal design of the grain requires a reasonable understanding of the mechanical behavior of the propellant, including the effect of an imposed hydrostatic pressure on ultimate properties.

The aim of this work is to provide both experimental information and a corresponding formalization in order to elucidate structural propellant grain safety during ignition. The experimental data were obtained from uniaxial tensile tests and simple shear tests performed with an imposed hydrostatic pressure varying from atmospheric pressure to 15 MPa. It is well established that the materials studied exhibit time–temperature and pressure-sensitive properties. The ultimate properties reported here are formalized in a proposed stress-failure criterion capable of including the pressure effect.

Solid-Propellant Grains

Solid-propellant rocket engines are attractive because of their simple propulsion principle, which avoids alimentation systems needed to produce combustion. Both oxidizer and fuel elements are present in a chemically stable material (the propellant), which is shaped to fit the motor itself. This feature, in particular, represents an obvious advantage for military applications such as nuclear ballistic missiles, which require a long-term storage capability and high service safety (3).

Main Parts of a Solid-Propellant Engine.   The *case* is a thin, high-tensile-strength metal or fiber-composite envelope in which the propellant is directly cast and bonded. The mechanical requirements for this part are that it offer sufficient resistance to service pressure and loading conditions during flight, such as vibrations and acceleration, and have optimum weight.

The *nozzle* provides the guidance capability of the motor through the orientation of combustion-gas ejection. In this part, combustion-gas pressure is transformed into high-velocity flow, which is the source of movement.

The *propellant grain* is the energetic source of the motor. Ballistic requirements are defined as a pressure versus time envelope with which are associated mass and safety figures. To provide maximum thrust during the ignition phase, the combustion chamber is star-shaped all along the central axis. This complicated structural architecture is obtained directly from the casting process. Combustion is initiated in this chamber and it propagates parallel to the axis until the flame front reaches the motor case, where it is stopped by an elastomeric insulator protecting the case.

The *solid propellant* is the energetic material involved in the combustion process. For a homogeneous propellant (or a double-base propellant), the oxidizing and reducing elements involved in the creation of energy through combustion are combined in the same molecule. Composite propellants, on the other hand, are typically composed of a more reducing polymeric matrix (polybutadiene or polyurethane) loaded with a solid powder oxidizer (ammonium perchlorate), and possibly with a metal powder (aluminum) that plays the role of a secondary fuel component. The material studied here belongs to the latter family of propellants.

Pressure Loading During the Ignition Phase

During the ignition phase of a propellant grain, the pressure in the chamber due to gases generated by combustion increases, in a few milliseconds, from atmospheric pressure, P_{atm}, at $t = 0$ s to the service pressure, P_{serv} (5–15 MPa for most applications) at time t_{serv} (50–100 ms). The pressure versus time history is quasi-linear during the early ignition phase (i.e., up to t_{serv}) and then the pressure reaches a constant value, which is maintained until the end of burning time, t_f. At the end of combustion, the pressure drops rapidly from the service pressure to ambient pressure without a controlled time dependence. A representative pressure–time curve is shown in Figure 2.

During the ignition phase, as the pressure increases, the propellant is loaded by hydrostatic pressure imposed on a biaxial tensile stress field. Because the propellant is incompressible in the ignition condition, the pressure is transmitted entirely to the case, which, being thin because of the weight requirement, presents significant hoop deformations. Therefore, a tensile strain

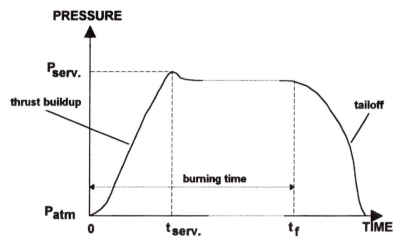

Figure 2. Pressure in the combustion chamber versus time.

field is generated in the grain with a maximum value at the surface of the combustion chamber. A schematic representation of this stress state is shown in Figure 3.

The Tensile Stress Field. The main tool used in industry to assess the structural integrity of the propellant grain is a commercially available finite-element code that solves problems of finite elastic deformations. From numerical investigation performed with the code, one of the maximum strain locations (the critical point) is identified. This point (M) is located at the surface of the combustion chamber near the middle of the symmetry axis.

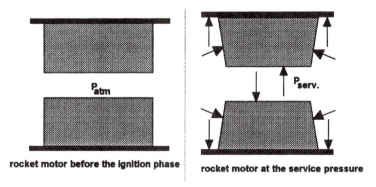

Figure 3. Schematic representation of pressurization during the ignition phase.

In the simplified case of a long cylinder (plane strain), as shown in Figure 4, the Cauchy stress tensor at point M can be written as:

$$\sigma = \begin{bmatrix} \sigma_r = -P_{serv} & 0 & 0 \\ 0 & \sigma_\theta = \sigma - P_{serv} & 0 \\ 0 & 0 & \sigma_z = \dfrac{\sigma}{2} - P_{serv} \end{bmatrix} \quad (1)$$

where σ is the stress due to the strain. From this particular result, it becomes clear that hydrostatic-pressure effects on the material properties of the propellant have to be investigated.

Tensile Behavior of Composite Solid Propellants

Overview of Mechanical Behavior. The current understanding of the mechanics and mechanisms of fracture in filled rubbers, including composite propellants, is highly advanced in some areas but surprisingly limited in others. The highly heterogeneous structure of this material leads to specific phenomena that occur mainly at the particle–matrix interface and produce a particular response to mechanical loading.

Tensile properties of composite propellants depend on the tensile properties of the matrix, concentration of the components, particle size, particle-size distribution, particle shape, quality of the interface between fillers and polymeric binder, and, obviously, experimental conditions (strain rate, temperature, and environmental pressure). Many authors (2, 3) have explained the effect of fillers on the mechanical properties of composites, the importance of the filler–matrix interface on physical properties, and the mechanism of reinforcement of the material. Other efforts have examined the effect of experimental conditions on the failure properties of filled elastomers. Landel and

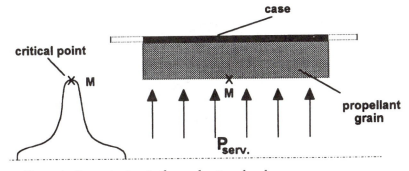

Figure 4. Pressurization in the combustion chamber.

Smith (4) have shown that tensile strength and ultimate elongation depend on both temperature and experimental time scale.

The first step of the failure process in filled rubber is the formation of vacuoles within the binder or at the binder–filler interface. Depending on the ratio of the strength of the adhesive bond between filler and binder to the cohesive strength of the binder, different phenomena may be observed at the yield point.

When rubber is not sufficiently well bonded to the inclusion, the filler–binder adhesion bond fails first and vacuoles are formed around the filler particles. Once debonding commences, it propagates to the vicinity of neighboring particles, thereby promoting failure of the material. This debonding process (see Figure 5) occurs at an applied stress that obeys a Griffith fracture energy criterion, but now the work of detachment of the elastomer from the inclusion is required in place of the work of tearing rubber apart (6). Gent and Park (6) demonstrated that the work of detachment greatly depends on the dissipative properties of the elastomer, increasing for more dissipative materials.

In composite propellants, a relatively strong bond between binder and filler is important for obtaining high tensile strength. However, the edges and surfaces of the inclusions serve as sites of dangerously high stress concentrations, which could cause internal failure of the softer matrix material. Most studies (1, 2, 6–9) have dealt with the problem of matrix fracture induced by the inclusions. Here, we present the mode of matrix fracture called cavitation,

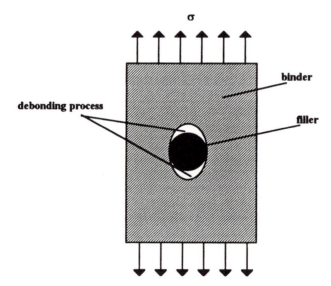

Figure 5. Debonding at the filler–binder interface during a tensile test.

which is characterized by the sudden appearance of voids within the elastomeric matrix in well-bonded filled elastomers. A schematic representation of this phenomenon is shown in Figure 6. The initial microvoid occurs near the inclusion by internal fracture of the elastomer under triaxial tension (or negative hydrostatic pressure; P), when P reaches a critical value, denoted P_c. Gent and co-workers (*1, 6–8*) and Oberth and Bruenner (*2*) have observed a direct proportionality between the critical tensile stress, σ_c, for vacuole expansion and the elastic modulus, E, of the elastomer. As we shall see, this phenomenon could explain the effect of a positive imposed hydrostatic pressure on the ultimate properties during a tensile test.

In composite propellants, the cavitation (or debonding) process, which has been shown to take place near (or at) the particle–matrix interface, is dependent on pressure, deformation, and additional viscoelastic and dissipative considerations (*10*).

Each type of propellant has specific mechanical characteristics, but the influence of test parameters (temperature, strain rate, and pressure) is the same for all propellants (*11*). Tensile tests are widely used to analyze propellant behavior as well as examine the manufacturing controls of the propellants. Because their behavior is not linear-elastic, it is necessary to define several parameters that allow a better representation of the experimental tensile curve. The stylistic experimental stress–strain response at a constant strain rate from a uniaxial tensile test is shown in Figure 7, where E is the elastic modulus (initial slope), $S_{1,P}$ is the tensile strength (used later for a failure criterion), and $e_{1,P}$ is the strain at tensile strength.

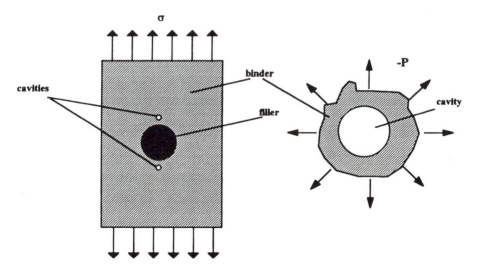

Figure 6. The cavitation process near the filler–binder interface during a tensile test.

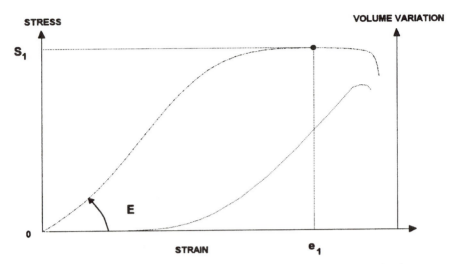

Figure 7. Representation of a uniaxial tensile test with T, V_c, and P fixed.

Moreover, during the tensile test, the physical nature of the propellant (i.e., the fillers and binder) changes, resulting in an increase in specimen volume (see Figure 7). This particular process is caused by the occurrence of vacuoles around solid fillers or an increase of microcracks in the binder. Knowledge of such phenomena contributes greatly to the determination of the mechanical behavior of propellants. The simultaneous measurement of volume dilation during tensile testing is done with a gas dilatometer developed by Farris (*11*). This method measures the pressure variation in an enclosure in which the specimen is placed. The change of pressure is directly linked to volume expansion of the specimen.

Qualitatively, the effects of the loading conditions may be considered as follows:

1. A decrease in test temperature, T, leads to an increase in rigidity (i.e., modulus) and ultimate strength, but a decrease in fracture strain.
2. An increase in strain rate (or crosshead velocity, V_c) leads to effects similar to those just described, with a possible time–temperature equivalence.
3. An increase in test pressure (hydrostatic pressure, P) does not affect the modulus but strongly increases the ultimate properties (e.g., stress and strain at failure).

The Hydrostatic Pressure Effect.   The particular hydroxy-terminated polybutadiene (HTPB) solid propellant employed here is a polybutadi-

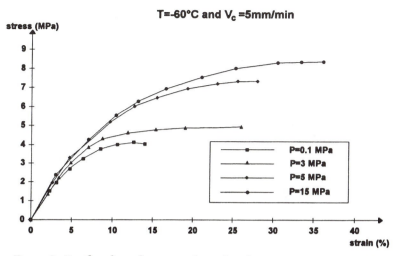

Figure 8. Results of tensile tests performed under different imposed hydrostatic pressures for T = –60 °C *and* V_c = 5 *mm/min.*

ene matrix material that contains 85 wt% of composition. The experimental setup offers a range of pressures, from ambient to 15 MPa; the test temperature can be varied from –60 to +60 °C; and the crosshead velocity, V_c, of the dynamometer can be varied from 5 to 50,000 mm/min (corresponding to a range in strain rate of 0.06 to 600 min^{-1}). Figures 8 and 9 show the effects of pressure on the stress–strain response for a fixed strain rate and fixed temperature conditions. An increase of pressure leads to an increase in the ultimate

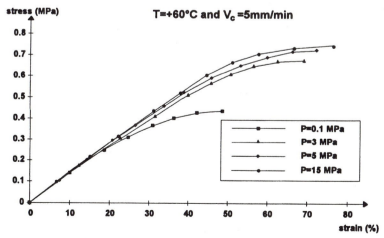

Figure 9. Results of tensile tests performed under different imposed hydrostatic pressures for T = 60 °C *and* V_c = 5 *mm/min.*

properties (stress and strain) (3, 12–15). It must be emphasized that depending on the experimental conditions, a further increase in pressure does not lead to an increase in ultimate properties once a limit pressure is exceeded. This pressure is referred to as the saturation pressure, P_{sat}, and it depends on the strain rate and temperature.

The saturation pressure seems to be proportional to the elastic modulus of the propellant (Figure 10). Although there is no solid experimental evidence, it is suspected that the positive hydrostatic pressure acts as a retardation parameter in cavitation or in debonding of particles from the polymeric matrix, as described by Gent and co-workers. Qualitatively, the effect of applied pressure during a tensile test is believed to delay the occurrence of vacuoles and to decrease their number. This assumption may be sustained by simultaneous volume-expansion measurements taken during tensile tests under different pressures. For an increase in pressure, the relative measured volume decreases (see Figure 11).

A Failure Criterion for Solid Propellants

A relationship expressing the general state of stress at which failure occurs in terms of one or more material parameters is termed a failure criterion. Knowledge of the failure criterion, combined with knowledge of the material parameters and their temperature and time dependence, completely determines the conditions under which the material will fail (3, 5, 15, 16).

Stress analysis is a fundamental requirement for determining rocket-engine structural capabilities (3, 5). Under a mechanical load, like ignition, there

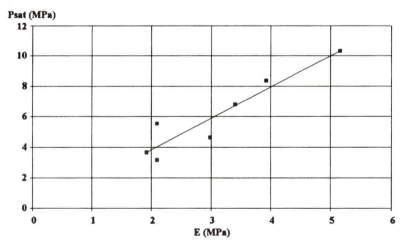

Figure 10. P_{sat} versus E of the propellant for different tensile test conditions (temperature and strain rate).

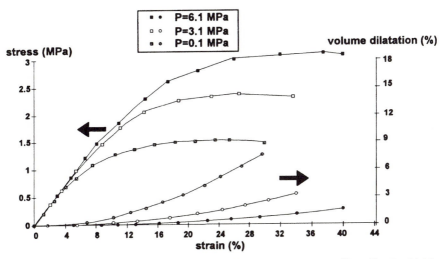

Figure 11. Results of tensile tests of an HTPB composite propellant (for T = 20 °C and V_c = 500 mm/min).

is at each point a particular three-dimensional state of stress. Consequently, it is not possible to perform a direct comparison between the capability obtained by monodimensional tests and a three-dimensional stress field. Because a propellant grain is subjected to a spectrum of three-dimensional stress states, the stress capability of a uniaxial tensile test is not sufficient for a complete failure analysis.

The stress tensor for each point of the grain can be represented by one point in principal stress space. In that space, there exists a volume where the propellant is made worthless by significant damage, even possibly a crack. A major difficulty of such a representation is the fact that the failure properties of propellant depend strongly on loading conditions (temperature and strain rate). So in this paper, for each loading condition (one strain rate and one temperature), we construct a failure surface based on experimental data for several multiaxial stress states.

As a convention, the failure point is hereafter defined by the maximum stress point on the stress–strain curves. This definition may lead to confusion, but in the industrial context, it is assumed that the propellant structure presents an extended irreversible degradation beyond this point which is incompatible with safe application.

Failure Criterion. In the design of a propellant structure, the output of the calculation is a tensor quantity in terms of either stress or strain, which must be compared with some experimentally available failure properties. These properties are generally obtained from uniaxial and multiaxial extension tests performed with or without an imposed hydrostatic pressure. In fact, the complicated experimental techniques needed to achieve multiaxial experiments limit the investigation of pressure dependence to uniaxial and simple shear tests.

As mentioned above, the stress state of the material at a particular point of the structure is represented by a tensor quantity that can be reduced to its principal stresses: σ_1, σ_2 and ϑ_3. The failure criterion is then a relationship of the type

$$F(\sigma_1, \sigma_2, \sigma_3) = 0 \qquad (2)$$

which can be graphically represented by a three-dimensional surface envelope in principal stress space, as shown in Figure 12. The inner volume defined by this envelope corresponds to a low failure probability, while the outer space corresponds to a high failure probability. The evaluation of a safety coefficient will therefore be any convenient measure of the distance from the representative point of the stress state to the failure envelope.

The corresponding experimental investigation should then focus on collecting values of the failure stress for any principal stress combination in order to fully describe the envelope. It is obvious that such a detailed description represents an unrealistic task, especially in an industrial environment. Some basic assumptions described in the following sections allow us to reduce the parameter field of such an investigation.

Isotropy Assumption.   Isotropy of material properties implies radial symmetry of the envelope around the $\sigma_1 = \sigma_2 = \sigma_3$ axis in principal stress space. The surface of the envelope is then entirely defined by its projection in a plane containing this particular axis and one of the principal stress axes. By

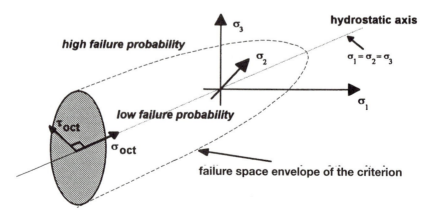

Figure 12. "Pressure effect" criterion in principal stress space.

convention, one may choose this plane as the octahedral plane where the abscissae and ordinates are respectively defined by

$$\sigma_{oct} = \frac{\sigma_1 + \sigma_2 + \sigma_3}{3} \tag{3}$$

and

$$\tau_{oct} = \frac{\sqrt{(\sigma_1 - \sigma_2)^2 + (\sigma_1 - \sigma_3)^2 + (\sigma_2 - \sigma_3)^2}}{3} \tag{4}$$

The average stress, σ_{oct}, is the trace of the stress tensor corresponding to the volumetric part of the stress field, while the octahedral shear stress, τ_{oct}, involves only the shear components of the same field.

A Failure Criterion for Solid Propellant Including the "Pressure Effect".   From a set of uniaxial tensile test results and simple shear results with and without superimposed hydrostatic pressures, a failure criterion is proposed for any given set of strain rate and temperature. The essential parameters in the definition of this criterion are (1) the tensile strength in uniaxial extension at atmospheric pressure, $S_{1,P=0.1MPa}$, and (2) the ratio of the tensile strength under saturation pressure in uniaxial tension to the tensile strength under atmospheric pressure, $F_{S,Psat} = S_{1,P=Psat}/S_{1,P=0.1MPa}$.

This stress criterion represents the progressive effect of pressure from atmospheric pressure up to the saturation pressure and satisfies the following experimental evidence:

1. For any given experimental set of strain rate and temperature, there exists a pressure over which no further influence of the testing pressure is observed. The saturation pressure therefore depends on the testing conditions.
2. The failure stress for an equitriaxial extension approximately equals the failure stress for a uniaxial extension, as we see in Table I. (The equitriaxial extension is performed using a thin disk of propellant bonded between two specially prepared plates of metal.)

Within these assumptions, the failure criterion can be modeled as

$$\tau_{oct} = \frac{\sqrt{2}}{3} F_{S,P=P_{sat}} \frac{\sigma_{oct} - S_{1,P_{atm}}}{\sigma_{oct} - \frac{S_{1,P_{atm}}}{3}(2F_{S,P=P_{sat}} + 1)} S_{1,P_{atm}} \tag{5}$$

It can be easily verified that this relation satisfies the experimental limiting cases:

**Table I. Comparison of Maximal Stresses for Uniaxial and
Equitriaxial Tensile Tests (HTPB Propellant)**

Temperature (°C)	Strain rate (min⁻¹)	$S_{1,P_{atm}}$: maximal tensile stress (MPa)	S_{eq}: maximal equitriaxial stress (MPa)	$\dfrac{S_{1,P_{atm}}}{S_{eq}}$
−40	1	1.90	1.83	1.04
−40	0.1	1.39	1.54	0.90
−20	1	1.16	1.18	0.98
−20	0.1	0.92	1.08	0.85
−5	1	0.93	0.99	0.94
−5	0.1	0.80	0.83	0.96
20	1	0.72	0.75	0.96
20	0.1	0.67	0.66	1.02
40	1	0.65	0.68	0.96
40	0.1	0.60	0.56	1.07
60	1	0.60	0.58	1.03
60	0.1	0.54	0.48	1.12

for uniaxial extension $(P = P_{atm})$: $\sigma_{oct} = \dfrac{S_{1,P_{atm}}}{3}$, $\tau_{oct} = \dfrac{\sqrt{2}}{3} S_{1,P_{atm}}$ (6)

for equitriaxial extension $(P = P_{atm})$: $\sigma_{oct} = \dfrac{S_{1,P_{atm}}}{3}$, $\tau_{oct} = 0$ (7)

for uniaxial extension under saturation pressure: if $P \rightarrow P_{sat}$, then $S_{1,P_{sat}} = S_{1,P_{atm}} F_{S,P_{sat}}$

and $\sigma_{oct} = \dfrac{S_{1,P_{atm}} F_{S,P_{sat}}}{3}$, $\tau_{oct} = \dfrac{\sqrt{2}}{3} S_{1,P_{atm}} F_{S,P_{sat}}$ (8)

As an illustration, Figure 13 presents results for three different strain-rate and temperature conditions. The ultimate tensile properties are used to construct the failure criterion, and the ultimate simple shear properties verify the validity of the failure criterion.

Conclusions

The ultimate properties of an HTPB propellant were investigated for various conditions of strain rate, temperature, and imposed hydrostatic pressure. In experiments, the pressure increased both the failure strain and stress until a maximum was reached at a saturation pressure. This saturation pressure was dependent on experimental conditions and seemed to be proportional to the elastic modulus of the propellant.

An original failure criterion was introduced to unify the experimental results obtained here in terms of ultimate stress and to include the effects of

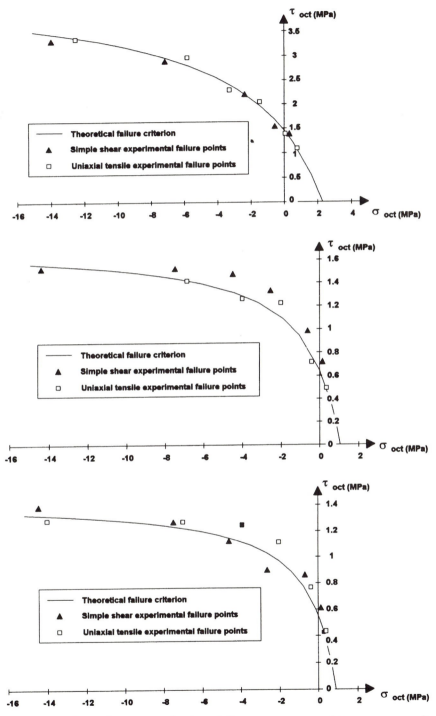

Figure 13. Failure criteria for three different strain-rate and temperature conditions. Top: T = –30 °C and $\dot{\epsilon}$ = 10 min⁻¹. Middle: T = 0 °C and $\dot{\epsilon}$ = 1 min⁻¹. Bottom: T = 20 °C and $\dot{\epsilon}$ = 10 min⁻¹.

pressure, time, and temperature. Excellent agreement between the proposed failure criterion and data was obtained.

References

1. Gent, A. N.; Lindley, P. B. *Proc. R. Soc. London Ser. A* **1958,** *249,* 195.
2. Oberth, A. E.; Bruenner, R. S. *Trans. Soc. Rheol.* **1965,** 9:2, 165.
3. Davenas, A. *Solid Rocket Propulsion Technology*; Pergamon: Oxford, England, 1993.
4. Landel, R. F.; Smith, T. L. *ARS J.* **1961,** *May,* 599.
5. Jones, J. W.; Knauss, W. G. *Proceedings of the 6th Solid Propellant Rocket Conference;* American Institute of Aeronautics and Astronautics: Washington, DC, 1965.
6. Gent, A. N.; Park, B. *J. Mater. Sci.* **1984,** *19,* 1947.
7. Cho, K.; Gent, A. N.; Lam, P. S. *J. Mater. Sci.* **1987,** *22,* 2899.
8. Gent, A. N. Presented at International Symposium Moffis-93 Mineral and Organic Functional Fillers in Polymers, Namur, Belgium, April 13–16, 1993.
9. Kinlock, A. J.; Young, R. J. *Fracture Behavior of Polymers*; Applied Science: London, 1983.
10. Heuillet, P.; Neviere, R.; Nottin, J. P.; Deneuville, P. *Damage Characterization for a Solid Propellant by an Energetic Approach*; Masters, J. E., Ed.; STP 1128; 1992; p 121.
11. Farris, R. J. *Development of a Solid Rocket Propellant Non-linear Viscoelastic Constitutive Theory*; Aerojet Solid Propulsion Company: prepared for Air Force Rocket Propulsion, (NTIS): June 1965.
12. Hazelton, I. G. *Appl. Polym. Symp.* **1965,** Interscience-John Wiley: New York, 217–228.
13. Ninous, J.; Besson, J. M.; Traissac Y.; Pouyet, J. *Composit. (Plastiques renforcés Fibre de verre textile)* **1993,** 16.
14. Ninous, J.; Traissac, Y.; Pouyet, J. Presented at the International Symposium Moffis-93 Mineral and Organic Functional Fillers in Polymers, Namur, Belgium, April 13–16, 1993.
15. Traissac, Y.; Neveire, R.; Stankiewicz, F. Presented at International Conference of Genie Mécanique des Caoutchoucs, Nancy, France, October 25–26, 1994.
16. Tschoegl, N. W. *Failure Surfaces in Principal Stress Space*; Polymer Science Symposium 32; John Wiley & Sons: New York, 1971; p 239

Mechanical Properties and Deformation Micromechanics of Rubber-Toughened Acrylic Polymers

P. A. Lovell, M. M. Sherratt, and R. J. Young

Polymer Science and Technology Group, Manchester Materials Science Centre, University of Manchester Institute of Science and Technology, Grosvenor Street, Manchester, M1 7HS, United Kingdom

A range of rubber-toughened acrylic polymers was prepared by dispersing preformed, multiphase toughening particles (1) directly into an acrylic polymer by extrusion, and (2) into methyl methacrylate monomer, which was then polymerized to form the acrylic matrix. The tensile and impact properties of the materials are reported together with the results from studies of deformation micromechanics performed using optical microscopy, transmission electron microscopy, and real-time, small-angle X-ray scattering. The effects of acrylic-matrix properties and toughening-particle composition, morphology, size, and level of inclusion on mechanical properties and fracture behavior are discussed.

T HE MOST COMMON METHOD for the preparation of rubber-toughened plastics involves in situ phase separation of rubbery toughening particles from the rigid plastic matrix as it is formed (*1–3*). In contrast, modern rubber-toughened acrylic materials utilize preformed toughening particles prepared by emulsion polymerization (*4–7*). The particles can be dispersed either directly into an acrylic polymer or into acrylic monomers that are then polymerized to form the acrylic matrix. Because the particles are cross-linked during their preparation, they retain their morphology and size in the blends so obtained. This route, therefore, has the advantage of allowing independent control of the properties of the acrylic matrix, the composition, morphology, and size of the toughening particles, and the level of inclusion of the toughening particles. This chapter gives an overview of some results from an extensive study of rubber-toughened acrylic materials. It describes the effects of systematic changes in the acrylic matrix and in toughening-particle composition, size, and morphology on tensile and impact properties, and the associated deformation mechanisms.

Experimental Work

Preparation of the Toughening Particles.

A range of two-, three- and four-layer (i.e., 2L, 3L, and 4L) toughening particles was prepared using sequential emulsion-polymerization processes (6, 7). Their sizes and morphologies are represented schematically in Figure 1. In order to retain a high percentage of visible-light transmission in the rubber-toughened acrylic materials, the compositions of the rubbery and glassy phases of the particles were chosen such that their refractive indices matched that of the acrylic matrix. The rubbery layers of the particles consisted of cross-linked poly[(n-butyl acrylate)-co-styrene] (78.2 mol% n-butyl acrylate), and the glassy layers consisted of poly[(methyl methacrylate)-co-(ethyl acrylate)] (94.9 mol% methyl methacrylate), the inner glassy layers being cross-linked. Each layer was graft-linked to the other layers with which it was in contact. With the exception of the 3LAI and 3LAII particles, allyl methacrylate (ALMA) was used at constant levels for both cross-linking and graft-linking. In the preparation of the 3LAI particles, ALMA was used at higher-than-normal levels in the interfacial regions to increase interfacial strength. The 3LAII particles were prepared with ALMA at normal levels in the interfacial regions, but for cross-linking the bulk of the glassy and rubbery phases ALMA was replaced by identical molar quantities of 1,2-ethanediol dimethacrylate and 1,6-hexanediol diacrylate, respectively.

For each type of toughening particle, the latex obtained from emulsion polymerization was coagulated by addition to magnesium sulphate solution to yield loose aggregates of the particles. These aggregates were isolated by filtration, washed thoroughly with water, and then dried at 70 °C.

Rubber-Toughened Acrylic Molding Materials.

The materials were prepared by blending the toughening particles with Diakon LG156 (supplied by ICI Acrylics), which is poly[(methyl methacrylate)-co-(n-butyl acrylate)] (92

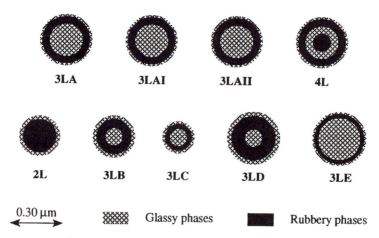

Figure 1. Schematics of sections through the equators of the toughening particles showing their sizes and internal structures. The compositions of the particle layers are given in the text.

mol% methyl methacrylate). This substance has a number-average molar mass (M_n) of 47 kg/mol, and M_W/M_n, where M_W is the weight-average molar mass, of 1.7 (gel permeation chromatography, polystyrene calibration).

The dried aggregates of toughening particles were blended with Diakon LG156 at 220 °C by either a single pass through a Werner Pfleiderer 30-mm twin-screw extruder or two passes through a Francis Shaw 40-mm single-screw extruder. This procedure ensured complete disruption of the aggregates to give uniform dispersions of the toughening particles, as evidenced by transmission electron micrographs of ultramicrotomed sections of the blends. For each type of particle, four blends containing different weight fractions (w_p) of particles were produced. The nomenclature used to define the blends is illustrated by 3LA22, which specifies that the blend contains 3LA particles with $w_p = 0.22$.

Diakon LG156 and each of the blends were compression-molded at 190 °C into plaques 3 mm thick (for tensile testing) and 6 mm thick (for impact testing).

Rubber-Toughened Acrylic Sheet Materials.

The sheet materials were specially prepared by ICI Acrylics for this research program by dispersing 3LA toughening particles into methyl methacrylate, which was then polymerized in sheet molds. A series of sheets 3 mm thick (for tensile testing) and 6 mm thick (for impact testing) were produced under identical conditions using 3LA toughening particles at 0, 1, 2, 4, and 8% by weight to methyl methacrylate. The acrylic sheet materials thus obtained were designated AS, RTAS-1, RTAS-2, RTAS-4, and RTAS-8 respectively.

Mechanical Testing.

Tensile stress–strain data were obtained in accordance with ASTM D638-84, employing a crosshead displacement rate of 5 mm/min, which corresponds to a nominal strain rate of 2×10^{-3} s^{-1}. Three-point-bend impact tests were carried out on freshly sharpened, single edge-notched specimens according to ASTM D5045-91, using an instrumented impact tester (Ceast Universal Pendulum) with loading times to failure in the range 0.8–2 ms; thus the values of critical strain-energy release rate obtained can be considered to be free from dynamic effects. All tests were conducted at 20 °C.

Studies of Deformation Micromechanics.

Polishing and sectioning techniques (*8–10*) for observing deformation zones were used to elucidate toughening mechanisms. Tensile specimens were deformed to fracture before polishing. Double edge-notched impact specimens were subjected to subcritical crack growth in three-point bend mode prior to polishing. The specimens were either (1) polished from two faces to obtain thin sections for optical microscopy, or (2) polished from one face, then ultramicrotomed, and then stained with ruthenium tetroxide to obtain sections for examination by transmission electron microscopy (TEM).

Tensile testing was carried out simultaneously with small-angle X-ray scattering (SAXS) using the synchrotron source at the Engineering and Physical Sciences Research Council's Daresbury Laboratory in the United Kingdom. The experiments were performed on beamline 8.2 employing monochromatic radiation of wavelength 1.52 Å and a 4.5 m camera length. The X-ray beam had a cross section of approximately 0.3 × 4 mm and a typical flux of 4×10^{10} photons/s. The specimens (overall dimensions 60 × 20 × 3 mm) had a continuously curved constricted region, which ensured that yielding would occur in the center of the gauge length where the specimen was approximately 5 mm wide. The specimens were mounted in a Polymer Laboratories Minimat tensometer, and the beam was positioned in the center of the gauge length with its long axis perpendicular to the tensile axis; the specimens were then deformed using a crosshead displacement rate of 3 mm/min. The voltage signal from the load cell was recorded continuously during deformation and, simultaneously, up to 20 consecutive SAXS patterns were recorded over 10 s time intervals using an area detector. The SAXS data were corrected for background scattering from the camera and undeformed specimen, and for the changes in specimen attenuation factor that arose from the changes in specimen thickness during deformation.

Results and Discussion

Tensile Deformation of the Rubber-Toughened Acrylic Molding Materials. Tensile testing revealed that, with the exception of the 3LAII particles, each type of toughening particle can induce extensive yielding in the matrix, leading to much greater fracture strains (up to 30–50%, compared with 3% for the unmodified matrix) and correspondingly higher energies to fracture. The poor performance of the 3LAII materials (~8% ultimate elongation for $w_p > 0.20$) results from the change in the chemistry of cross-linking in the 3LAII particles, which leads to a substantial increase in cross-link density (11).

The values of Young's modulus (E), yield stress (σ_y), and fracture stress (σ_f) determined from tensile testing are shown in Figures 2 through 4. In each case, the data are plotted against both the volume fraction of particles (V_p), which excludes the outer glassy layer of the particles, and the volume fraction of rubber (V_r), which excludes both the outer and internal glassy layers of the particles. Inspection of Figure 2a shows that the values of E are scattered when plotted against V_p, with the 2L and 3LE materials at the lower and upper bounds of the modulus ranges. However, when plotted against V_r (Figure 2b), the values of E for all the materials fall within a relatively narrow band, indicating that Young's modulus is controlled by the volume fraction of rubber. Similar observations can be made for the values of σ_y from Figure 3. These observations are in accord with finite-element analysis of the 2L, 3LA, 3LB, and 3LD materials, which predicts that Young's modulus and von Mises stress correlate with V_r rather than with V_p (12). In general, however, the predicted values of E and σ_y are higher than those measured experimentally. The measured E and σ_y data are consistent with results from other work on rubber-toughened poly(methyl methacrylate) materials; (13, 14) hence the relatively low values of these quantities remain an anomaly and are the subject of further investigations. Although the σ_f data plotted in Figure 4 are somewhat more scattered, they also show better correlation with V_r than with V_p. The greater scatter in the σ_f data is not unexpected because, as an ultimate property, σ_f will depend not only on elastic properties but also on the level of flaws in the different materials. In addition, because the values of tensile properties were calculated from the cross-sectional areas of the undeformed specimens, the errors in values of σ_f will inevitably be greater and more variable than those in E and σ_y.

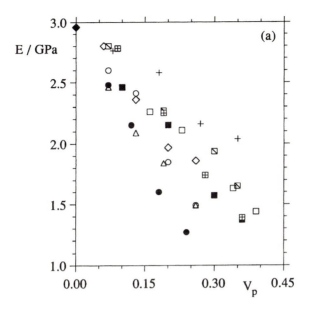

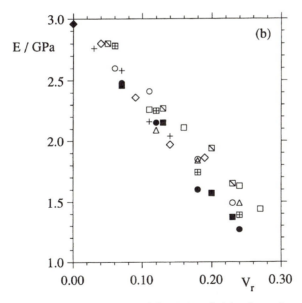

Figure 2. Variation of Young's modulus (E) with (a) volume fraction of particles (V_p), and (b) volume fraction of rubber (V_r), for rubber-toughened acrylic molding materials. Key: •, 2L; □, 4L; ■, 3LA; ◨, 3LAI; ⊞, 3LAII; ○, 3LB; ◇, 3LC; △, 3LD; +, 3LE; and ◆, Diakon LG156.

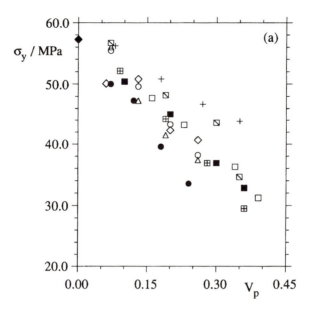

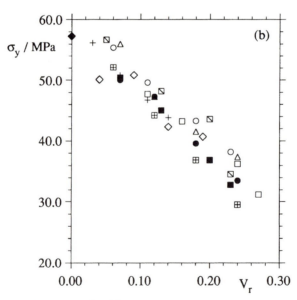

Figure 3. Variation of yield stress (σ_y) with (a) volume fraction of particles (V_p), and (b) volume fraction of rubber (V_r), for rubber-toughened acrylic molding materials. Key: •, 2L; □, 4L; ■, 3LA; ◨, 3LAI; ⊞, 3LAII; ○, 3LB; ◇, 3LC; △, 3LD; +, 3LE; and ◆, Diakon LG156. Diakon LG156, the 3LAII materials, 2L12, and 3LC10 did not yield, and so fracture stresses are plotted for these materials.

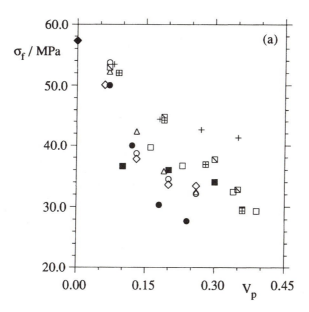

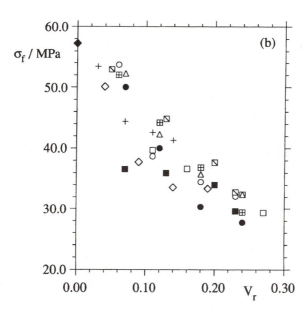

Figure 4. Variation of fracture stress (σ_f) with (a) volume fraction of particles (V_p), and (b) volume fraction of rubber (V_r), for rubber-toughened acrylic molding materials. Key: ●, 2L; □, 4L; ■, 3LA; ◨, 3LAI; ⊞, 3LAII; ○, 3LB; ◇, 3LC; △, 3LD; +, 3LE; and ◆, Diakon LG156.

With the exception of the 3LAII materials, yielding was accompanied by stress-whitening and the formation of shear bands. Although the observation of stress-whitening indicates the operation of voiding processes, more sophisticated techniques are required to elucidate its origin. Previous scanning electron microscopy (SEM) of tensile fracture surfaces showed the presence of numerous holes and domelike features with diameters similar to, and also smaller than, those of the toughening particles, suggesting that cavitation and/or debonding of the toughening particles contributes to stress-whitening (6, 15, 16). In the present work, optical microscopy and TEM of sections taken from stress-whitened material, and simultaneous tensile testing and real-time SAXS, were used to obtain more detailed information on the mechanism of deformation.

Optical microscopy of specimens from tensile tests shows significant differences in deformation behavior. The 2L, 3LA, 3LAI, 3LE, and 4L materials develop diffuse shear bands and stress-whiten at, and beyond, the yield point. The 3LB and 3LD materials stress-whiten with coarse shear bands emanating from diamond-shaped features that form just prior to yield; they also develop an undulating surface texture. The 3LC materials show similar coarse shear bands emanating from diamond-shaped features and have surface texture, but they stress-whiten only faintly. Examples of the different types of shear banding are shown in Figure 5.

The voids that gave rise to the stress-whitening were too small to be resolved in the optical microscope, and so TEM had to be employed to gain insight into the voiding processes. However, the observation of voided particles by TEM requires great care because, when exposed to electron radiation, the sections can undergo substantial relaxation, which results in either partial or complete closure of the voids. Although staining of the sections using ruthenium tetroxide greatly improves their stability in the electron beam, it also obscures the voids and so is of no benefit. With the exception of the 3LC, 3LE, and 3LAII materials, TEM micrographs of unstained ultramicrotomed sections taken from just below fracture surfaces show the presence of particles containing voided rubbery layers, which appear to have formed by failure of the rubber within the rubbery layer or at the interfaces between the glassy and rubbery layers. Representative micrographs are shown in Figure 6. The 3LC and 3LE materials have thin rubbery layers and, therefore, require examination at higher magnifications, exacerbating the problem of specimen relaxation. For these materials then, the TEM evidence for voided particles is inconclusive. The 3LAII materials, however, were examined in the same way as the 3LA and 3LAI materials but, in contrast, they show no evidence of voided particles. Thus, for the 3LAII materials, which are comparatively brittle, the voiding processes appear to be suppressed and unstable yielding is rapidly succeeded by fracture. By comparison with observations of the other materials, this result suggests that particle cavitation is advantageous in promoting tensile ductility in the rubber-toughened acrylic molding materials.

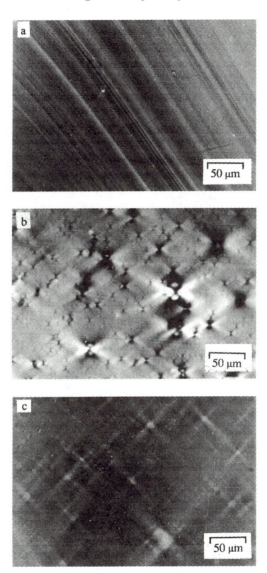

Figure 5. Optical micrographs of fractured tensile specimens showing shear-deformation bands viewed between crossed, polarizing filters to enhance contrast between the birefringent bands and the undeformed matrix; (a) bands typical of those observed in 2L, 3LA, 3LAI, 3LE, and 4L materials; (b) bands typical of those observed in 3LB and 3LD materials; and (c) bands observed in 3LC materials.

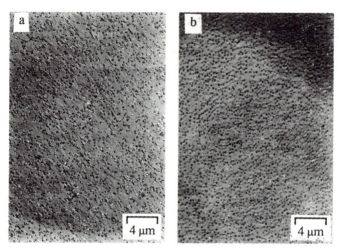

Figure 6. TEM images of unstained sections taken from just below the fracture surface in the stress-whitened regions of tensile specimens of 3LA34 (a) and 3LD31 (b). Cavities are evident inside the toughening particles.

The poor tensile properties of the 3LAII materials can be interpreted in terms of the change in cross-linking chemistry used for the preparation of the 3LAII particles. On the basis of studies of the polymerization of allyl methacrylate (11), the degree of cross-linking in the rubbery layer of 3LAII particles is estimated to be an order of magnitude greater than for the other types of particles. This difference has two obvious consequences. First, the stress required for cavitation in the rubbery layer is higher than for the other types of particles and, second, the glass-transition temperature of the rubbery phase is increased. The absence of cavitation on tensile deformation of the 3LAII materials, therefore, can be explained on the basis that both of these effects would be expected to delay cavitation and make the cavitation process more difficult.

The evidence from microscopy leads to the conclusion that particle cavitation is responsible for the stress-whitening observed in tensile specimens and that the voiding process is important for stable shear yielding of the materials. However, although the microscopic techniques provide a considerable amount of information about the deformation mechanisms operating in the materials, they are limited by being postmortem methods. Because the voids form within the rubbery layers of the particles, and stress-whitening is manifest only when the voids have grown large enough to scatter visible light, the point at which whitening is first observed cannot, with certainty, be taken to indicate the onset of voiding.

In an attempt to identify the sequence of deformation events, tensile testing was performed simultaneously with real-time SAXS (17, 18). Using SAXS, it is possible to detect voids only a few nanometers in size, thereby facilitating detection of the onset of void formation, even for voids formed within the thin rubbery layers of the 3LC and 3LE particles. When performed in real time, SAXS enables the onset of voiding to be located in relation to the stress–strain curve of a material.

The 3LC32, 3LD31, and 3LE31 materials were selected for examination because they contain particles with significant differences in size and morphology, and represent the three types of shear-band formation. Despite these differences, however, the observations are essentially the same. Oval-shaped, two-dimensional SAXS patterns evolve during tensile deformation and have intensity profiles that indicate the presence of voids that become increasingly elongated in the tensile direction as the strain increases (e.g., see Figure 7). The absence of the cross-shaped SAXS pattern characteristic of crazes (*17, 18*) confirms that crazing does not contribute to tensile ductility in these materials. The data obtained are exemplified by those given in Figure 7, which shows the evolution of the two-dimensional SAXS pattern with tensile deformation for 3LE31. As for the 3LC32 and 3LD31 materials, the results demonstrate that voiding coincides with yielding and that the onset of the increase in scattering intensity (due to voiding) occurs just prior to the yield point. The total scattering intensity increases markedly in the early stages of yielding, again showing

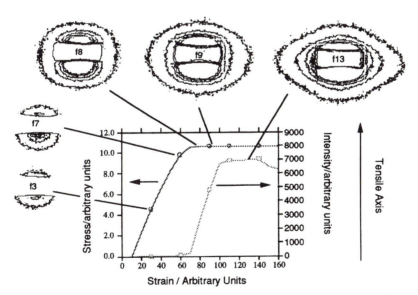

Figure 7. Results of simultaneous SAXS and tensile testing of 3LE31, showing that the onset of voiding occurs just prior to the yield point and that the extent of voiding increases dramatically during the early stages of plastic flow. The contour scattering patterns obtained at the points indicated suggest that the voids become increasingly elongated in the tensile direction as the strain increases.

that particle cavitation is important for stable macroscopic yielding of the matrix poly(methyl methacrylate). The principal difference between the results for the three materials is that the increase in scattering intensity on yielding for 3LC32 is approximately two orders of magnitude smaller than for the 3LD31 and 3LE31 materials, showing that the volume fraction of voids is much lower. It is apparent that the small size of the 3LC particles limits the voiding process, both in terms of the number of voids formed and their ability to grow (the voids remain too small to scatter visible light), and that this results in less control of yielding and lower fracture strains.

Impact Properties of the Rubber-Toughened Acrylic Molding Materials.

Figure 8 shows plots of critical strain-energy release rate (G_{Ic}) against V_r and V_p for each of the materials. The plots clearly show that under impact loading, G_{Ic} is strongly dependent on the size, morphology, and chemistry of the toughening particles.

The 3LAII and 3LC materials perform poorly under impact. For the 3LC materials, this poor performance appears to arise from the small particle size (152 nm outer diameter in the rubbery layer). The poor performance of the 3LAII materials can again be attributed to the relatively high level of cross-linking within the rubbery layers of the 3LAII particles.

An increase in G_{Ic} with increasing particle size is evident. The effect is most clearly seen by comparing the data for the 3LC, 3LB, and 3LD materials, which contain particles with 100 nm diameter glassy cores but with rubbery-layer outer diameters of 152, 202, and 255 nm, respectively. The results suggest that particles with rubbery-layer outer diameters greater than 200 nm are required to significantly improve toughness. Since the effect of particle size is small in going from the 3LB to the 3LD materials, the marked increase in G_{Ic} for the 3LA materials as compared with the 3LD materials can be assigned principally to the change in particle morphology rather than the further increase in size (from 255 nm to 284 nm).

The small differences in the G_{Ic} data for equivalent 3LA and 3LAI materials show that the increased level of interfacial graft-linking in the 3LAI particles has no significant effect on impact properties.

Comparing the data for the 2L and 3LB materials suggests that introduction of a glassy core into a homogeneous rubber particle leads to a small increase in G_{Ic} for a given value of V_r. However, the difference becomes less significant when the G_{Ic} data are plotted against V_p, implying that the effect is due to the corresponding increase in V_p at a given value of V_r for the 3LB materials as compared with the 2L materials.

The effect of the glassy core in three-layer particles is further demonstrated by the data for the 3LA and 3LE materials. These materials contain particles with essentially identical rubbery-layer outer diameters (284 and 290 nm, respectively), but with glassy-core diameters of 200 and 244 nm, respectively. The values of G_{Ic} for these materials are similar when plotted against V_p, but,

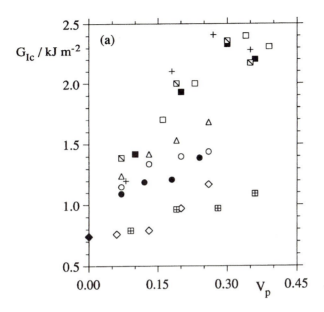

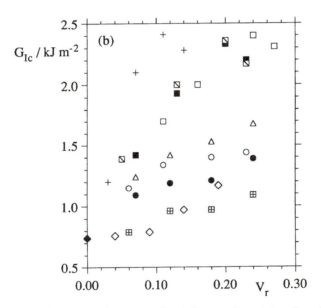

Figure 8. Variation of impact values of G_{1c} with V_p (a) and V_r (b), for rubber-toughened acrylic molding materials. Key: ●, 2L; □, 4L; ■, 3LA; ◨, 3LAI; ⊞, 3LAII; ○, 3LB; ◇, 3LC; △, 3LD; +, 3LE; and ◆, Diakon LG156.

because the 3LE particles have a much smaller rubber content, when considered at equivalent values of V_r the G_{Ic} values for the 3LE materials are considerably higher than for the 3LA materials. Because Young's modulus and yield stress correlate with V_r rather than V_p, the results show one major advantage of using three-layer particles, namely, that materials with improved tensile properties can be obtained without loss of impact performance by increasing the size of the glassy core.

Finally, comparison of G_{Ic} values for the 3LA materials with those for the equivalent 4L materials demonstrates that, contrary to patent claims (5), the introduction of a 100 nm diameter rubbery core into the glassy core of the 3LA particle has no significant effect on fracture resistance.

In order to investigate the deformation mechanisms operating under impact, double edge-notched specimens were subjected to impact loading in three-point bend geometry (8–10). The plastic zones formed around cracks grown subcritically in this way were sampled at different points by ultramicrotoming sections for examination by TEM.

The TEM images of 3LE31 shown in Figure 9 are typical of the deformation observed around and ahead of the subcritically grown cracks. Figure 9a is a micrograph taken at relatively low magnification in the plane normal to the crack and shows the presence of craze-like planar yield zones adjacent to the fracture surface. At higher magnification (Figure 9b), the yield zones can be seen to have propagated not only through the matrix between particles, but also through the particles. There is, however, no evidence of particle cavitation. Similar unstained sections also show no evidence of voids in particles, confirming that cavitation does not contribute to the deformation process. Hence, fracture under impact conditions is dominated by the formation of a multiplicity of craze-like planar yield zones.

Close inspections of high-magnification TEM images of stained sections reveals particles at the fracture surface that have either no glassy core or no rubbery layer at their exposed surface, and also that in most instances the yield zones propagate through particles close to the internal interfaces (see Figure 9b). Both observations suggest that the particles undergo failure at their internal interfaces.

Comparison of TEM micrographs of plastic deformation around subcritically grown cracks for a range of the materials shows that the values of G_{Ic} correlate with formation of the craze-like planar yield zones. Low values of G_{Ic} correspond to lower extents and densities of planar yield zone formation. For example, comparison of the TEM micrographs of stained sections from 3LE31 (Figure 9a) and 3LC32 (Figure 10a) shows that, for the latter material, the planar yield zones are finer and extend a shorter distance away from the fracture surface. TEM images of stained regions around subcritically grown cracks in 3LAII34 (Figure 10b) show a low density of planar yield zones, with no deformation beyond about three particle diameters away from the fracture surface.

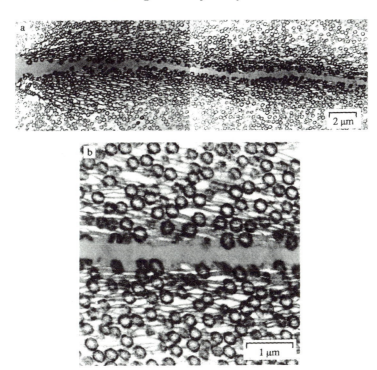

Figure 9. TEM images of stained sections showing deformation in the vicinity of a crack subcritically grown under impact in a double edge-notched, three-point bend specimen of 3LE31: (a) low-magnification micrograph showing the presence of planar deformation zones adjacent to the crack; and (b) the zones in (a) shown at a higher magnification.

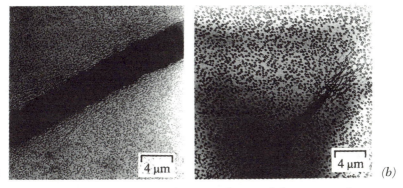

Figure 10. TEM images of stained sections showing deformation in the vicinity of a crack subcritically grown under impact in double edge-notched, three-point bend specimens of 3LC32 (a) and 3LAII34 (b).

Tensile Deformation and Impact Properties of the Rubber-Toughened Acrylic Sheet Materials. The rubber-toughened acrylic sheet (RTAS) materials were prepared by polymerization of methyl methacrylate into which 3LA toughening particles had been dispersed. Thus the properties of the RTAS materials can be compared with those of the 3LA materials described in the previous sections. The RTAS materials were prepared to investigate the effect of a substantial increase in the molar mass of the acrylic matrix (M_n for the acrylic sheet matrix, designated AS, was estimated by gel permeation chromatography to be 1,400 kg/mol, which may be compared with 47 kg/mol for Diakon LG156). However, TEM images of ultramicrotomed sections show that, unlike the 3LA materials, the particles are not completely dispersed in the RTAS materials and exist in clusters typically 5 to 10 μm in size. Hence, the differences in particle dispersion also need to be taken into account when interpreting the properties of the RTAS materials and comparing them with those of the 3LA materials.

Tensile testing showed that the 3LA particles induce yielding in the acrylic matrix leading to moderate increases in fracture strain (from 5% for the AS material to 8% for RTAS-8). The extent of yielding was similar to that of 3LA11, which has a fracture strain of 10%. The effects of increasing the weight fraction (w_p) of toughening particles on Young's modulus (E), yield stress (σ_y), and fracture stress (σ_f) are shown in Figure 11. As expected from the results for the rubber-toughened acrylic molding materials, both E and σ_y decrease approximately linearly with w_p. Comparing the values of σ_y and σ_f shows that yielding with strain softening is only pronounced for RTAS-8. The higher values of E for the RTAS materials as compared with 3LA11 is a reflection of differences in the values of E for the two acrylic matrix materials, which in turn can be attributed to differences in chemical composition: Diakon LG156 is a copolymer of methyl methacrylate that contains 8 mol% of n-butyl acrylate repeat units, whereas the AS material is a homopolymer, poly(methyl methacrylate).

During tensile testing, the RTAS materials stress-whiten, the extent of whitening increasing with strain and with w_p. In contrast, crazes can be seen prior to fracture of the AS material. So far, these visual observations have been further investigated only with real-time SAXS experiments. Preliminary results are shown in Figure 12. Whereas Diakon LG156 fractures by rapid, uncontrolled growth of a single craze and gives no SAXS information during tensile deformation, the AS material gives the characteristic cross-shaped SAXS pattern of crazes, (17, 18) which becomes more intense as the strain increases (Figure 12a). The formation of stable crazes in the AS material is responsible for its higher fracture strain (5% compared with 3% for Diakon LG156), and may be attributed to its much higher molar mass (i.e., greater number of entanglements per molecule), which enables the fibril-drawing process to proceed in a stable manner (1–3). The results of simultaneous SAXS and tensile testing of RTAS-8 are quite different from those of the AS

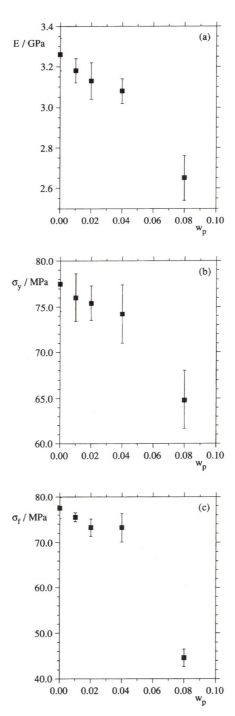

Figure 11. Plots of (a) Young's modulus (E), (b) yield stress (σ_y), and (c) fracture stress (σ_f) against weight fraction (w_p) of 3LA particles for the RTAS materials. The AS material and RTAS-1 did not yield, and so fracture stresses are plotted for them in (b).

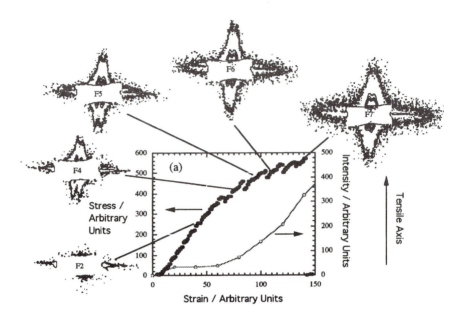

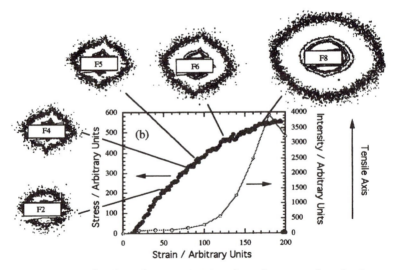

Figure 12. Results of simultaneous SAXS and tensile testing of acrylic sheet materials. (a) Results for the AS material show the formation of crazes during deformation. (b) Results for RTAS-8 show that the deformation processes that induce yielding are dominated by particle cavitation. The contour scattering patterns were obtained at the points indicated on the stress–strain curves.

material (compare Figures 12a and 12b). Although there is evidence that crazes are formed in RTAS-8 (as indicated by the vertical spikes in the SAXS patterns in Figure 12b), the SAXS patterns are otherwise similar to those obtained for the rubber-toughened acrylic molding materials and suggest that particle cavitation–debonding is the major deformation mechanism. Together with visual observation of shear bands, this result shows that incorporation of 8% by weight of the 3LA particles into the AS material causes the dominant mechanism operating under tensile deformation to switch from crazing to shear yielding.

Figure 13 shows that impact values of G_{Ic} increase significantly with w_p in the range investigated. The increase is slightly greater than that achieved by dispersing 3LA particles at the same level in Diakon LG156; for example, 3LA11 has a G_{Ic} value of 1.42 kJ/m², which may be compared with the value for RTAS-8 of 1.54 kJ/m². In order to interpret these observations, it is important to establish whether there are differences in the mechanism of fracture under impact loading for the two types of rubber-toughened acrylic material. At present, however, only preliminary studies of the mechanisms of deformation in the RTAS materials have been carried out, and the results are not sufficiently complete to be definitive. SEM images of fracture surfaces from both tensile and impact test specimens are markedly different from those reported (*6*, *15*, *16*) for the 3LA materials and show no evidence of particle cavitation. Instead, the SEM images show what appear to be complete particles sitting in

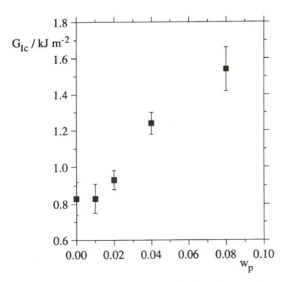

Figure 13. Variation of impact values of G_{Ic} with w_p of 3LA particles for the RTAS materials.

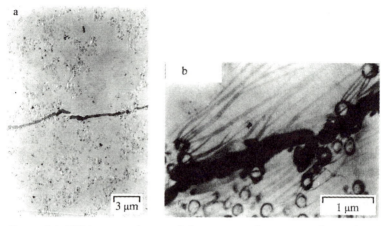

Figure 14. TEM images showing deformation in the vicinity of a crack subcritically grown under impact in a double edge-notched, three-point bend specimen of RTAS-8: (a) low-magnification micrograph of an unstained section showing voided particles and faint deformation zones in the plane normal to the crack; (b) micrograph of a stained section showing the deformation zones in (a) at a higher magnification.

wells, suggesting that particle debonding may be an important process in deformation of the RTAS materials.

More revealing information has been obtained from TEM examination of the deformation adjacent to a crack grown subcritically under impact in a double edge-notched specimen of RTAS-8 (see Figure 14). Figure 14a shows voids that appear to be located in the outer regions of the 3LA particles, and also reveals faint deformation zones in the matrix approximately parallel to the plane of the crack. The deformation zones can be seen more clearly in the stained section shown in Figure 14b, but they are not as distinct as those observed in micrographs of stained sections of the rubber-toughened acrylic molding materials. If the reasonable assumption is made that the stain highlights the deformation zones principally by filling void space (i.e., by highlighting density differences), this observation suggests differences in the nature of the deformation zones for the two types of material, with the void content of the zones being lower for the RTAS materials.

Summary and Conclusions

The Young's modulus and yield stress of rubber-toughened acrylic materials are effectively determined by the volume fraction of rubber and are insensitive to toughening-particle size and morphology. Tensile yielding in the ductile materials results from shear yielding in the matrix, which is stabilized and enhanced by the formation and growth of voids in the particles. For the RTAS materials, there is evidence that crazing may also make a small contribution.

The toughness of the rubber-toughened acrylic molding materials under impact is strongly dependent on the size, morphology, and chemistry of the toughening particles. For toughening particles that contain a glassy polymer core within the rubbery particle, it is possible to increase Young's modulus and yield stress without compromising toughness by increasing the size of the core.

The mechanism of deformation under three-point bend impact loading of edge-notched specimens is quite different from that operating during loading of tensile specimens. The dominant process under impact loading is the formation of planar yield zones ahead of, and approximately parallel to, the crack plane. For the rubber-toughened acrylic molding materials, toughness correlates approximately with the extent and density of formation of these planar yield zones. Although particle cavitation does not occur under impact loading in the rubber-toughened acrylic molding materials, examination of a similarly loaded specimen of an RTAS material revealed the formation of planar yield zones together with a small degree of particle cavitation.

Initial investigations of the RTAS materials suggest that the change in matrix material results in subtle changes in the deformation mechanisms operating under both tensile and impact loading. Further experimental work is necessary to probe these differences more thoroughly.

Acknowledgments

We express our thanks to the Engineering and Physical Sciences Research Council and ICI Acrylics for funding the research reported here. We also thank Martyn Murray of ICI for preparing the sheet materials and Tony Ryan of the Manchester Materials Science Centre for his invaluable help and guidance with the SAXS work. The assistance of Mike Chisholm and Bill Jung of ICI is gratefully acknowledged. One of the authors (R. J. Young) is grateful to the Royal Society for support in the form of the Wolfson Research Professorship in Materials Science.

References

1. Bucknall, C. B. *Toughened Plastics;* Applied Science: London, 1977.
2. Kinloch, A. J.; Young, R. J. *Fracture Behaviour of Polymers;* Applied Science: London, 1983.
3. Bucknall, C. B. In *Comprehensive Polymer Science*; Allen, G.; Bevington, J. C., Eds.; Pergamon: Oxford, England, 1989; Vol. 7, p 27.
4. Rohm and Haas Company, Br. Patent 1 414 187, 1975, 1 340 025, 1973.
5. E. I. Du Pont de Nemours and Company, Br. Patent 2 039 496A, 1979.
6. Lovell, P. A.; McDonald, J.; Saunders, D. E. J.; Sherratt, M. N.; Young, R. J. In *Toughened Plastics I: Science and Engineering*; Riew, C. K.; Kinloch, A. J., Eds.;

Advances in Chemistry 233; American Chemical Society: Washington, DC, 1993; pp. 61–77.

7. Lovell, P. A.; McDonald, J.; Saunders, D. E. J.; Young, R. J. *Polymer* **1993**, *34*, 61.
8. Pearson, R. A.; Yee, A. F. *J. Mater. Sci.* **1986**, *21*, 2475.
9. Sue, H-J.; Yee, A. F. *J. Mater. Sci.* **1989**, *24*, 1447.
10. Pearson, R. A.; Yee, A. F. *J. Mater. Sci.* **1991**, *26*, 3828.
11. Heatley, F.; Lovell, P. A.; McDonald, *J. Eur. Polym. J.* **1993**, *29*, 255.
12. Guild, F. unpublished results
13. Bucknall, C. B.; Partridge, I. K.; Ward, M. V. *J. Mater. Sci.* **1984**, *19*, 2064.
14. Milios, J.; Papanicolaou, G. C.; Young, R. J. *J. Mater. Sci.* **1986**, *21*, 4281.
15. Saunders, D. E. J. PhD Thesis, Victoria University of Manchester, England, 1990.
16. Lovell, P. A.; McDonald, J.; Saunders, D. E. J.; Sherratt, M. N.; Young, R. J. *Plast. Rubb. Comp. Proc. Appl.* **1991**, *16*, 37.
17. Kramer, E. J.; Brown, H. R. *J. Macromol. Sci., Phys. Ed.* **1981**, *B19*, 487.
18. Bubeck, R. A.; Buckley, D. J.; Kramer, E. J.; Brown, H. R. *J. Mater. Sci.* **1991**, *26*, 6249.

16

Loading-Rate Dependence of the Fracture Behavior of Rubber-Modified Poly(methyl methacrylate)

O. Julien[1], Ph. Béguelin[2,*], L. Monnerie[3], and H. H. Kausch[2]

[1]Elf Atochem, Cerdato, 27470 Serquigny, France
[2]Ecole Polytechnic Fédérale de Lausanne, Laboratoire de Polymères, MX-D, CH 1015 Lausanne, Switzerland
[3]Ecole Supérieure de Physique et Chimie Industrielles, Laboratoire de PCSM, 75231 Paris Cedex 05, France

The fracture behavior of neat and rubber-toughened poly(methyl methacrylate) (PMMA) was investigated at room temperature over five decades of testing velocities (from 10^{-4} to 14 m/s). The toughened PMMA materials underwent three successive transitions, whereas the neat material remained brittle over the entire range of testing velocities. The transitions seen in the toughened PMMA correspond to stable to partially unstable fracture, partially unstable to fully unstable fracture, and disappearance of toughening. The PMMA matrix was modified by incorporating core–shell latex particles (15, 30, and 45 vol%) and three-layer particles (30 vol%). Although materials modified by two-layer particles showed all three transitions (shifted to higher test speeds when the particle content was increased), three-layer PMMA had a higher fracture toughness and showed only the first transition. Together with an energetic approach, a fractographic study was used to investigate these transitions. The interdependence of toughening mechanisms promoted by the particles and relaxation behavior of the PMMA matrix is discussed.

T HE IMPACT PROPERTIES OF POLY(METHYL METHACRYLATE) (PMMA) may be improved by mechanical blending of small rubber particles with the matrix. Several authors have investigated the toughening mechanisms of these materials, which at present are only partially understood. In 1989, Mauzac and Schirrer (1) studied the effect of the rubber particles at the crack tip. They looked at particle volume fractions that ranged over five decades of concentra-

*Corresponding author.

tion, and found a progressive perturbation of the single craze at the crack tip. For particle contents above about 4%, they proposed the existence of craze bundles giving rise to an extended damage zone. Improvement of the toughness is associated with the extension of this damage zone.

The influence of the volume fraction of modifying particles was also studied by several other authors (2–4). Although the sizes and the particle morphology differed between studies, the most efficient toughening was observed for volume fractions between 25 and 40%. In creep experiments, Bucknall and co-workers (5) measured only a small volume change, even in highly-stress-whitened material. They postulated that shear yielding is the prevailing mechanism in creep in rubber-toughened PMMA (RTPMMA). In their study of the influence of the strain rate on the tensile volume strain, Frank and Lehmann (6) demonstrated the strong dependence of the strain rate on the cavitational process occurring for longitudinal strains greater than 3%. They saw a considerably higher volume strain at high strain rates than at low strain rates. In one of the first studies on RTPMMA, Hooley et al. (7) observed improved toughness in three-layer particle systems compared with two-layer systems. More recently, systematic studies have been conducted on several morphologies of toughening particles (8–10). The effect of a glassy core in a homogeneous rubbery particle increased the modulus, E, the yield stress, σ_y, and the critical-strain-energy release rate, G_{Ic}, for a given fraction of particles. They showed that increasing the size of the glassy core increased G_{Ic}.

Most of the authors cited observed a strong dependence of the toughening mechanisms on the testing rate. However, the loading rates used for fracture testing in these studies were generally one static rate and one impact rate.

In a previous study (11), we investigated the brittle–ductile transition temperature in the fracture of several fractions of modifying particles (0, 15, 30, and 45%) in RTPMMA. As the aim of the present work is to study the influence of the loading rate on the same materials as used in reference 11, it was important to cover a wide range of testing rates. The use of different test methods to cover this range often fails to provide consistent data, and at impact-testing rates the analysis of data by fracture mechanics is complicated by the dynamic nature of the tests (12–13). Here, a new testing procedure based on the reduction of the initial acceleration during the test by damping the contact stiffness was applied (14–16). Rather than determining the impact behavior of polymeric materials, this test procedure allows the measurement of properties at high loading rates, under quasi-static stress conditions. Using this technique, the fracture behavior of neat and rubber-toughened PMMA was investigated over five decades of test speed, from 10^{-4} to 14 m/s.

For two different morphologies of modifying rubber particles, the mechanical transitions were identified by means of an energetic approach together with postmortem surface analysis on a macroscopic scale. The dependence of these transitions on the content of modifying particles is examined.

Experimental Details

Test Equipment. Fracture tests in mode I opening were performed with a Schenck high-speed servohydraulic testing apparatus. To avoid loss of contact between the test specimen and the adjacent parts at high testing rates, a compact tension geometry was used (see Figure 1). The load was measured by means of a reduced-weight piezoelectric load cell, and the specimen fixture had a high natural frequency (50 kHz). The specimen displacements were measured by a contact-free optical device based on laser light and optical-fiber technology. Further details on the experimental procedure can be found in references 15 and 16.

Materials. A PMMA matrix with a molecular weight, M_w, of 130,000 g/mol was toughened with two-layer, rubbery, core–glassy shell particles. The materials were synthesized by Atochem (6). The core component is a copolymer based on poly(*n*-butyl acrylate-*co*-styrene), prepared by emulsion polymerization. The shell consists of PMMA grafted to the core to ensure good adhesion between the particle and the matrix. The average external particle diameter is 160 nm, while that of the rubbery core is 140 nm. Toughened PMMA granulates were prepared by mechanical blending of the neat matrix with 15, 30, and 45% volume fractions of latex. The neat PMMA used as the matrix of those systems was also investigated. For convenience, these systems are named here 2 L0 for the neat matrix, and 2 L15, 2 L30, and 2 L45.

For comparison, a commercially available RTPMMA (Altuglas ARS, named here 3 L30), containing a 30% volume fraction of three-layer particles, was tested under the same conditions. The modifying particles have a rubbery inner shell about 200 nm in diameter surrounding a glassy core with rubbery inclusions. The third layer is a glassy outer shell of 250 nm. The morphology of the particles is shown schematically in Figure 2.

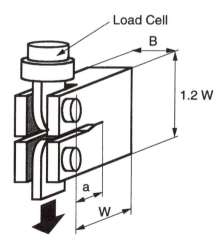

Figure 1. Geometry of the compact-tension test. W = 20 mm, B = ~6 mm, *and* a = 9–11 mm.

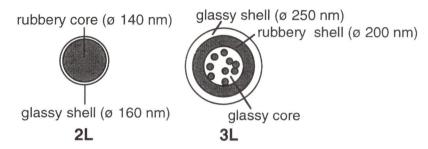

Figure 2. Schematics showing the morphology of two-layer particles (left) and three-layer particles (right). The diameter is given by φ.

Notching. Owing to the wide range of toughness of our materials, great care was taken in the notching procedure. For the 2 L0 (neat matrix), a fresh razor blade was tapped in the machined notch and a "natural" crack was propagated as a prenotch. Using the same method for 2 L15, "naturally" prenotched specimens by brittle failure of the matrix were also obtained; comparison of the fracture toughness measured on specimens prenotched by drawing a fresh razor blade along the machined notch showed lower values as well as lower scatter with this latter technique. Further checks by photoelasticity and the method of caustics confirmed that tapping a razor blade produces residual deformation ahead of the prenotch in RTPMMA. Therefore, the prenotch used for the 2 L15, 2 L30, 2 L45, and 3 L30 specimens was produced by drawing a razor blade along the notch tip.

Testing Conditions and Analysis. The fracture behavior was investigated at room temperature at nominal piston velocities, from 10^{-4} m/s to 10 m/s. For test speeds higher than 10^{-1} m/s, the damped test procedure described in reference 15 was used. Quasi-static stress conditions therefore prevailed in the specimen, even at high loading rates. This fact allowed the analysis of fracture-mechanics parameters to be performed using a static approach.

As discussed in the next section, the maximum stress intensity factors (K_{Imax}) were calculated for the maximum force. The calculations of the strain-energy release rate (G_{Imax}) were based on the integration of the load versus displacement curves. When stable fracture occurred after the maximum force was reached, the energies for crack initiation and propagation were calculated separately. The integration of the load-displacement curve up to the maximum force (F_{max}) is associated with the initiation process. The energy after peak load is associated with the propagation of the fracture.

Crack speeds for unstable propagation were measured for certain specimens using a graphite gauge (17).

For the postmortem analysis of stress-whitening resulting from the slow growth of a crack, a numerical image of each fracture surface was made under identical lighting conditions using a video camera. Computer analysis of the images allowed us to quantify these surfaces. For each concentration of latex, a constant gray-level threshold was used to obtain a binary image of the stress-whitened zone of the fractured ligament. The relative surface of this zone was then measured.

Results

Load Displacement and Fracture Behavior Over a Range of Test Speeds.

As discussed at the beginning of this chapter, many authors have reported an apparent correlation between the toughening mechanisms and the development of a stress-whitened zone that diffracts light in the material. The occurrence of stress-whitening at 3–4% strain coincides with the appearance of both cavitation (6) and shear strain in the matrix (8).

A typical force versus displacement curve is shown in Figure 3. The initial regime, **a**, of the force-displacement diagram is linear. It is followed by a nonlinear regime, **b**, where the damage zone, which is stress-whitened, is developed together with some subcritical crack extension. This damage zone is fully developed when the force reaches its maximum (F_{max}). Then the force starts to drop and the third regime, **c**, corresponds to crack extension, the crack growing either in a stable or unstable manner. When subcritical (stable) crack extension occurs, it proceeds together with a damaged zone surrounding it. If the crack extension is critical (unstable), the force drops rapidly at F_{max}. In this case, the size of the stress-whitened zone remains what it was when unstable growth started, and it does not increase as the crack continues to propagate. Intermediate behavior is also seen where crack propagation is stable immediately after F_{max} but subsequently unstable.

Because the existence and extent of the nonlinear regime, **b**, is highly dependent on the fraction of the modifying phase and the test speed, it is believed to play an important role in the kinetics of toughening and thus is considered for fracture characterization. In order to apply the same analysis over the entire range of test speeds, the standard linear elastic fracture mechanics

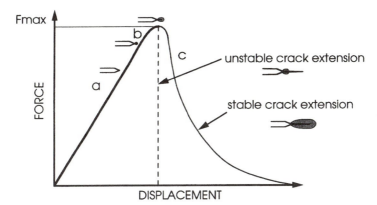

Figure 3. Typical force versus displacement curve associated with growth of the damaged zone. Key: a, initial, linear regime; b, nonlinear regime; and c, the third regime, corresponding to crack extension.

(LEFM) analysis was performed. Instead of rigorously applying the European Structural Integrity Society TC4 protocol (*18*), which requires linearity criteria to be respected, the maximum force, F_{max}, and the length, *a*, of the prenotch prior to the loading of the specimen were used for the analysis. The length of the prenotch, *a*, was the same for all specimens, which gave constant ratio, *a/W*, of 0.5 (where *W* is the width of the specimen). Thus, the results are expressed with respect to the test speed instead of the crack-tip loading rate *dK/dt*, which was used in previous studies (*15–16*).

Results of Fracture-Toughness Tests over the Range of Test Speeds.

Figure 4 shows the stress intensity factor, K_{Imax}, versus test speed for all the materials tested. The test speed is the true opening displacement rate of the specimen when fracture occurs. It is measured by the optical device, and it can be up to 14 m/s. Typical times to fracture vary from ~10 s at the lowest speeds to ~150 µs at the highest.

The neat PMMA (2 L0) behaves in a brittle manner over the entire range investigated. At the lowest test speed, stable or partially stable crack propagation can occur. At higher test speeds, crack propagation is always fully unstable. The apparent rise in the fracture toughness, K_{Ic}, with test speed may be associated with a reduced contribution of thermal activation resulting from the reduced time-to-fracture when the test rate is increased. Such an effect is consistently observed in the fracture of thermoplastic polymers (*19*). Because chain slip is involved, this effect depends on molecular weight; thus a previous

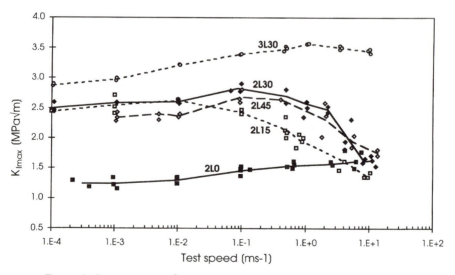

Figure 4. Stress intensity factor, K_{Imax}, versus the logarithm of the test speed, for all systems tested.

study of neat PMMA of high molecular weight ($M_w = 4 \times 10^6$) (*13*), with identical experimental conditions, did not show the same increase in K_{Ic} with test speed.

Considering the fracture toughness of the products toughened with two-layer particles (2 L), there are obvious transitions with increasing the loading rate at room temperature. The fracture toughness of all 2 L RTPMMA materials remained high over three decades of test speed; and the materials exhibited decreasing fracture resistance at rates higher than about 1 m/s. This ductile-to-brittle transition appears to be especially noticeable for 2 L15, where it occurred at lower test speeds than in materials with a higher rubber content. Increasing the rubber content improved fracture toughness, particularly in the high-speed regime, where loss of toughness occurs. Under the same conditions, K_{Imax} remained constant and even increased with test speed for the material toughened with three-layer modifying particles (3 L30). For all the test velocities investigated, the fracture toughness and fracture energy (which is not reported here) of 3 L30 remained higher than those of 2 L. No ductile-to-brittle transition was observed for 3 L30.

We observed exactly the same range of variation of K_{Imax} values in the 2L materials, and thus the same trend of fracture behavior, by lowering the temperature (*11*). The main difference was that the transition zones appeared more extended in speed than in temperature. Furthermore, the testing technique applied here permitted a better splitting of the contributions of initiation and propagation to the overall process of fracture than the Charpy testing described in reference 11.

Nevertheless, such a fracture-mechanics approach deals only with the initiation stage of the fracture process. In order to consider the whole fracture process (initiation and propagation, as described in the preceding discussion), an energetic approach is proposed here, which allows us to characterize the transitions in a precise and reproducible manner.

Determination of Transitions in the Fracture Process.   To determine the transitions, different fracture parameters must be considered together: K_{Imax}, G_{Imax}, initiation energy, U_{init}, and propagation energy, U_{prop}, as well as the macroscopic stress-whitened area of the fracture surface.

Fracture Behavior and Transitions.   The evolution of typical fracture behavior with test speed is summarized in Table I. At low test speed, propagation is fully stable and the whitened zone covers the entire fracture surface. Then as the test speed increases, the whitened area of the fractured ligament (which corresponds to the area of stable crack growth) and the energy of propagation decrease continuously, and the fracture becomes partially unstable. Such a transition from stable to unstable crack propagation depends on the geometry and size of the specimen used, especially the ligament length, $W - a$. This transition is called Transition 0.

Table I. Transitions and Fracture Behavior of 2 L RTPMMA over the Range of Testing Speeds Investigated

Upon increasing the test speed further, the crack changes beyond F_{max} from partially stable to fully unstable, and the additional energy required to fracture the specimen drops sharply. In fact, above a certain speed, if rapid crack propagation occurs at or below F_{max}, the propagation energy cannot be separated from the initiation energy because the driving force of the crack is the elastic energy stored in the specimen. At these speeds, the crack velocity is high and the whitened zone is restricted to the process zone developed ahead of the precrack during the initiation process. This transition to unstable fracture is called Transition I.

The final transition, Transition II, occurs at a test speed about two decades above that at which Transition I is seen. The unstable fracture already occurs in the linear domain of the force–displacement curve (**a** in Figure 3). This transition is related to the total disappearance of toughening effects. The absence of a whitened zone on the specimen is noticeable. Under these conditions, the $K_{Imax} = K_{Ic}$ and $G_{Imax} = G_{Ic}$ values measured for the 2 L15 system are even lower at high test speeds than those measured for the neat PMMA.

The Method Used to Define the Transitions.   In determining these three transitions, the most important parameters are the relative size of the whitened area on the fracture surface (normalized by the entire surface of the ligament), and the decrease in propagation energy. These parameters define the transitions marked by arrows in Figure 5. The parameters are listed in Table II. Figure 5 shows an example of how the transitions were defined using the data for 2 L15. The results obtained for all materials, as well as the crack velocities measured, are summarized in Table III.

It is important to note that for all 2 L RTPMMAs, the whitened area of the fracture surface decreases continuously between transitions 0 and II when the test speed is raised, and it does not present the abrupt variation that can be observed when the temperature is varied.

Discussion

Although the toughening mechanisms may be different in RTPMMA, the transitions in fracture behavior observed here at high rates are qualitatively similar to those described by Bucknall (*20*) for the impact of high-impact polystyrene containing different fractions of modifier. Here, the main result is that all the transitions are shifted simultaneously when the 2 L rubber content is increased.

Because the highest testing speed reached in our experiment setup was 14 m/s, it is not clear whether Transition II for 30% and 45% RTPMMA was fully completed at the highest velocities investigated. If it was not, a further decrease in fracture resistance may occur at higher velocities. Transitions 0 and I of 2 L15 occur at speeds close to those of the transitions in the neat ma-

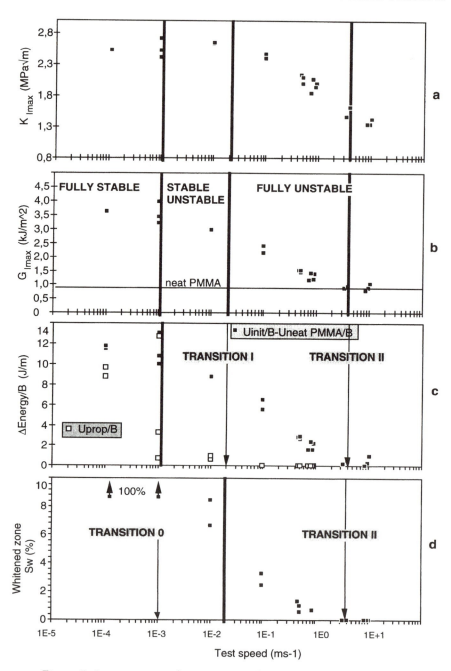

Figure 5. Stress intensity factor, a: K_{Imax}. b: Strain-energy release rate, G_{Imax}. c: Normalized initiation energy, U_{init}, and normalized energy for propagation, U_{prop}. U_{neat} PMMA is the energy to fracture neat PMMA (2 LØ) d: Stress-whitened area of the fracture surface, Sw, used to determine the transitions in 2 L15.

Table II. Criteria Used to Locate the Transitions

Transition	Criterion
Transition 0: fully stable to partially stable fracture	Below this transition, the whitened zone covers 100% of the fracture surface (see Figure 5d)
Transition I: partially stable to fully unstable fracture	Energy for initiation and propagation cannot be separated (Figure 5c) (drastic drop in force at F_{max})
Transition II: total disappearance of toughening mechanisms	Fracture energy reaches values equal to or below that of the neat matrix (Figure 5c); total disappearance of the stress-whitened zone (Figure 5d)

NOTE: Transitions are in order of increasing speed.

trix. Nevertheless, at low rates this system shows a higher fracture resistance than the neat PMMA. The lower values measured at high speed for 2 L15 compared with the neat matrix can be explained by the lower amount of PMMA present at the crack tip.

From the crack-velocity measurements performed at a piston speed of 1 m/s, it appears (in the case of unstable propagation) that the crack speeds in the toughened materials are higher than those measured in the neat material. This result is probably due to the higher elastic energy stored in the specimen subsequent to crack blunting in the toughened material. These crack velocities are close to the maximum speeds generally measured in polymers.

Transitions and the Whitened Zone. The variation of K_{Imax} with the whitened area of the fracture surface has been plotted for all 2 L rubber contents. The results fit on a single "master" curve (Figure 6). A universal trend is observed, although the kinetics of growth of the whitened zone are highly dependent on rubber content, which is probably determined by the matrix. This observation suggests that the same mechanisms are involved in de-

Table III. Test Speeds at Which Transitions Occur, and Crack Speed for Neat PMMA and 2 L RTPMMAs

Material 1	Content of modifier (vol%)	Transition 0 (m/s)	Transition I (m/s)	Transition II (m/s)	Maximum crack speed at a piston velocity of 1 m/s
2 L	0	0.001	0.01	—	120
2 L	15	0.001	0.02	4	540
2 L	30	0.009	0.2	~13	550
2 L	45	0.05	0.8	Not reached	Not measured
3 L	30	0.1	Not reached	Not reached	Not measured

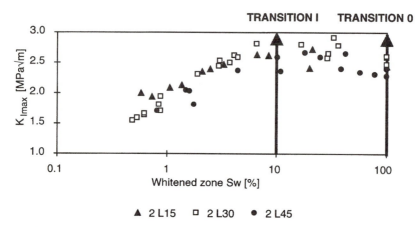

Figure 6. Variation of K_{Imax} with the length of the relative area of the ligament covered by stable crack growth (the stress-whitened surface area).

formation and fracture for all the 2 L products. As a consequence, the break in the slope marking Transition I appears at the same stress-whitened area for all rubber contents: it covers roughly 10% of the surface of the total fracture area. In other words, the highest test speed at which the plastic zone can confine a stable fracture process (Transition I) is reached at roughly the same crack length, whatever the particle content.

Contribution of the Whitened Zone to Crack Stability and Fracture Toughness. A higher toughness at high speeds is achieved when the observed transitions are shifted to higher rates. Therefore, a knowledge of the mechanisms that affect the rates at which transitions occur is important. The ability of the material to balance the energy supplied to the specimen and the energy dissipated in the growth of the plastic zone (stress-whitened) together with the blunt crack is needed to prevent unstable fracture, and hence to retain a high toughness at high rates.

Conditions of Growth of the Whitened Zone. The nature of the volumic plastic zone was investigated by Lovell and co-workers (10) with in situ small X-ray scattering. In tensile experiments, they found that cavitation inside the particles precedes yielding. Although they did not observe interparticular crazes at low rates in tension, they found that subcritical crack growth in impact was dominated by microcrazing. A time-dependent process, related to relaxation times of the matrix, is involved in propagating the micromechanisms of deformation between the particles. Cavitation of the particles, which activates the shear processes of the matrix, is believed to be a determinant of toughening. It cannot occur at high rates because of early brittle failure of the matrix by crazing. This inhibits the development of the macroscopic volumic plastic zone enhanced by the particles at lower rates. In other words, high values of initiation and propagation energies are achieved by large-scale plasticity around the crack. The kinetics of such large-scale plasticity are highly time-dependent. For the toughening mechanism to be efficient, two conditions must be satisfied:

1. The rubber particles must be in the rubbery state to initiate the toughening mechanism (i.e., cavitation)
2. Once initiated, matrix shearing must have time to grow and propagate throughout interparticular PMMA to be effective.

Unlike time, temperature acts equally on both matrix and rubber particles. In our previous study (11), where the temperature was varied, the lower transition temperatures were identical (~–35 °C) for 2 L30 and 2 L45. Identical temperatures occurred because below this temperature the particles are in the glassy state (21).

Here, the test rate is varied at room temperature, which is a more sensitive way of studying the interactions between the particles and the properties of the matrix (condition 2). For example, it is likely that in our experiments, the particles remain in the rubbery state (T = 23 °C) while the speed varies over the five decades. Corroborating this point is the fact that scanning electron micrographic (SEM) fractographic observations show that particle bridging and localized plastic shear deformation of the matrix still occur at the interface between particles and matrix in the fracture plane for any tested system. Figure 7 shows the fracture surface created by an unstable crack running at a

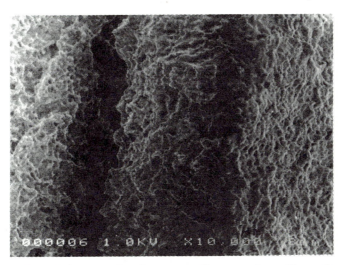

Figure 7. SEM fractographic observation of an area of unstable crack growth, in a 2 L45 sample tested at 2 m/s.

speed of several hundred meters per second. In such a case, this small-scale plasticity is restricted to the surface plane, no whitened zone is visible in this particular part of the ligament, and the fracture energy is rather low. However, although the crack velocities of the unstable fracture of 2 L15 and 2 L30 are very similar (see Table III), the test speeds at which Transition I occurs are different. Thus the shift of this transition cannot be explained by a change in fracture velocity.

Crack Stability.   At low test speeds, stable crack growth with an extended stress-whitened plastic zone and crack blunting occur by the same mechanisms as those involved in the kinetics of the plastic zone, namely, rubber cavitation followed by shear deformation of the matrix. The ability of the matrix to shear is controlled by its relaxation behavior, which therefore determines its plasticity and the deformation imposed on rubbery particles distant from the notch.

At high speeds, the volumic dilatation of the particles is too low when failure of the matrix occurs at the notch tip. Therefore, no cavitation can occur in an extended volume around the notch, eliminating the possibility of toughening by shearing of the matrix. Decreasing the interparticular distance by raising the content of particles shifts the disappearance of toughening mechanisms to higher speeds.

However, when stable–unstable fracture occurs (between Transitions 0 and I), evidence of secondary crack nucleation has been obtained by SEM on postmortem fracture surfaces. These cracks lead to parabolic markings ahead of the main blunt crack propagating in the plastic zone. Figure 8a shows an example of such secondary cracks developed at the boundary between the stable growth region and the unstable fracture surface in a 2 L45 specimen tested at 0.1 m/s.

The orientation of the parabolic markings provides an indication of the relative direction of the crack. In Figure 8b it is clear that several secondary cracks have grown rapidly in region 3, moving toward the stable crack, which has propagated together with the plastic zone in the nonlinear region of the load-displacement trace (region 1). Secondary cracks have also grown in region 2, where they are moving toward the edge of the specimen.

The higher toughness, at all speeds, of PMMA modified with three-layer particles is related to the ability of these particles to cavitate (or debound) more easily than the two-layer ones. This ability is mainly due to (1) an increase in shear stress at the interface of the rubbery shell, and (2) a higher local strain resulting from their larger diameter. These two effects occur because the rubber lies between two glassy parts: the inner PMMA core and the outer PMMA shell. This statement means once more that we assume that the cavitation process is the main event triggering the shearing mechanisms of the matrix.

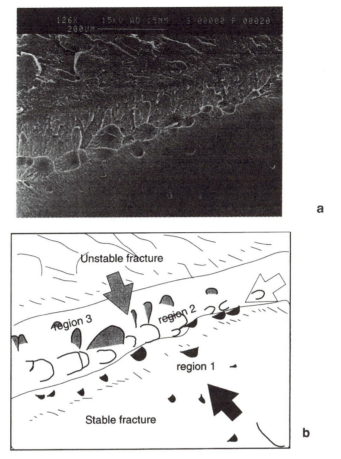

Figure 8. 2 L45 tested at 0.1 m/s: (a) SEM fractographic observation of the intermediate region between stable and unstable crack propagation; and (b) direction of the main and secondary cracks, indicated by arrows, according to the three regions (see text) containing parabolic marks.

Conclusions

The three mechanical transitions observed in the RTPMMA are shifted to higher test speeds when the particle content is increased. None of these shifts to higher test speeds can be explained by a change in crack-propagation velocity because similar velocities have been measured for 15 and 30% volume fractions of modifier. Thus it is postulated that the shift is related to the change in the interparticular distance. As suggested by Bucknall (*20*), the growth kinetics of the micromechanisms of toughening in the interparticular matrix are involved.

Our investigations over a wide range of test speeds have shown that the fracture behavior is a fully time-dependent process, which involves the glassy matrix as well as the rubbery particles. The transitions result from the progressive disappearance of the ability of the material to develop large-scale plasticity triggered by the presence of the particles. They appear then to be mainly controlled by matrix-relaxation behavior together with the nature of the particles.

PMMA toughened by three-layer particles, such as the 3 L30 system studied here, has a higher fracture toughness over the entire range of test rates than PMMA toughened with two-layer particles. In our three-layer toughening system, only Transition 0 was observed. Transitions I and II may exist, however, at higher test rates than those investigated here.

Acknowledgments

We gratefully acknowledge the financial support of Elf Atochem. We also thank B. Senior of the Institute of Microscopy, Ecole Polytechnique Fédérale de Lausanne, for assistance with scanning electron microscopy.

References

1. Mauzac, O.; Schirrer, R. *J. Appl. Polym. Sci.* **1989,** *38,* 2289–2302.
2. Mauzac, O.; Schirrer, R. *J. Mater. Sci.* **1987,** *25,* 5125–5133.
3. Wrotecki, C.; Heim, Ph.; Gaillard, P. *Polym. Eng. Sci.* **1991,** *31(4),* 213–222.
4. Gloaguen, J. M.; Steer, P.; Gaillard, P.; Wrotecki, C.: Lefebvre, J. M. *Polym. Eng. Sci.* **1993,** *33(12),* 748–753.
5. Bucknall, C. B.; Partridge, I. K.; Ward, M. V. *J. Mater. Sci.* **1984,** *19,* 2064–2072.
6. Frank, O.; Lehmann, J. *Colloid Polym. Sci.* **1986,** *264,* 473–481.
7. Hooley, C. J.; Moore, D. R.; Whale, M.; Williams, M. J. *Plast. Rubber Process. Appl.* **1981,** *1(4),* 345–349.
8. Lovell, P. A.; McDonald, J.; Saunders, D. E. J.; Sherratt, M. N.; Young, R. J. *Toughened Plastics I: Science and Engineering;* Riew, C. K.; Kinloch, A. J., Eds. American Chemical Society: Washington, DC, 1993; pp. 61–77.
9. Archer, A. C.; Lovell, P. A.; McDonald, J.; Sherratt, M. N.; Young, R. J. *Polym. Prepr. (Am. Chem. Soc. Div. Polym. Mater. Sci. Eng.)* **1994,** *70,* 153–154.
10. Lovell, P. A.; Ryan, A. J.; Sherratt, M. N.; Young, R. J. *Polym. Prepr. (Am. Chem. Soc. Div. Polym. Mater. Sci. Eng.)* **1994.**
11. Julien, O. M. Phil. Thesis, Cranfield Institute of Technology, Cranfield, UK, 1991.
12. Böhme, W.; Kalthoff, J. F. *J. Phys. Colloq.* **1985,** *C5,* (Suppl. 8, No. 46), 213–218.
13. Williams, J. G.; Adams, G. C. *Int. J. Fract.* **1987,** *33,* 209–222.
14. Béguelin, Ph.; Barbezat, M. *J. Phys. III* **1991,** *1,* 1867–1880.
15. Béguelin, Ph.; Kausch, H. H. *J. Mater. Sci.* **1994,** *29,* 91–98.
16. Béguelin, Ph.; Kausch, H. H. In *ESIS on Impact and Dynamic Fracture of Polymers and Composites;* ESIS 19, Williams, J. G.; Pavan, A., Eds.; Mechanical Engineering Publications: London, 1996; pp 3–19.

17. Stalder, B.; Béguelin, Ph.; Roulin Moloney, A.-C.; Kausch, H. H. *J. Mater. Sci.* **1987,** *24*, 2262.
18. Williams, J. G.; Cawood, M. J. *Polymer Testing* **1990,** *9*, 15–26.
19. Kausch, H. H. *Polymer Fracture,* 2nd ed.; Springer Verlag: Berlin, Germany, 1987.
20. Bucknall, C. B. *Makromol. Chem. Macromol. Symp.* **1988,** *16*, 209.
21. Bucknall, C. B. In *Toughened Plastics,* Applied Science Publishers: London, 1977.

17

Investigations of Micromechanical and Failure Mechanisms of Toughened Thermoplastics by Electron Microscopy

G. H. Michler and J.-U. Starke

Department of Materials Science, Martin-Luther-University Halle-Wittenburg, D–06217 Merseburg, Germany

In different toughened thermoplastics modified with particles, the competitive influence of particle diameter and interparticle distance on the toughening mechanism was studied. Using the techniques of transmission and scanning electron microscopy, including in situ techniques, the morphology and the micromechanical processes of deformation and fracture were investigated. The toughening is the result of different mechanisms, which depend on the micromechanical behavior of the base polymer. In toughened high-impact polystyrene and acrylonitrile–butadiene–styrene, the enhanced toughness is based mainly on the formation of crazes, and the particle diameter, D, is the parameter of primary importance. When toughening is based mainly on shear deformation, the interparticle distance, A, is the parameter of most importance. In both cases there are critical sizes of D and A.

B Y ADDING RELATIVELY SMALL AMOUNTS of elastomers to polymers, it is possible to increase the fracture toughness up to one order of magnitude. Toughened polymers may be distinguished by using different criteria, including composition, morphology, and the toughening mechanism (1–3).

Composition (type of polymeric components). The base polymer (which is to be modified) may be an amorphous polymer [e.g., polystyrene (PS), styrene–acrylonitrile copolymer, polycarbonate, or poly(vinyl chloride)], a semi-crystalline polymer [e.g., polyamide (PA) or polypropylene (PP)], or a thermoset resin (e.g., epoxy resin). The modifier may be a rubber-like elastomer (e.g., polybutadiene, ethylene–vinyl acetate copolymer, ethylene–propylene copolymer, or ethylene–propylene–diene copolymer), a core–shell modifier, or another polymer. Even smaller amounts of a compatibilizer, such as a copolymer, are sometimes added as a third component to control the morphology.

Morphology or phase structure (type of arrangement of the different polymer phases). The base polymer is the matrix in which the rubber or modifier phase is dispersed in particles (dispersed systems). The base polymer may be present in particle form and surrounded by thin elastomer layers like a honeycomb or network (network systems). The shape and size of the different phases (particles) and the volume content of the phases are important parameters.

Toughening mechanism (the micromechanical mechanism that enhances toughness). In dispersed systems there are two categories of mechanisms: either the energy-absorbing step is the preferred formation of crazes at the rubber particles (multiple crazing), as in high-impact PS (HIPS) and numerous grades of acrylonitrile–butadiene–styrene (ABS) polymers (1–4), or the energy absorption mainly takes place through shear deformation between the modifier particles (multiple shearing), as in impact-modified PA or PP (3, 5–6). In the network of honeycomb systems a third mechanism is present whereby an intensive yielding of the thermoplastic particles occurs inside the meshes of the network (multiple particle yielding) (2, 7, 8).

Many parameters that are connected with one another are important for toughness in dispersed polymer systems. On the basis of the molecular parameters of the components (molecular structure, macromolecular arrangement, and compatibility), and depending on processing or manufacturing, two sets of parameters are of primary importance:

1. Morphological parameters, including
   - Phase structure
   - Supermolecular structure of the components
   - Volume content of the particles
   - Shape and size of the particles
   - Interparticle distance
   - Interfacial structure.
2. Micromechanical parameters, including
   - Local stress concentration in the matrix at and between particles (depending on particle shape, particle content, Young's modulus, and Poisson's ratio of matrix polymer and particles)
   - Fracture strength of the matrix, particle, and interface
   - Yield stress of the matrix polymer
   - Deformation characteristic of the matrix (crazing or shear deformation).

Two parameters are of particular interest: the diameter, D, of modifier particles and the distance, A, between particles (interparticle distance). Until now, the literature has given conflicting accounts of the influence of these parameters on different systems. HIPS and ABS have often served as model systems for producing new materials, but the fact that the morphologies as well as the micromechanical mechanisms of toughness enhancement can be different has not always been considered.

The aim of this work is to study the influence of particle size, interparticle distance, particle volume content, and local stress state on the toughening mechanism in several dispersed systems. The systems consist of a matrix of an amorphous or semicrystalline thermoplastic (see Figure 1). It is necessary to determine whether the particle diameter or the interparticle distance is of primary importance. But it is difficult to check the influence of both parameters because there is an interrelation between D, the average minimum value of A, and the particle volume content, v_P:

$$A = \left(\sqrt[3]{\frac{\pi}{6v_P}} - 1 \right) D \tag{1}$$

Figure 2 shows the variation of A with v_P for several values of the D.

To study the influence of these structural parameters on the micromechanical mechanisms of toughening, several techniques of electron microscopy were used. Electron microscopic techniques allow investigations not only of detailed morphology but also of the micromechanical processes of deformation and fracture (*3, 9–11*).

Experiments

Materials Studied.   The materials investigated are listed in Table I. Different amorphous and semicrystalline thermoplastics were used as matrix materials, which were modified with several types of rubber. The volume content of the rubber and the rubber-particle diameter were changed to enable A

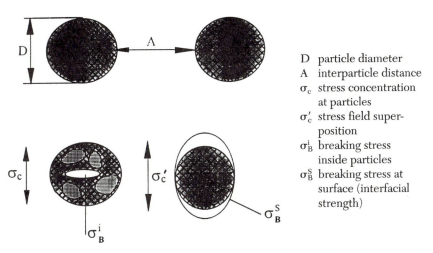

D particle diameter
A interparticle distance
σ_c stress concentration at particles
σ_c' stress field superposition
σ_B^i breaking stress inside particles
σ_B^S breaking stress at surface (interfacial strength)

Figure 1. Schematic of important concepts used to study toughening mechanisms.

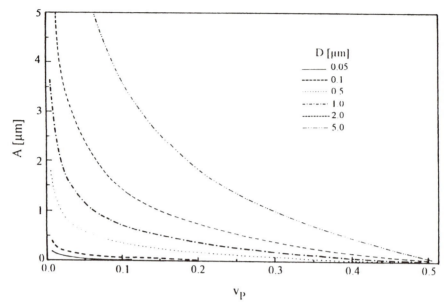

Figure 2. Variation of interparticle distance, A, with particle volume content, v_P, and particle diameter, D.

Table I. Materials Investigated

	Modifier		
Matrix	*Type*	*Average Particle Diameter, $\bar{D}$ (μm)*	*Content (vol%)*
Polystyrene	Polybutadiene	1.0	20
Styrene–acrylonitrile	Polybutadiene	0.5	0.6
copolymer		0.05	2.4
			21
			25
Polyamide 6	Ethylene–vinyl acetate copolymer	0.5–1.5	5–15
Polyamide 6	Ethylene–propylene copolymer	0.15	
		0.6	20
		1.0	
Polyamide 6	Ethylene–propylene–diene copolymer	0.1	20
Polypropylene	Ethylene–propylene copolymer	1–3	20
Polypropylene	Ethylene–propylene–diene copolymer	1–3	20

to vary within a large range: from less than 10 nm up to a few micrometers (see Figure 2). The usual manufacturing techniques were applied to produce these materials, namely, graft polymerization for HIPS, separate polymerization of modifier particles and mixing for ABS, and blending techniques for high-impact PA and PP. Compression-molded as well as injection-molded samples were investigated.

Investigation of the Morphology.   Three preparation techniques and electron microscopic investigations were used to study the morphology of the polymer systems:

1. Selective chemical staining of the rubber phase of the samples using chlorosulphonic acid and osmium tetroxide, preparation of ultrathin sections (about 0.1 μm thick) in a cryoultramicrotome, and investigation of the sections by conventional transmission electron microscopy (TEM).
2. Preparation of semithin sections (up to a few micrometers thick) by ultramicrotomy, and investigation of these sections in a 1000-kV, high-voltage electron microscope (HVEM) to reveal larger particles more precisely and, thus, the true particle-diameter distribution.
3. Preparation of brittle (low-temperature) fracture surfaces of the unmodified materials, and investigation in a scanning electron microscope (SEM) to preferentially reveal the largest particles.

Investigation of Micromechanical Processes.   The micromechanical mechanisms of deformation and fracture were studied by several electron microscopic techniques, including in situ electron microscopy (3, 9) (see Figure 3). When passing from technique (a) to techniques (b) and (c) in Figure 3, larger magnifications with better resolutions can be applied to reveal more and more details of the micromechanical processes. A combination of these techniques allows the study of deformation processes in several modes, including at different strain rates and stress rates (e.g., plane stress or plane strain state).

Results and Discussion

Toughening by the Multiple-Crazing Mechanism

Electron Microscopic Results.   The fundamental deformation step is the formation of crazes at the rubber particles (Figure 4). The crazes start directly at the interface between rubber particles and matrix in the equatorial zones around the particles, that is, in the zones of highest stress concentration. The structure of the amorphous material is transformed by local plastic defor-

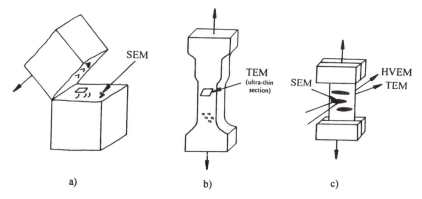

Figure 3. Electron microscopic techniques used to study micromechanical processes in polymers: (a) investigation of fracture surfaces by SEM; (b) investigation by TEM of ultrathin sections prepared from deformed and selectively stained bulk material; and (c) deformation of samples of different thicknesses (bulk, semithin, and ultrathin), using special tensile stages with SEM, HVEM, and TEM. The technique in (c) shows the possibility of conducting in situ deformation tests in the electron microscope.

mation into the structure of the crazes, which contain fibrils and elongated microvoids. After initiation, the crazes propagate into the adjacent matrix and end there, or—if the distance between the particles is small enough—they propagate from one particle to neighboring ones located near the propagation of the crazes.

In many cases the large rubber particles in HIPS, which possess a salami

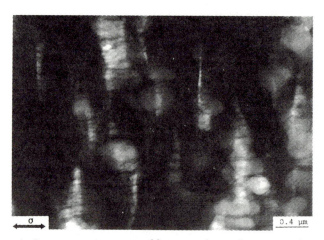

Figure 4. Crazes originating at rubber particles, with some partly ending in the matrix (HVEM image). The deformation direction is horizontal.

or network structure, reveal a cavitation and fibrillation of the rubber network between the PS inclusions (Figure 5). This effect increases the stress concentration at the particles and enables them to deform more easily, leading to the formation of numerous and larger crazes. The strength at the rubber–matrix interface is high enough to prevent cavitation or debonding.

The same mechanism can appear in ABS polymers. Besides the formation of the fibrillated crazes, and depending on the matrix and local stress state, a homogeneous plastic deformation between particles, comparable to the appearance of homogeneous crazes in SAN (*12, 13*), is also possible (Figure 6). The homogeneous deformation in ABS is associated with cavitation inside the rubber particles. In general, this mechanism precedes the formation of the fibrillated crazes.

In the past, several ABS grades were produced with a broad or bimodal particle-diameter distribution. Particles of different sizes can contribute to crazing in different ways. Figure 7 shows such a material and clearly reveals the greater capability of the largest particles to initiate crazes. Beside these large particles, the smaller ones show only a small effect *on craze initiation*.

Characteristics of the Craze Mechanism.   HIPS, as well as the ABS grades studied here, deform mainly by the formation of crazes. The reason is the strong tendency of matrix material to form crazes under load (*3, 12, 13*). Details of the toughening mechanism have been reported recently (*1–4*). Therefore, only a brief review of the main points is given here, to clarify the difference between this mechanism and the shear mechanism. The processes

Figure 5. Rubber particles with crazes in HIPS. The rubber particles are strongly elongated by cavitation and fibrillation in the rubber network around PS inclusions (HVEM image). The deformation direction is vertical.

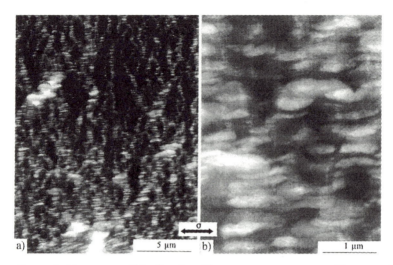

Figure 6. Formation of fibrillated and homogeneous crazes at rubber particles in ABS (HVEM image). The deformation direction is shown by the arrow.

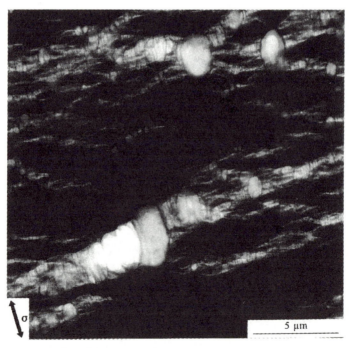

Figure 7. Preferred formation of crazes at the largest particles in an ABS polymer (HVEM image). The deformation direction is shown by the arrow.

of deformation and fracture have been arranged in a three-stage mechanism (Figure 8).

Stage 1 (craze initiation). Each rubber particle concentrates the stress in the surrounding matrix in an area whose size is strongly related to the particle diameter. The highest concentration of stress is along the equatorial zones of the rubber particles. The criteria for craze initiation are related to this stress concentration; therefore, crazes start from these zones and propagate perpendicularly to the direction of tension. This separate initiation of crazes around the particles is the primary mechanism (see Figure 4). It is noteworthy that the stress concentration at rubber particles is sufficient to create crazes; the often-detected cavitation inside particles (as seen in Figure 5) increases stress con-

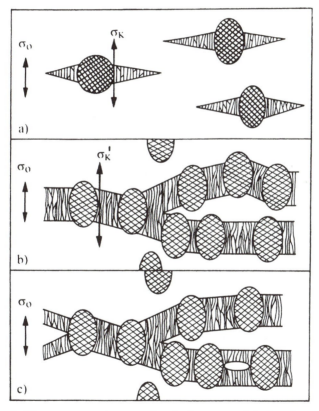

Figure 8. Three-stage mechanism of multiple crazing: (a) stage 1: stress concentration and craze initiation at rubber particles; (b) stage 2: superposition of stress fields (small interparticle distance, high rubber volume content) and formation of broad craze bands; and (c) stage 3: limitation of crack length and crack stopping at rubber particles.

centration and makes the formation of crazes easier, but it is not considered to be a precondition of craze initiation (see the section "Summary of Different Toughening Mechanisms").

Stage 2 (superposition effect). At a rubber-particle content of more than 15 vol%, the stress concentration directly at the particle surface is increased by the stress fields of neighboring particles (Figure 9). This fact favors the formation of broad crazes and long craze bands.

Stage 3 (Crack Propagation). After the formation of cracks inside the crazes, the cracks are stopped by weak rubber particles. This mechanism of stopping crack and restricting crack lengths delays the propagation of cracks and the ultimate fracture.

Important Parameters

Stress Rate at Particles. The stress component, $\sigma_{\nu\nu}$, acting parallel to the boundary between rubber particles and matrix is important for the initiation of crazes. It reaches a maximum value (which can be about twice the outer stress, σ_o) at the equatorial regions of the particles. Besides depending on the shape of the particles and Poisson's ratio, the elastic-stress concentration at the rubber particles depends mainly on the ratio $\chi = G_P/G_M$, where G_P and G_M are the Young's modulus of the particles and the matrix, respectively. This ratio has been calculated by Michler (*14*) on the basis of the solution obtained by Goodier for an isolated particle embedded in a matrix and subjected to uniaxial tension (*15*) (see Figure 9).

In many practical cases, the ratio G_P/G_M is smaller than 0.1 (*16*), yielding a stress concentration, $\sigma_{\theta\theta}$, of 1.8 or more. This value is more than 85% of the maximum possible stress concentration (of 2.04, as for voids). In the case of stiffer particles, the stress concentration can be increased by cavitation of the particles (a decrease of the ratio G_P/G_M). Relief of the initial thermal stress and of the hydrostatic stress inside the particles can be considered another advantage of particle cavitation (*17*). The effect of particle cavitation is discussed in more detail by Lazzeri and Bucknall (*18*).

The initiation point of crazes is directly at the surface of the rubber particles. Along with the formation of crazes, there is the creation of microvoids, which increase the stress concentration at the craze tip. By this stress concentration, the polymeric material at the craze tip is transformed into the craze. Thus, with craze propagation the stress state necessary for craze initiation is reproduced continuously at the craze tip in the matrix (Figure 10).

The stress concentration at rubber particles starts to increase by the superposition effect if the interparticle distance, A, is smaller than about $0.5D$, that is, if the rubber volume content is higher than 15 vol% (Figure 11). There is an increasing tendency of craze formation with increasing stress concentration, which means with decreasing particle modulus and increasing rubber volume content.

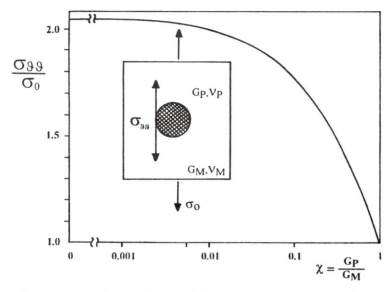

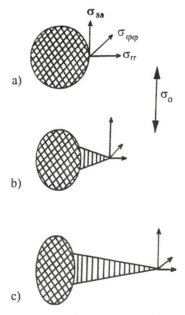

Figure 9. Dependence of the size of the stress component, $\sigma_{\theta\theta}$ at rubber particles on the ratio G_P/G_M, where G_P and G_M are the Young's modulus of the particles and the matrix, respectively.

Figure 10. Propagation scheme of the craze and of the triaxial stress state in front of the craze tip, which is necessary to transform the matrix material into crazes.

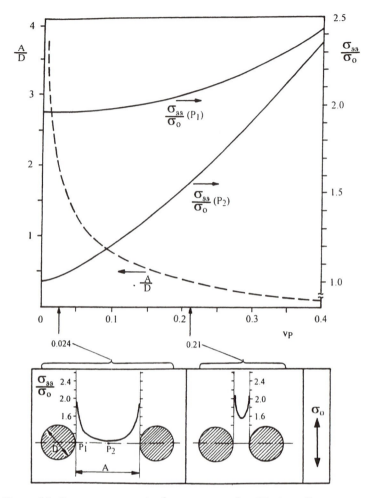

Figure 11. Stress concentration between particles: (Top) smallest average inter-particle distance, A, and stress concentration, $\sigma_{\theta\theta}$, at point P_1 and between the particles (point P_2), as a function of rubber-particle volume content, v_P. (Bottom) stress concentration between particles for two different rubber volume contents.

Particle Diameter.   The increased stress around the equator of particles is larger than the outer stress, σ_o, in a zone within an angle range of $\pm 35°$ (*12*). So the size, D_z, of this stress-concentration zone has the same magnitude as the particle radius, $D/2$:

$$D_z \approx D/2 \qquad (2)$$

In this zone, the craze-initiation zone, there is preferred initiation of crazes. The initiation of crazes is effective if D_z is comparable to the characteristic size or thickness of the crazes in the matrix material (*3, 19*). This effect defines an optimum particle diameter, D_{opt}.

Particles whose diameters are significantly greater than D_{opt} are less effective for craze initiation. The stress-concentration zone is much broader than the optimum thickness of crazes. Therefore, several thinner crazes having a total volume smaller than the volume of a craze with a thickness equal to D_z are initiated (*3, 19*).

Particles with small diameters are ineffective for another reason. The level of stress concentration is independent of particle diameter, but the dimensions of the stress-concentration zone decrease with decreasing particle diameter. By studying craze formation in glassy polymers (*20*), it is known that there is a minimum size or thickness of crazes. That is, crazes smaller than a critical size, $2g_{min}$, where g_{min} is the thickness of the minimum boundary layer between the highly deformed material inside the craze and the undeformed surroundings, cannot be produced in glassy polymers. The size of the stress-concentration zone, D_z, has to be larger than this minimum critical size. Therefore, for the smallest effective rubber particles a diameter, D_{min}, follows (*19*):

$$D_{min} = 2D_z = 4g_{min} \tag{3}$$

For PS and SAN, a value of 10 nm was determined for g_{min}, yielding a minimum effective particle diameter, D_{min}, of 40 nm. This calculation correlates well with experimental results (*1, 21*).

The experimental results clearly reveal that the size of the stress-concentration zone, that is, the diameter of the rubber particles, is the primary factor for the initiation of crazes. Therefore, the effectiveness of the different-sized rubber particles can be shown in an "effectiveness curve," which gives the probability of the particles to initiate craze (*19, 22*). Starting from D_{min} for craze initiation, the effectiveness of particles of different sizes increases up to a maximum and then drops at larger diameters.

Particle-Diameter Distribution. The dependence of the effectiveness of different-sized particles is valid when all particles possess the same modulus and, therefore, yield the same stress concentration. In many cases, however, this is not given, particularly in composed (salami) particles and if grafting reactions exist (*2*). Figure 12a shows the results of measurements of the ratio of the length and width of rubber particles (with a network or salami structure) in a HIPS. With increasing particle size the ratio increases too, indicating an increasing "deformability" or decreasing stiffness or modulus of the particles. This decrease of modulus is shown in Figure 12b as relative modulus compared with the modulus of large particles, $G3_{\mu m}$. The easier deformability of the large particles can be explained by the greater number of inclusions inside them and their easier internal movement. On the other hand, small particles can possess a larger modulus that is due to either a hardening effect of surface grafting (the existence of a stiff, grafted surface layer

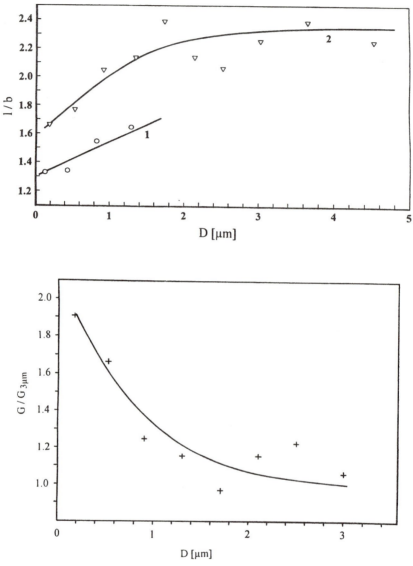

Figure 12. Dependence of the modulus of rubber particles on particle diameter in HIPS: (a) increase of the length-to-width ratio of particles with increasing diameter; (b) decrease of the modulus, G, of particles with increasing diameter. The modulus is shown as the ratio between G and $G_{3\mu m}$, the value of G of large particles.

of the matrix polymer) or a higher degree of internal grafting, as discussed in reference 21.

The result of the diameter-dependent modulus is an increased stress concentration and increased tendency to craze initiation with increasing particle diameter. This result is demonstrated in the micrograph of an ABS polymer in Figure 7, which shows the preferred craze formation at the largest rubber particles. Therefore, toughened materials with a broad particle-diameter distribution or a bimodal diameter distribution often show preferred craze initiation at the largest particles, which has the disadvantage of reduced effectiveness (*23*). The maximum formation of crazes appears in a material with rubber particles of optimum diameter, D_{opt} and a small diameter distribution.

Another role of large particles in ABS with a bimodal rubber mixture has been shown in reference 11 by direct investigation of deformed specimens by TEM. The larger particles (diameters between 0.2 and 1.0 μm) initiate shear deformations, which propagate through the material along aligned bands of cavitated smaller particles (diameter 0.1 μm). These bands of shear deformation and cavitated small particles look like coarser crazes (*4, 11*). The propagation direction of these bands depends on the nearest neighbors. The interparticle distances between nearest neighbors are smaller than the average interparticle distances according to equation 1 and are calculated by computer modeling in reference 24. Each formation of homogeneous deformation bands or homogeneous crazes is associated with cavitation inside the rubber particles. The sequence and type of the deformation events were studied in ABS and also in HIPS by real-time, small-angle X-ray scattering in reference 25. In this work it was also demonstrated that the homogeneous mechanism and rubber-particle cavitation occur before crazing.

Toughening by Multiple Shear Deformation

Electron Microscopic Results.   Electron micrographs were obtained from deformation tests of toughened PP and PA. The micrographs of toughened PP shown in Figure 13 reveal ruptured particles and plastically deformed matrix material between these voids. The cavitation step inside the particles, with the subsequent deformation and fibrillation of the adjacent material, can also be clearly seen in Figure 14.

The micrograph in Figure 15a shows deformed and elongated rubber particles as bright particles in the PA matrix. The material between the particles is highly plastically deformed and visible in the form of bright, diffuse zones. The whole material between particles is involved in plastic deformation, but no internal structure or cavitation (as in the case of crazes in HIPS) is visible. Cavitation inside the particles is clearly visible in Figure 15b, which shows higher magnification. In this material, the plastically deformed areas are spread over a large volume of the sample, which indicates high toughness. The polymer con-

Figure 13. Highly deformed, toughened PP, showing cavitation inside the particles and intense plastic deformation of the matrix strands (HVEM image). The deformation direction is vertical.

tains about 20 vol% rubber particles with diameters between 0.1 and 0.4 μm and an average diameter of about 0.2 μm. According to equation 1, the average interparticle distance, A, is about 0.1 μm.

The deformation behavior of toughened PA is compared for larger and smaller particles in reference 26. In large particles with an average diameter, D, of about 1 μm and an average minimum interparticle distance, A, of about 0.5 μm, an intense plastic deformation appears in only a few bands. In the case

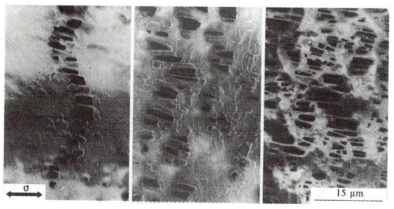

Figure 14. Deformation of toughened PP, showing an increasing number of cavitated particles with increasing elongation (SEM image produced in an in situ deformation test). The deformation direction is horizontal.

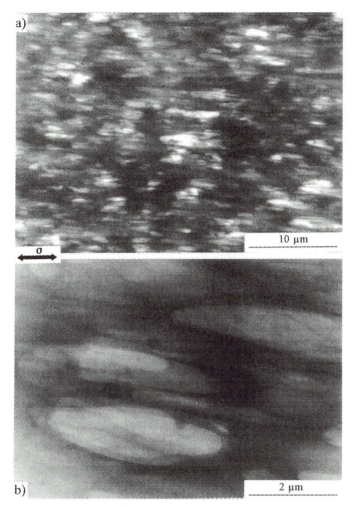

Figure 15. Strained semithin sections of rubber-modified PA (HVEM images). The deformation direction is horizontal. (a) highly deformed particles with plastically deformed matrix material in between; and (b) cavitation or microvoids inside particles.

of very small particles, with D less than 0.1 μm and A less than 0.05 μm, there is a nearly complete absence of any plastic deformation. With increasing load the sample deforms like unmodified PA without any toughening effect. A material with slightly larger particles ($\bar{D} \approx 0.15$ μm) and interparticle distances ($\bar{A} \approx 0.08$ μm) shows a beginning shear deformation between particles, visible mainly in the deformed and elongated particles (Figure 16). Instead of an intense shear yielding, there is only a formation of angled shear bands [see the

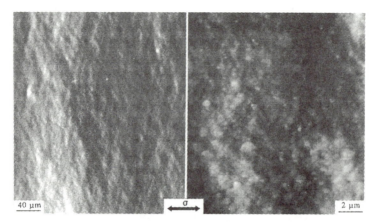

Figure 16. Modified PA with small particles of diameter D *(D̄ = 0.15 μm) and small interparticle distances* A *(Ā = 0.08 mm), showing beginning plastic deformation (HVEM image). The deformation direction is horizontal.*

lower-magnification image in Figure 16 (left)]. This case corresponds to the first stage in Figure 17 and defines a transition stage.

Characteristics of the Shear Mechanism. The results of the investigation of the micromechanical processes can also be summarized in a three-stage-mechanism (Figure 17). In stage 1, under an external stress, stress concentrations, σ_K, or stresses increased by superposition of local stress fields are built up at, and between, the rubber particles (as in the systems described in the preceding discussion). At places with a maximum shear-stress component, weak shear bands are formed between the particles at an angle of about 45° to the load direction (see Figures 16 and 17a).

In stage 2 (void formation), larger hydrostatic stresses are built up inside the particles owing to stress concentration and the formation of shear bands. These larger stresses give rise to cracking of the particles and the formation of microvoids inside. The result is a higher local stress concentration, σ'_K, between the particles (Figure 17b).

In stage 3 (induced shear deformation), shear processes are initiated in the matrix strands (matrix ligaments or bridges) between the particles (or voids) owing to high local stress. The shear-deformation process proceeds simultaneously at numerous adjacent matrix bridges and thus takes place in a fairly large polymeric volume (Figure 17c; see also Figure 15).

As in the systems characterized by craze formation, there is also an influencing effect on crack propagation:

1. With increased stretching of the adjacent matrix ligaments, the microvoids in the particles are elongated and give rise to slight-

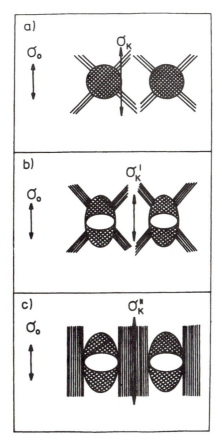

Figure 17. Schematic representation of the three-stage mechanism of toughening in PA6 blends: (a) elastic stress concentration, σ_K, between particles and the formation of weak shear bands at the particles (σ_o is applied stress); (b) void formation inside the particles, and increased stress concentration between particles and voids; and (c) plastic deformation of matrix strands between particles and voids.

ly reduced stress levels. The strain hardening of the plastically stretched matrix bridges prevents elongation and widening of the voids. The formation of cracks with overcritical length and premature crack propagation are thus avoided.

2. Cracks that enter the elongated microvoids are rounded at their tips (crack-tip blunting) and prevented from premature propagation.

The deformation step during which energy is mainly absorbed is the plastic stretching of the matrix bridges between the rubber particles.

As a result of the toughening mechanism, the transition temperature, T_{BT}, at which the fracture type changes from brittle to tough shifts to lower values (from 60 °C for unmodified PA6 to about –30 °C). The dependence of T_{BT} on rubber-particle content, particle diameter, and type of rubber has been extensively studied by Gaymans, Borggreve, and co-workers (6, 31, 33, 34).

The deformation mechanism at decreasing temperature has been studied by Cieslinski in toughened PP directly in TEM (11). At room temperature, PP and impact-modified PP deform by shear yielding. Below T_{BT}, crazing dominates. This is consistent with the result that shear yielding gradually gives way to crazing as the temperature decreases, and with the idea that molecular chain movement is restricted and chain scission is the dominant mechanism (followed by crazing) at lower temperatures (3).

Important Parameters

Stress State at Particles.   If the modifier particles consist of rubber-like material, they act as stress concentrators as in HIPS and ABS. Whereas in HIPS and related polymers the maximum stress component, $\sigma_{\theta\theta}$, at the equatorial regions around the particles is responsible for initiating crazes, in polymers with a tendency to shear deformation the maximum shear stress at the particles, yielding the formation of shear bands, must be considered (see Figure 17a). But in contrast to crazes, which have a stress-concentrating ability, shear bands to not increase the stress between particles as effectively. Therefore, the formation of microvoids inside the particles is necessary as an additional mechanism to increase the stress at and between particles. To make the polymeric material between particles yield, not only is the stress concentration at particles necessary (as in craze formation), but so is the stress field between particles.

To deform the matrix material between particles by shear processes, a more uniaxial stress state (i.e., a stress state with one major component) is more effective than a triaxial stress state. Such a uniaxial stress state only becomes possible when the distances between particles are not too large, and thus relief of the plastic constraint in the interparticle matrix strands appears. This effect is schematically illustrated in Figure 18 (26). Cases (b) and (c) demonstrate stress concentrations between particles after cavitation inside the particles. When the matrix strands between particles are too thick, the matrix material is under triaxial stress, including a strong hydrostatic component. This stress state is not effective in initiating local plastic deformations (Figure 18c). If the thickness of the matrix strands is small enough, a plastic response of the matrix material can appear, relieving the plastic constraint (Figure 18b). As a result, the hydrostatic stress component in the matrix is reduced in favor of an increased uniaxial stress component in a direction parallel to the loading. An intense plastic yielding of the matrix strands is made possible in a large area.

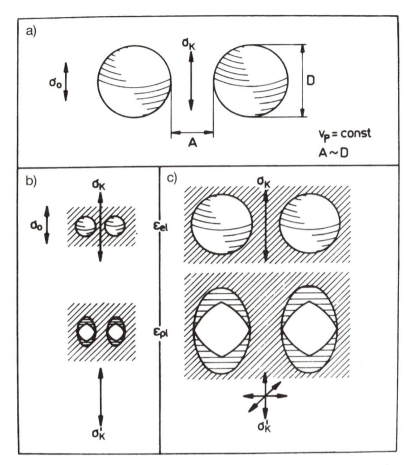

Figure 18. Effect of interparticle distance, A, on plastic deformation of matrix strands between particles: (a) definitions of the size parameters: D = particle diameter, v_p = particle volume content, σ_o = applied stress, and σ_K = stress concentration; (b) with a small interparticle distance, a uniaxial stress state is dominant between the particles and microvoids after cracking of the particles, and plastic yielding can be obtained; and (c) with a large interparticle distance, thick matrix strands favor a triaxial stress state between the particles and microvoids, and plastic yielding is hindered.

The same effect is possible when the interfacial strength is low (reduced coupling). Therefore, from the micromechanical point of view there is no difference between void formation by cavitation and by debonding.

The effect necessary for stress transformation from a triaxial stress state into a more uniaxial stress state requires relatively small interparticle distances because the stress state must be changed in the whole polymeric material between particles. This fact may explain the experimental results for critical interparticle distances, although the effect has not been well understood until now (see the discussion later in this section).

The section "Toughening by the Multiple-Crazing Mechanism" mentioned the effect of superposition of the local fields of stress concentration, which must also be considered. As shown in Figure 8, there is a remarkable increase of the local stress between particles by superposition if the interparticle distance is smaller than the particle diameter ($A/D < 1$, which corresponds to particle volume contents above 5%).

Particle Diameter.   Because plastic deformation depends on stress concentration in the whole volume between particles and not directly on the stress at the particles, D is not of primary importance. The only precondition is that for a given particle volume content, the particles must be small enough to ensure that A is less than the critical value according to equation 1. Therefore, the most important function of particles is to produce a dense pattern of microvoids. Recently, an increase of the toughness of modified PA with increasing tendency to form microvoids inside the particles (with decreasing stress to crack the particles or with decreasing cavitation strain) was found (6). The cavitation stress of an elastomer is dependent on its modulus (27).

Interparticle Distance.   As shown in the preceding discussion, A must be small enough to create the effective uniaxial stress state. This means that a critical value of A, A_c, exists, which describes a brittle-to-tough transition of the material. The decisive role of A in contrast to D was explored first by Wu (28, 29) and later by Borggreve and co-workers (6, 30). The findings on the dependence of the brittle-to-tough transition temperature, T_{BT}, on rubber content and particle size can be summarized in a dependence of T_{BT} on A. Borggreve et al. (31) reported a decrease of T_{BT} from –10 °C to –30 °C with decreasing A from 0.3 μm to 0.1 μm. The value of A_c in the PA used by Wu shows the same decrease: from about 0.3 μm at room temperature to about 0.1 μm at –30 °C (32).

The reason for the existence of A_c is not given by the increased stress concentration between closely adjacent particles; it has been found that for the same ratios A/D (i.e., for the same effect of stress superposition), an improved toughness appears only if the absolute value of A is below A (33). Results given in reference 34 show that the use of the average distance between rubber particles as a criterion for the toughness of nylon–rubber blends is too simple because A_c also appears to increase with increasing rubber cavitation stress. A possible explanation is that what is critical is not the average distance between all rubber particles (according to equation 1), but the average distance between cavitated particles. When the elastomer has a high cavitation stress, not all the particles will cavitate, and the interparticle distance will be larger than that calculated on the basis of rubber concentration and particle size (according to equation 1).

The experimental results also indicate that there is a critical minimum interparticle distance, A_{min}. In reference 35 it was shown that blends containing small particles with diameters below 200 nm yield a reduced toughness (or increased T_{BT}). The critical particle size was reported to increase with increasing temperature. This was confirmed in reference 36. It is assumed that a modified stress state exists in the small interparticle ligaments between the small particles (see reference 22). As another explanation, it was stated in reference 34 that small particles cavitate at higher stresses than larger ones and, therefore, do not participate in the toughening mechanism. An up-to-date summary of the influence of different structural parameters on the toughness of nylon–rubber blends is given in reference 34.

Comparison of the Craze and Shear Mechanisms

Summary of Different Toughening Mechanisms.   In toughened polymers with a dispersed modifier phase (i.e., in the dispersed systems), the three mechanisms sketched in Figure 19 may, in general, be distinguished. The characteristics of these different mechanisms are as follows.

Case a: stress-induced formation of fibrillated crazes. The weak rubber particles act as stress concentrators. Crazes are formed starting from the particle–matrix interface around the equatorial region of particles. The voids inside the crazes initiate a stress concentration at the craze tip, which propagates together with the propagating craze; therefore, the crazes reproduce the stress state necessary for their propagation. Cavitation inside the rubber particles is not necessary, but it enables a higher stress concentration and easier deformation of the particles.

Case b: stress-induced formation of homogeneous crazes. The stress concentration at the particles causes homogeneous crazes to start at the particle–matrix interfaces. Propagation of these crazes into the matrix is accomplished by an increase of volume, which arises from cavitation inside the particles (the possible mechanism of cavitation inside the originally homogeneous crazes is unlikely). Therefore, these crazes are closely connected to the cavitated rubber particles—they cannot propagate for distances as long as those of the fibrillated crazes—and appear mainly between particles.

Case c: stress-induced formation of shear deformation. The stress concentration of the modifier particles, usually too small, is increased by the formation of cavities inside the particles. By these cavities, and if A is small enough, the originally triaxial stress state between particles is transformed into a more uniaxial stress state. Then the matrix strands between particles can be plastically deformed where the necessary volume increase, which arises from the cavitated particles, appears.

This classification is valid for a given temperature. With variation in temperature there can be a transition from one mechanism to another. For instance, decreasing the temperature of toughened PP shifts a shear yielding mech-

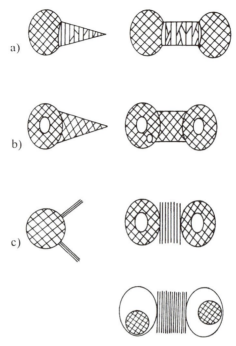

*Figure 19. Schematic representation of the three different toughening mecha-
nisms in dispersed systems, where the assumed loading direction is vertical: (a)
induced formation of fibrillated crazes (i.e., with microvoids in them) at the
equatorial zones of rubber particles; (b) induced formation of homogeneous
crazes at cavitated particles; and (c) induced formation of shear deformation between cavitated particles.*

anism according to case c into a crazing mechanism according to case a (*11*).

If fibrillated crazes (case a) coexist with homogeneous deformation (cases
b or c), the homogeneous mechanisms and rubber-particle cavitation precede
the formation of crazes.

Comparison of Important Parameters.   Several parameters important
for toughening are compared in Table II.

Conclusions

If the toughening effect is based mainly on the formation of crazes, the
particle diameter, D, is of primary importance. Particles act as stress concentrators and directly initiate the crazes. There are critical particle diameters,
which depend only on the ability of the matrix to form crazes (if all particles
possess the same modulus, i.e., give the same stress concentration). The opti-

**Table II. Comparison of Several Parameters Important
for Toughening Mechanisms**

Parameter	Craze mechanism	Shear mechanism
Important region of stress concentration	At particles	Between particles
Condition needed for stress-field superposition to start	$A < 0.5D$	$A < 1.0D$
Stress state	$\sigma_1 \geq \sigma_2, \sigma_3$ (triaxial) $\sigma_1 > \sigma_y$	$\sigma_1 \gg \sigma_2, \sigma_3$ (uniaxial)
Relief of stress constraint at particles	By crazes	By particle cavitation
Matrix yielding	With cavitation	Without cavitation
Particle cavitation necessary?	No	Yes
Where the increase of volume by cavitation mainly occurs	Crazes	Particles
Important structural size parameter	Particle diameter, D	Interparticle distance, A
Structural limits	$D_{min} < D < D_{max}$ $D \approx D_{opt}$	$A_{min} < A < A_{max}$

mum particle diameter corresponds to the typical thickness of crazes in the matrix material. The critical minimum particle diameter, D_{min}, acting as a material constant, is defined by the size of the smallest possible crazes. The value of D_{min} is about 50 nm for HIPS as well as for ABS. The ratio A/D (i.e., the rubber volume content) is of secondary importance.

If the toughening effect is based mainly on shear deformation, the interparticle distance, A, is of primary importance. The particles cavitate, enabling a change of the stress state from a triaxial to a more uniaxial one. Yielding is made possible if the distance between cavitated particles is small enough. For critical interparticle distance, there is a minimum, A_{min}, and a maximum, A_{max} (where A is the distance between cavitated particles if the particles do not possess the same modulus). Both act as real material constants. Only if the average value of A is between A_{min} and A_{max} can intense local plastic yielding be obtained. The values of A_{min} and A_{max} are about 50 and 300 nm, respectively, for toughened PA6. The particle diameter is of secondary importance.

The different basic mechanisms shown in Figure 19 are valid for other toughened polymers as well. For example, the mechanism in Figure 19 is decisive for rubber-toughened poly(vinyl chloride) (37), polycarbonate, (38, 39), poly(methyl methacrylate) (40, 41), and rubber-toughened epoxies (42, 43).

Until now, these critical values could be determined only experimentally. When they are known, criteria can be defined to optimize the morphology of modified polymers in order to realize maximum toughness. Therefore, the investigation of micromechanical processes is helpful in choosing the appropriate polymeric components and manufacturing procedures.

Acknowledgments

We thank J. Heydenreich for giving us the opportunity to carry out the deformation tests in the 1000-kV high-voltage electron microscope of the Max-Planck-Institute of Microstructure-Physics. We gratefully acknowledge the financial support provided to J.-U. Starke by the Max-Buchner-Forschungsstiftung and the Deutsche Forschungsgemeinschaft.

References

1. Bucknall, C. B. *Toughened Plastics;* Applied Science: London, 1977.
2. Michler, G. H.; Gruber K. *Plaste Kautsch.* **1976,** *23,* 346; 496.
3. Michler, G. H. *Kunststoff-Mikromechanik: Morphologie, Deformations- und Bruchmechanismen;* Carl Hanser Verlag: Munich, Germany, 1992.
4. Michler, G. H. *Acta Polym.* **1985,** *36,* 285.
5. Wu, S. *J. Appl. Polym. Sci.* **1988,** *35,* 549.
6. Borggreve, R. J. M.; Gaymans R. J.; Eichenwald H. M. *Polymer* **1989,** *30,* 78.
7. Michler, G. H. Plaste Kautsch. **1981,** *28,* 191.
8. Michler, G. H. *Kunststoff-Mikromechanik: Morphologie, Deformations- und Bruchmechanismen;* Carl Hanser Verlag: Munich, Germany, 1992; Section 9.2.3., pp. 277–285.
9. Michler, G. H. *Morphology of Polymers;* Bethge H.; Heydenreich J., Eds.; Electron Microscopy in Solid State Physics; Elsevier: Amsterdam, Netherlands, 1987; p. 386.
10. *Electron Microcopy in Plasticity and Fracture Research of Materials;* Messerschmidt, U.; Appel F.; Heydenreich J.; Schmidt, V., Eds.; Physical Research Series; Akademia-Verlag: Berlin, Germany, 1990; Vol. 14.
11. Cieslinski, R. C. *J. Mater. Sci. Lett.* **1992,** *11,* 813; *Proceedings of the 9th International Conference on Deformation, Yield and Fracture of Polymers;* 1994; pp. 5/1–4.
12. Michler, G. H. *Plaste Kautsch.* **1979,** *26,* 497.
13. Michler, G. H. *J. Mater. Sci.* **1990,** *25,* 2321.
14. Michler, G. H. *Acta Polym.* **1985,** *36,* 325.
15. Goodier, J. N. *Trans. ASME* **1933,** *55,* 39.
16. Michler, G. H.; Hamann, B.; Runge. *J. Angew. Makromol. Chem.* **1990,** *180,* 169.
17. Boyce, M. E.; Argon, A. S.; Parks, D. M. *Polymer* **1987,** *28,* 1680.
18. Lazzeri, A.; Bucknall, C. B. *Mater. Sci.* **1993,** *28,* 6799.
19. Michler, G. H. *Plaste Kautsch.* **1988,** *35,* 347.
20. Michler, G. H. *Colloid. Polym. Sci.* **1986,** *264,* 522.
21. Michler, G. H. *Plaste Kautsch.* **1979,** *26,* 680.
22. Michler, G. H. *Kunststoffe* **1991,** *81,* 449, 548.
23. Gust, H.; Starke, J.-U.; Michler, G. H. *Polymerphysik* **1994,** P12.12.
24. Hall, R. A. *J. Mater. Sci.* **1991,** *26,* 5631.
25. Bubeck, R. A.; Buckley, D. J., Jr.; Kramer, E. J.; Brown, H. R. *J. Mater. Sci.* **1991,** *26,* 6249.
26. Michler, G. H. *Acta Polym.* **1993,** *44,* 113.
27. Ball, J. M. *Philos. Trans. Roy. Soc.* London, Ser. A **1982,** 306.
28. Wu, S. *Polymer* **1985,** *26,* 1855.
29. Wu, S. *Polym. Eng. Sci.* **1990,** *30(13),* 753.
30. Borggreve, R. J. M.; Gaymans, R. J. *Polymer* **1988,** *29,* 1441.

31. Borggreve, R. J. M.; Gaymans, R. J.; Schuijer, J.; Ingen Housz, J. F. *Polymer* **1987,** *28,* 1489.
32. Margolina, A. *Polym. Commun.* **1990,** *31,* 95.
33. Borggreve, R. J. M., Gaymans, R. J.; Schuijer, J. *Polymer* **1989,** *30,* 71.
34. Dijkstra, K. Ph.D. Thesis, University of Twente, Netherlands, 1993.
35. Ostenbrink, A. J.; Dijkstra, K.; Wiegersma, A.; Wal, A. v.d.; Gaymans, R. J. *Proceedings of the 8th International Conference on Deformation, Yield and Fracture of Polymers*; 1991; pp. 50/1–4.
36. Oshinski, A. J.; Keskulla, H.; Paul, D. R. *Polymer* **1992,** *33,* 268.
37. Haaf, F.; Breuer, H.; Stabenow, J. *Angew. Makromol. Chem.* **1977,** 58/59, 95.
38. Chihmin, C.; Hiltner, A.; Baer, E.; Soskey, P. R.; Mylonakis, S. G. *J. Appl. Polym. Sci.* **1994,** *52,* 177.
39. Hourston, D. J.; Lane, S. In *Rubber Toughened Engineering Plastics*; Collyer, A. A., Ed.; Chapman & Hall: London, 1994; p. 243.
40. Bucknall, C. B.; Partridge, I. K.; Ward, M. V. *J. Mater. Sci.* **1984,** *19,* 2064.
41. Lovell, P. A.; McDonald, J.; Saunders, D. E. J.; Sherratt, M. N.; Young, R. J. In *Toughened Plastics I: Science and Engineering;* Riew, C. K.; Kinloch, A. J., Eds.; Advances in Chemistry 233; American Chemical Society: Washington, DC, 1993; pp. 61–77.
42. Kinloch, A. J. In *Rubber-Toughened Plastics;* Riew, C. K., Ed.; Advances in Chemistry 222; American Chemical Society: Washington, DC, 1989, pp. 67–91.
43. *Toughened Plastics I: Science and Engineering*; Riew, C. K.; Kinloch, A. J., Eds.; Advances in Chemistry 233; American Chemical Society: Washington, DC, 1993.

Rubber-Toughening of Polycarbonate–Nylon Blends

Feng-Chih Chang and Dah-Cheng Chou

Institute of Applied Chemistry, National Chiao-Tung University, Hsinchu, Taiwan, Republic of China

Incompatible and brittle polymer blends of polycarbonate (PC) and nylon-6 [polyamide (PA)] have been compatibilized with high-molecular-weight bisphenol A epoxy resin and toughened by reactive and unreactive core–shell rubbers. The epoxy resin has proved to be an effective, reactive, in situ compatibilizer for the incompatible PC–PA blends. Rubber particles in the PC–PA blends can be controlled so as to be distributed in either PC, PA, or both phases. Unreactive methyl methacrylate–butadiene–styrene (MBS) rubber is more compatible with PC than with PA and tends to reside in the PC phase and near the PC–PA interface. Reactive MBS rubber containing maleic anhydride and methacrylic acid (MBS-MA) can react with nylon end groups, and the resulting nylon-linked rubber particles are retained within the nylon phase. Rubber distributed in the PC phase is more effective in toughening the PC–PA blends than that distributed in the nylon phase. Rubber-toughening of blends with PC as a matrix is more efficient than in blends with nylon as a matrix.

THE DEVELOPMENT OF RUBBER-TOUGHENED THERMOPLASTIC RESINS is an important contribution to the commercial polymer industry. Toughening a single-polymer matrix by the incorporation of a second rubber phase is relatively simple because the added rubber can only be distributed in this matrix. Rubber toughening of a binary polymer blend is much more complex, for the third rubber phase may reside in either matrix, along the interface, or in any combination of those places. Rubber-phase distribution is critical in determining the mechanical toughness of the resulting blend products, and this fact has long been recognized by polymer researchers. It is important to understand the rubber-toughening mechanisms and to know whether rubber distributed in the more ductile component, the more brittle component, or along the interface is most effective in rubber-toughening a binary blend.

In practice, controlling rubber-phase distribution is extremely difficult in any binary blend, and it has rarely been achieved because few rubbers that are compatible with matrices are available for this purpose. Only recently has literature started to appear on identifying the rubber-phase distribution in rubber-toughened binary blends, mainly through straining the rubber phase and transmission electron microscopy (TEM). Hobbs et al. first reported that core–shell methacrylate–butadiene–styrene (MBS) rubber particles were distributed only in the polycarbonate (PC) phase in melt-blended PC–poly(butylene terephthalate)–MBS (PC–PBT–MBS) blends (1, 2). Similarly, we found that MBS or butyl acrylate core–shell rubber particles were distributed exclusively in the PC phase in PC–poly(ethylene terephthalate)–rubber blends, regardless of blending sequence (3, 4). To further prove such preferential distribution, Hobbs et al. later reported that the PC phase enveloped poly(methyl methacrylate) (PMMA) domains in PMMA–PC–PBT blends (5). Inoue and co-workers studied the ternary alloy of PBT–PC–MBS and reported a three-phase structure in which the fine rubber particles were covered with a PC shell and distributed in the PBT matrix (6). We reported that MBS particles were distributed in the styrene–maleic anhydride (SMA) phase in PC–SMA–MBS blends (7), and in the acrylonitrile–butadiene–styrene (ABS) phase in PC–ABS–MBS blends (8). Chang et al. also found that the thermoplastic polyurethane elastomer (TPE) was located at the PC–polyacetal interface in PC–polyacetal–TPE blends (9).

Paul and co-workers reported that MBS particles were located along the interfacial region of a blend of high-impact polystyrene (HIPS) and ABS (10). Fowler et al. (11) later demonstrated that different blending methods can change the rubber distribution, and they reported that MBS rubber has a strong tendency to be located at the interface of polystyrene (PS) and styrene–acrylonitrile (SAN) by one-step blending of PS–SAN–MBS; and the rubber is preferably distributed in the SAN phase if the MBS is preblended with SAN and then blended with PS. Cheung et al. also reported that MBS rubber particles were distributed preferentially at or near the interface between polysulfone (PSF) and poly(phenylene sulfide) (PPS) in PSF–PPS–MBS blends (12). The MBS rubber particles were found to be distributed in the SAN phase or at the PC–SAN interface, depending on the acrylonitrile (AN) content in the SAN of the PC–SAN–MBS blends (13). MBS rubber was located in PMMA in PC–PMMA–MBS blends, as would be expected (13). Recently, Chang and co-workers also found that MBS particles were distributed in the PMMA phase in PBT–PMMA–MBS blends (14) and in the PBT phase in PBT–poly(phenyl oxide)–MBS blends (15).

Compatibilities between the rubber phase and the blend components play the most important role in determining the final destination of the rubber, although, in a few cases, processing conditions such as blending sequence may change the distribution. Matching solubility parameters between rubber and any blend pair may provide the first prediction of rubber-phase distribution, although such an approach is not always satisfactory. A more accurate approach is the spreading coefficient, which has been used to successfully explain observed rubber migration and the final distribution (*5, 13*).

Essentially all the literature reviewed in the preceding paragraph emphasized identification rather than control of rubber-phase distribution. Recently, our laboratory conducted a series of investigations to identify and control rubber-phase distribution in several binary blends by using functionalized core–shell rubber. We were able to control butyl acrylate core–shell rubber in PC, PET, or both phases in PC–PET–rubber blends by functionalizing the shell structure of the core–shell rubber with glycidyl methacylate monomer units (*16*).

Polymer blends of PC and various nylons [polyamides (PAs)] have been the subject of active investigations (*17–28*). Chemical reactions have been reported to take place between these two homopolymers under melt conditions, mainly through an aminolysis process, to form PC–PA copolymer (*18, 20, 21*). Several compatibilizers have been used to achieve better compatibility between PC and PA, including epoxy-containing compounds (*17*), poly(allylate-*co*-maleic anhydride) (*24*), PC–PA block copolymer (*25*), and polyurethane (*27*). In order to improve low-temperature impact properties, maleic anhydride–grafted polypropylene or ethylene–propylene rubber has been used as an impact modifier in the PC–PA blends (*27*). In this study, we chose high-molecular-weight bisphenol A epoxy resin (molecular weight about 5000) as an in situ reactive compatibilizer for blends of PC and PA (nylon-6). The epoxy end groups can react with PA end groups to form the phenoxy-PA block copolymer under melt conditions. This in situ–formed copolymer can function as a phase compatibilizer because the phenoxy segment of the copolymer is known to be nearly miscible with PC (*29*). In order to better understand the effect of rubber distribution on the resulting properties, two MBS rubbers, one with a regular PMMA shell and the other with PMMA containing a small fraction of methacrylic acid (MAA) and maleic anhydride (MA) monomer, have been employed to toughen the PC–PA blends. In this chapter, we emphasize the control of rubber-phase distribution in PC–nylon–rubber blends through functionalized core–shell rubber. Details of work on the epoxy-resin-compatibilized PC–PA blends will be reported elsewhere.

Experimental Details

The following materials were used:

    PC: caliber melt flow rate (MFR) = 15 natural grade (Dow Chemical Co.).
    PA: Novamid 1010C2, nylon-6 (Mitsubishi Chemical Industries).
    Compatibilizer: NPES-909, bisphenol A type solid epoxy resin, epoxide equivalent weight (EEW) = 2026 g/eq (Nan Ya Plastics Corp., Taiwan).

Catalyst: ethyltriphenylphosphonium bromide (Merck Co.).

MBS rubber: KCA–102, core–shell with PMMA shell (Kureha Chemical Co., Japan).

MBS–MA rubber: KCA–503, a similar type of core–shell rubber, except the shell contains methyl methacrylate (MMA), methacrylic acid (MAA), and maleic anhydride (MA) monomer units (Kureha Chemical Co., Japan).

The blends were prepared using a 2.0-cm twin-screw extruder (Welding Engineers) with length/diameter (L/D) = 48 and counter-rotating, intermeshing screws. Blended pellets were dried and molded into various 0.125-in. test specimens using a 3-oz. injection-molding machine (Arburg).

Tensile properties (ASTM-D638) were measured using a tensile tester (Instron Model 4201). Notched impact tests (ASTM-D256) were carried out, using a 10-mil notch radius, in an impacter (TMI). Instrumented falling-weight impact tests (ASTM-D3029) were performed on 0.125-in.-thick specimens under ambient conditions. Morphological characterization of the blends was conducted by scanning electron microscopy (SEM) and TEM. The fracture surfaces of the specimens were examined by SEM after etching with 30% aqueous KOH solution and coating with gold. The TEM thin-layer specimens were cut using a microtome and stained with 2% OsO_4 solution.

Results and Discussion

Figure 1 shows SEM micrographs of the uncompatibilized and compatibilized PC–PA (30/70) blends after etching away of the PC phase with dilute KOH so-

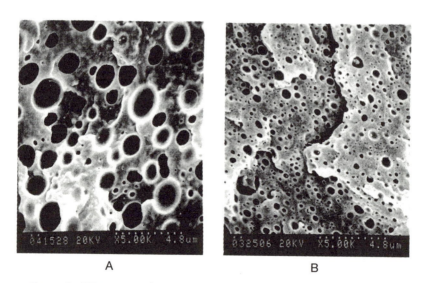

A   B

Figure 1. SEM images of uncompatibilized and compatibilized blends: (A) PC/PA = 30/70; and (B) PC/PA/compatibilizer = 70/30/5.

lution. The presence of 5 parts per hundred resin by weight (phr) epoxy resin can reduce phase domain significantly, mainly because of lower interfacial tension during melt blending. Figure 2 shows SEM images of the 10-phr rubber-modified blends. The presence of MBS in the blend results in a significantly larger PC domain than in the corresponding unmodified blend (Figure 2A vs Figure 1A). The MBS rubber particles with a PMMA shell structure have a strong tendency to be located in the PC phase and near the PC–PA interface in this PC–PA–MBS (30/70/10) blend (Figure 5B). Therefore, the PMMA shells of the MBS rubber particles, rather than the PC phase, actually provide most of the direct contact with the PA phase. PA is considered more compatible with PC than with PMMA because PA has a solubility parameter ($\delta = 13.6$) closer to that of PC ($\delta = 9.8$) than to that of PMMA ($\delta = 9.1$) (*30, 31*). Therefore, the larger observed PC domains in this PC–PA–MBS blend can also be interpreted as the result of less compatibility between PA and PMMA. Alternatively, this phenomenon can be interpreted as follows: the presence of MBS rubber changes the intrinsic nature of PC to make this PC phase less compatible with PA relative to the virgin PC.

Figure 2B shows the SEM image of the blend modified with MBS–MA, where the domain size of the PC phase is comparable to that of the unmodified one (Figure 1A). The MBS–MA rubber particles are evenly distributed within the PA phase (*see* Figure 6), and so the PC phase makes direct contact with the PA phase. Therefore, the interfacial tensions of this blend modified

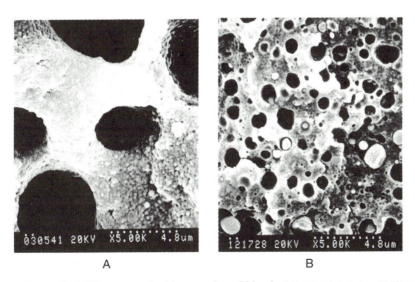

Figure 2. SEM images of rubber-toughened blends: (A) PC/PA/MBS = 30/70/10; and (B) PC/PA/MBS–MA = 30/70/10.

with MBS–MA, and the corresponding unmodified blends, are expected to be close, which will result in almost equal domain sizes of the PC phase. The corresponding compatibilized rubber-toughened blends (MBS and MBS–MA) show similar trends: the MBS-modified blend has larger PC phase domains than the blend modified with MBS–MA (Figures 3A and 3B vs. Figures 2A and 2B).

Characterization of Phase Morphologies by TEM.   Figure 4 shows TEM images of OsO_4-stained samples of the unmodified PC–PA (30/70) blends, where the light phase is PC and the darker phase is PA. Figure 5 shows micrographs of MBS-modified PC–PA (30/70) blends in which the MBS content ranges from 5 to 30 phr. The unreactive rubber particles are distributed in the PC phase and near the interface, even at as high as 30 phr of rubber.

On the other hand, the reactive MBS–MA rubber particles are evenly distributed only in the PA phase (Figure 6). The acid and MA groups of the MBS–MA rubber shell can react with the amine end groups of PA, and the resulting PA-linked rubber particles are therefore retained in the PA phase. It is feasible to have a PC--PA blend with rubber particles in both phases by blending both MBS and MBS–MA rubbers simultaneously (Table I, No. 18).

To better understand the ability of rubber particles to undergo interphase

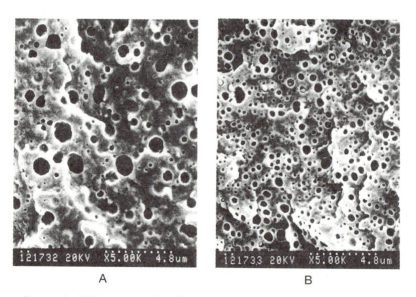

                    A   B

Figure 3. SEM images of rubber-toughened and compatibilized blends: (A) PC/PA/compatibilizer/MBS = 30/70/5/5; and (B) PC/PA/compatibilizer/MBS–MA = 30/70/5/5.

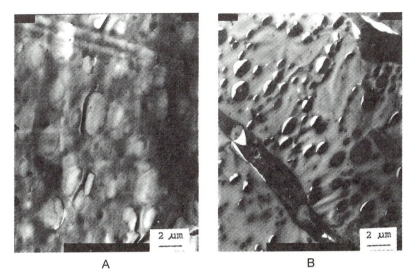

A   B

Figure 4. TEM images of unmodified blends: (A) PC/PA = 30/70; and (B) PC/PA = 70/30.

migration in typical melt mixing, various two-step sequential blendings were carried out. Figure 7A shows a micrograph of the PC–PA–MBS blend prepared by preblending PA with MBS and then blending with PC, where the MBS particles have migrated from the PA phase into the PC phase. Figure 7B shows that the MBS–MA rubber particles are distributed only in the PA phase in a blend prepared by preblending PC with MBS–MA and then blending with PA. This result indicates that the MBS–MA rubber particles have enough time to contact and react with the PA phase during second-stage melt blending. The PA-linked MBS–MA rubber particles are forced to remain in the PA phase. Figure 8 shows micrographs of blends with PC as the major component. Again, the MBS rubber particles are distributed exclusively in the PC phase (Figure 8A), whereas the MBS–MA particles are distributed only in the PA phase.

Mechanical Properties.   The mechanical properties of the selected blends studied, including the results of tensile, Izod impact, and falling-weight impact tests, are summarized in Table I. Generally, lower-than-expected toughening efficiency (especially for Izod impact) and data scattering were obtained because it is hard to maintain consistent moisture content of the specimens (especially for blends in which PA is the major component). The presence of the compatibilizer in all the blends resulted in consistent and substantial improvement in properties (Nos. 2, 3, and 4 vs No. 1, and No. 20 vs

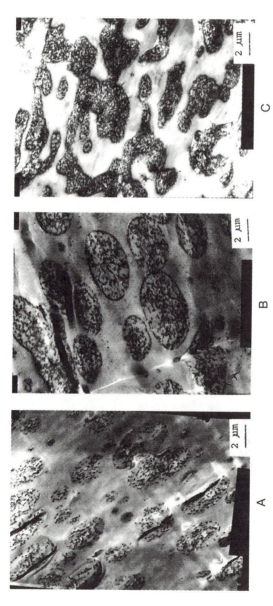

Figure 5. TEM images of MBS-modified blends: (A) PC/PA/MBS = 30/70/5; (B) PC/PA/MBS = 30/70/10; and (C) PC/PA/MBS = 30/70/30.

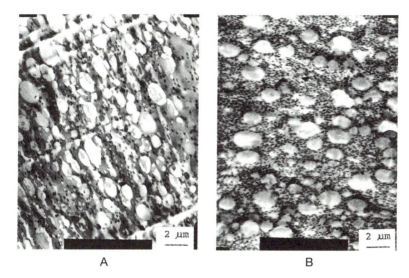

Figure 6. TEM images of blends modified with MBS–MA: (A) PC/PA/MBS–MA = 30/70/5; and (B) PC/PA/MBS–MA = 30/70/30.

19), mainly because of finer domains of the compatibilized blends relative to the uncompatibilized counterparts.

The blends modified with MBS rubber (without compatibilizer) show substantial improvement over the unmodified blend if PC is the major component (Nos. 21 and 22 vs No. 19), but they show minimum improvement when PA is the major component (Nos. 6, 7, and 8 vs No. 1). The blends modified with MBS–MA rubber also show significant improvement if PC is the major component (Nos. 23 and 24 vs No. 19), but they fail to show improvement when PA is the major component (Nos. 9, 10, and 11 vs No. 1).

In general, the combination of compatibilizer (with or without catalyst) and rubber in the blend shows a synergistic effect, but the effect is not substantial and is somewhat inconsistent. The blend containing both MBS and MBS–MA rubbers (No. 18) is expected to have the rubber particles distributed in both the PC and PA phases, but it fails to show any further improvement.

Generally, the blend with PC as the major component can be toughened by rubber more effectively than the blends with PA as the major component. Morphologically, the dispersed PA phase is present as an elongated lamellar structure in blends in which PC is the major component, whereas the dispersed PC phase is present in an elliptical form in blends where PA is the major component.

Table I. Summary of Mechanical Properties

No.	Composition[a]	Tensile Modulus (MPa)	Tensile Yield Strength (MPa)	Tensile Elongation (%)	Izod Impact Strength (J/M)	Falling-Weight Impact (J)
1	C30/N70	2370	—	9	38	1.6
2	C30/N70/Comp2	2600	60	112	50	3.8
3	C30/N70/Comp5	2580	54	95	72	4.7
4	C30/N70/Comp10	2550	58	109	66	5.7
5	C30/N70/Comp5/Cat	1890	54	59	85	3.9
6	C30/N70/RA5	2050	49	27	34	3.3
7	C30/N70/RA10	1870	40	26	40	2.3
8	C30/N70/RA30	1280	31	17	25	—
9	C30/N70/RB5	2140	—	9	16	2.4
10	C30/N70/RB10	1880	38	22	17	1.4
11	C30/N70/RB30	1450	28	22	18	—
12	C30/N70/Comp2/RA5	1940	46	92	89	4.6
13	C30/N70/Comp2/RA5/Cat	1980	42	237	87	5.8
14	C30/N70/Comp5/RA5/Cat	1810	46	49	30	29.7
15	C30/N70/Comp2/RB5	1940	46	—	41	4.3
16	C30/N70/Comp2/RB5/Cat	2170	47	57	31	4.9
17	C30/N70/Comp5/RB5/Cat	1870	46	25	17	1.6
18	C30/N70/Comp5/RA5/RB5/Ca	2160	50	28	88	—
19	C70/N30	2400	50	11	50	0.9
20	C70/N30/Comp5	2640	57	215	113	7.5
21	C70/N30/RA5	1960	45	228	137	2.6
22	C70/N30/RA10	2120	44	217	256	9.6
23	C70/N30/RB5	1860	42	193	138	2.3
24	C70/N30/RB10	2230	46	171	199	1.7
25	C70/N30/Comp5/RA5/Cat	—	—	—	530	10
26	C70/N30/Comp5/RA5/Cat	2510	52	26	57	3.8

[a]C: PC; N:PA; Comp: compatibilizer; RA: MBS; RB: MBS–MA; Cat: catalyst, 200 ppm.

Conclusions

This study demonstrated that the final destination of the added core–shell rubber particles, in PC, PA, or both, in the PC–PA binary blend can be controlled by properly selecting the chemical structure of the shell in the core–shell rubber. The unreactive MBS rubber tends to reside in the PC phase and near the vicinity of the PC–PA interface. The reactive MBS–MA rubber can have a chemical reaction with PA end groups and can therefore be retained within the PA phase. High-molecular-weight bisphenol A epoxy resin has proved to be an efficient compatibilizer for PC–PA blends. Rubber-toughening of the PC–PA blend in which PC is the matrix is much more effective than with blends in which PA is the matrix.

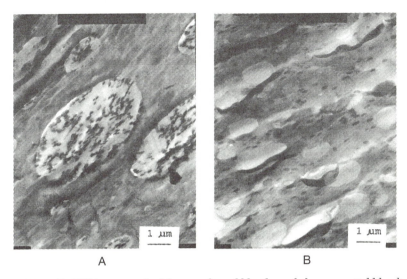

A   B

Figure 7. TEM images of rubber-toughened blends made by sequential blending: (A) preblending PA with MBS, then adding PC, where PC/PA/MBS = 30/70/5; and (B) preblending PC with MBS–MA, then adding PA, where PC/PA/MBS–MA = 30/70/5.

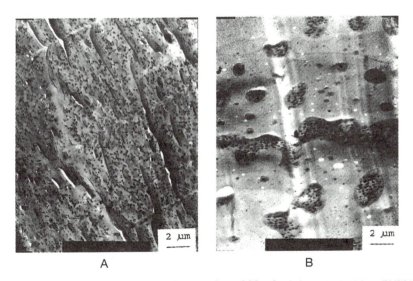

A   B

Figure 8. TEM images of rubber-toughened blends: (A) PA/PA/MBS = 70/30/5; and (B) PC/PA/MBS–MA = 70/30/5.

Acknowledgment

This study was financially supported by the National Science Council of the Republic of China under contract No. NSC82–0405–E–009–311.

References

1. Hobbs, S. Y.; Dekkers, M. E. J.; Watkins, V. H. *Polym. Bull.* **1987,** *17,* 341.
2. Hobbs, S. Y.; Dekkers, M. E. J., Watkins, V. H. *J. Mater. Sci.* **1988,** *23,* 1219.
3. Chang, F. C. Company Internal Report; Dow Chemical, Freeport, TX, 1985.
4. Liao, Z. L.; Chang, F. C. *J. Appl. Polym. Sci.* **1994,** *52,* 1115.
5. Hobbs, S. Y., Dekkers, M. E. J.; Watkins, V. H. *Polymer* **1988,** *29,* 1598.
6. Okamoto, M.; Shinoda, Y.; Kojima, T.; Inoue, T. *Polymer* **1993,** *34,* 4868.
7. Chen, R. C.; Chiang, F. C. *Polym. Networks Blends* **1993,** *3,* 107.
8. Wu, J. S.; Shen, S. C.; Chang, C. F. *Polym. J.* **1994,** *26,* 33.
9. Chang, C. F.; Yang, M. Y.; Wu, J. S. *Polymer* **1991,** *32,* 1394.
10. Keskkula, H.; Paul, D. R.; McCready, K. M.; Henton, D. E. *Polymer* **1987,** *28,* 2063.
11. Fowler, M. E.; Keskkula, H.; Paul, D. R. *J. Appl. Polym. Sci.* **1989,** *37,* 225.
12. Cheung, M. F.; Golovoy, A.; Mindroin, V. E.; Plummer, H. K., Jr.; Van Oene, H. *Polymer* **1993,** *34,* 3809.
13. Cheng, T. W.; Keskkula, H.; Paul, D. R. *Polymer* **1992,** *33,* 1606.
14. Fan, D. H.; Wu, J. S.; Chang, F. C. unpublished results.
15. Liu, W. B.; Chang, F. C. unpublished results.
16. Lin, K. P.; Chang, F. C. *Polym. Networks Blends* **1994,** *4,* 51.
17. Maresca, L. Eur. Patent 227053, 1987.
18. Eguiazabal, J. I.; Nazabal, J. *Makromol. Symp.* **1988,** *20/21,* 255.
19. Heggs, R. P.; Marcus, J. L.; Markbam, R. L.; Mangaraj, D. *Plast. Eng.* **1988,** *June,* 29.
20. Cortazar, M.; Eguiazabal, J. I.; Iruin, J. J. *Brit. Polym. J.* **1989,** *21,* 395.
21. Gattiglia, E.; LaMantia, F. P.; Turturro, A.; Valenza, A. *Polym. Bull.* **1989,** *21,* 47.
22. Gattiglia, E.; Turturro, A.; Pedemonte, E. *J. Appl. Polym. Sci.* **1989,** *38,* 1807.
23. Eguiazabal, J. I.; Nazabal, J. *Plast. Rub. Proc. Appl.* **1990,** *14,* 211.
24. Sato, M.; Akiyama, S.; Honda, S. *Kobunshi Ronbunshu* **1990,** *47,* 287.
25. DeRudder, J. L. U.S. Patent 4960836, 1990.
26. Gattiglia, E.; Turturro, A.; Pedemonte, E.; Dondero, E. *J. Appl. Polym. Sci.* **1990,** *41,* 1411.
27. Perros, P. J.; Bourbonals, E. A. U.S. Patent 5019625, 1991.
28. Gattiglia, E.; Turturro, A.; Lamantia, F. P.; Valenza, A. *J. Appl. Polym. Sci.* **1992,** *46,* 1887.
29. Chu, J. H.; Chang, F. C. *Proceedings of the 1990 Annual Conference of the Chinese Society for Material Science;* 1990; 1021.
30. Billmeyer, F. W. *Textbook of Polymer Science,* 3rd ed.; Wiley-Interscience: New York, 1984; p 153.
31. Bucknall, C. B. *Toughened Plastics;* Applied Science: London, 1977; p 13.

19

Toughened Polypropylene–Polyamide 6 Blends Prepared by Reactive Blending

J. Rösch and R. Mülhaupt

Freiburger Materialforschungszentrum und Institut für Makromolekulare Chemie der Albert-Ludwigs Universität, Stefan-Meier-Strasse 31, D-79104 Freiburg im Breisgau, Germany

Reinforced polypropylene (PP) was prepared by blending 70 vol% PP with 30 vol% polyamide 6 (PA6) in the presence of compatibilizers such as maleic anhydride–grafted polypropylene and maleic anhydride–grafted rubbers. The key to this reactive blending technology was the in situ formation of segmented polymers via covalent imide-coupling involving amino-terminated PA6 and succinic anhydride functional compatibilizers. Compatibilizer volume fraction and molecular architecture gave control of PA6 dispersion and interfacial adhesion between PP and PA6. In contrast to the simultaneous dispersion of separate PA6 and rubber microphases, the in situ formation of core–shell-type dispersed microphases, comprising a rigid PA6 core and a rubber shell, accounted for a substantially improved balance between the toughness and stiffness of the PP–PA6 (70/30) blend.

R ECENT BREAKTHROUGHS IN CATALYST and process development have greatly simplified polypropylene (PP) production and broadened the application spectrum of PP materials, which have economic, ecological, and recycling advantages over environmentally less-friendly polymers (*1, 2*). Among commodity thermoplastics, only PP exhibits heat-distortion temperatures above 100 °C. In order to compete successfully with traditional materials, such as metals and engineering thermoplastics, in higher-value-in-use engineering applications, an important challenge in PP development must be met. The challenge is to overcome the limitations of the properties of PP associated with its hydrocarbon nature, namely, poor dyeability, poor adhesion, low resistance to hydrocarbon permeation, and comparatively high chain flexibility. Moreover, another important R&D objective in the development of engineering resins is to improve toughness without sacrificing stiffness and strength. One of the widely applied approaches toward reinforced PP is to incorporate stiff isotropic or

0-8412-3151-6   © 1996 American Chemical Society

anisotropic dispersed phases, preferably inorganic fillers, into the continuous PP phase. Frequently, such stress-concentrating dispersed microphases reduce plastic deformation of the PP matrix, giving rise to drastically lower toughness (3, 4). Therefore, blend technologies are being developed to achieve property synergisms that do not reflect the properties of the blend components when the mixing ratio is taken into account. Especially in automotive applications, blends of PP with polyamides (PAs) are of particular interest for reducing polyamide water-uptake and improving the stiffness and paintability of PP (5).

Because PP and PA6 are mutually immiscible, polymeric dispersing agents, often referred to as blend compatibilizers, must be added to enhance both PA6 dispersion and PP–PA6 interfacial adhesion. The use of compatibilizers to improve dispersion and interfacial adhesion has been demonstrated successfully for a large variety of multiphase polymer blends (6–9). Typical PP blend compatibilizers are maleic anhydride–grafted PP (PP-g-MA) and maleic anhydride–grafted rubbers, for example, ethene–propene rubber (EPR-g-MA) or polystyrene-block-poly(ethene-co-1-butene)-block-polystyrene (SEBS-g-MA). During melt processing, succinic anhydride groups react with amino end groups of PA6 to produce segmented polymers.

The purpose of our research was to explore new strategies for controlled formation of multiphase PP–PA6 blends containing 30 vol% PA6 as dispersed reinforcing microphase, and to exploit the potential of reactive processing technology and tailor-made blend compatibilizers. PP–PA6 (70/30) and rubber-modified PP–PA6 (70/30) blends were examined to elucidate the basic parameters governing PA6 microphase dispersion, interfacial adhesion, stress transfer, and impact-energy dissipation. Moreover, the influence of reactive and unreactive compatibilizing rubbers was studied to correlate the molecular architecture and volume fraction of functionalized PP and rubbery blend compatibilizers with the mechanical and morphological properties of rubber-modified PP–PA6 (70/30) blends. Another objective of this research was to compare the blend architectures of rubber-modified PP–PA6 (70/30) blends comprising either (1) separately dispersed PA6 and rubber microphases, or (2) in situ formed core-shell-type microphases with a PA6 core and a covalently bonded rubber shell.

Experimental Details

Materials. All polymers were commercially available and used without further purification. PP was purchased from Hoechst AG (Hostalen PPN 1060, number-average molecular weight (M_n) = 63,000 g/mol, weight-average molecular weight (M_w) = 182,700 g/mol, as determined by size exclusion chromatography (SEC) in 1,2,4-trichlorobenzene at 135 °C using a polystyrene standard, melt flow index (MFI) (230/2,16) = 2 dg/min, and melting temperature (T_m) = 165 °C). Non-functionalized EPR (Exxelor VM22, MFI (230/2,16) = 5 dg/min, T_g = 56 °C), maleic anhydride–grafted PP (Exxelor PO2011, 0.031 mol anhydride/kg polymer, MFI (230/2,16) = 125 dg/min), and maleic anhydride–grafted EPR (Exxelor VA 1803,

60 mol anhydride/kg polymer, MFI (230/2,16) = 3 dg/min, glass-transition temperature (T_g) = –52 °C) were products of Exxon Chemicals. Maleic anhydride–grafted SEBS was supplied by Shell (Kraton G1901 X2, 208 mol anhydride/kg SEBS, MFI (230/2,16) = 3.2 dg/min). PA6 was obtained from Snia, Milano (Sniamid ASN 27, T_m = 222 °C, 0.031 mol amine end groups/kg PA6).

Reactive Blending and Characterization.   PA6 was dried for 6 h at 80 °C under oil-pump vacuum prior to use. Melt blending was performed using a Haake Rheomix 90 twin-screw kneader equipped with a 60-ml mixing chamber and on-line temperature and torque recording. In a typical run, after preheating the mixing chamber at 240 °C for 10 min, 40 g of the blend composed of PP, PA6, PP-*g*-MA, and 0.2 g stabilizer mixture (80 wt% Irganox 1010 and 20 wt% Irgafos 168) was charged. Blending was performed for 4 min at 240 °C; this time included 2 min required for melting the components. Afterwards, the blend was quickly recovered and quenched to room temperature between water-cooled brass plates.

For testing, sheets 1.5 mm in thickness were prepared by compression molding as follows: the samples were annealed for 10 min at 260 °C in a heated press (Schwabenthan Polystat 100) and then quenched between water-cooled metal plates. The cooling rate was monitored by means of a thermocouple and was 50 K/min between 230 and 110 °C and 20 K/min below 110 °C. For tensile testing, dumbbell-shaped tensile bars 18 mm in length were cut and machined as described in DIN 53544. After sample conditioning (3 days at 23 °C and 50% humidity), stress–strain measurements to determine Young's modulus (from the initial slope of the stress–strain curve) and yield stress were recorded at a 10-mm/min crosshead speed on a tensile tester (Instron 4204) at 23 °C. The average deviation of the modulus measurement was less than 15%. Notched Charpy impact strength was determined on five test specimens according to standard procedures (DIN 53453) using a Zwick 5102 pendulum impact tester equipped with a 2 J pendulum. Morphological studies were performed using a Zeiss CEM 902 transmission electron microscope (TEM). Thin sections suitable for TEM analysis were cut after staining and hardening the samples in ruthenium tetroxide vapors for 6 h. Microtoming of the samples into sections 80 to 100 nm thick was performed using a Reichert Jung Ultracut E device equipped with diamond knives.

Results and Discussion

PP-*g*-MA-Compatibilized PP–PA6 (70/30) Blends.   One of the traditional concepts for compatibilizing PP and PA, pioneered by Ide and Hasegawa (*10*) during the early 1970s, involves using PP-*g*-MA as a blend compatibilizer. During melt processing, the reaction of PA6 amine end groups with succinic anhydride–functional PP affords imide-coupled PP-*graft*-PA6, which represents efficient dispersing agents and interfacial-adhesion promoters for PP–PA6 blends. This basic principle is illustrated in Figure 1. Several research groups have investigated the morphological, rheological, and mechanical properties of such blends (*10–16*), emphasizing the influence of blend compatibilizers. However, it is important to note that most blend compatibilizers produced by reactive extrusion processes are ill-defined. For instance, free-radical grafting of PP with MA is accompanied by extensive PP degradation. Thermal PP degradation in the presence of maleic anhydride results in a

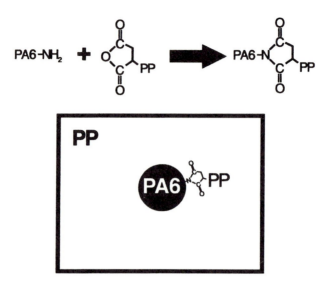

Figure 1. In situ formation of PP-block-PA6 derived from amino-terminated PA6 and mono(succinic anhydride)-terminated PP.

mixture of mono- and bis(succinic anhydride)-terminated propene oligomers via ene-type addition of maleic anhydride to olefin-terminated PP intermediates resulting from β-chain scission of the PP chain.

In our research, we used well-defined mono(succinic anhydride)-terminated PP-g-MA blend compatibilizers with independently varied molecular weights and stereoregularities. In the melt PP-*block*-PA6, diblock copolymers are formed at the PP–PA6 interface and compatibilize PP with PA6. As reported in more detail in a previous communication (*17, 18*), PP-g-MA model compatibilizers with one terminal succinic anhydride end group are readily available via ene-type addition of maleic anhydride to vinylidene-terminated oligopropenes, which are produced in high yields using metallocene-catalyzed propene oligomerization (*18–21*). In contrast to conventional melt-grafted PP-g-MA, this process yields completely atactic as well as highly isotactic mono(succinic anhydride)-terminated PPs with narrow molecular-weight distributions ($1.5 < M_w/M_n < 2.5$), and M_n varying between 400 and 30,000 g/mol. The morphological and mechanical properties of PP–PA6 (70/30), prepared in the presence of atactic PP-g-MA with $M_n = 5000$ and isotactic PP-g-MA with $M_n = 10,000$ g/mol as blend compatibilizers, are displayed in Figure 2 as a function of the PP-g-MA volume fraction.

Provided M_n was greater than 3000 g/mol, both atactic and isotactic PP-g-MA enhanced the dispersion of PA6 in the continuous PP matrix. Because of the miscibility of atactic and isotactic PP in the melt phase, stereoregularity did not influence PA6 dispersion, which was controlled primarily by the PP-g-

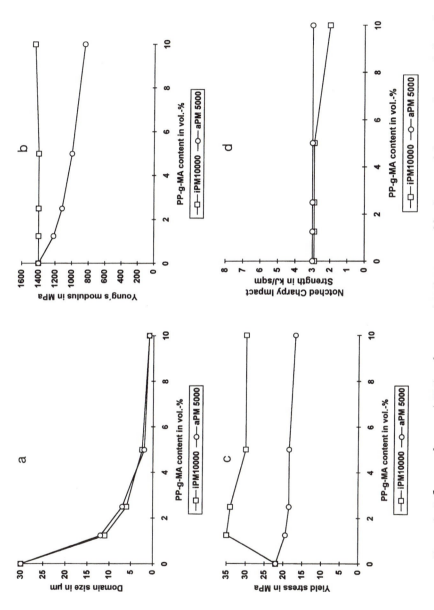

Figure 2. Influence of atactic (aPM) and isotactic (iPM) PP-g-MA blend compatibilizers on the morphological and mechanical properties of PP–PA6 (70/30): PA6 domain size (a), Young's modulus (b), yield stress (c), and notched Charpy impact strength (d).

MA volume fraction. In terms of solid state properties, as expressed by Young's modulus, yield stress, and notched Charpy impact strength, high-molecular-weight highly isotactic PP-*g*-MA compatibilizers were much more efficient than atactic PP-*g*-MA. In fact, as is apparent from Figure 2, yield stress and notched Charpy impact strength increase substantially with increasing stereo-regularities, molecular weights, and volume fractions. This behavior is closely associated with cocrystallization of the isotactic PP segment of the in situ formed PP-*block*-PA6, which is likely to accumulate at the PP–PA6 interface, although complete coupling is unlikely (*17*). In spite of a doubling in impact strength with respect to PP, this balance between the toughness, stiffness, and strength of PP–PA6 must be improved further to meet the demands of engineering applications. Therefore, ternary blends of PP, PA6, and rubber (i.e., EPR and SEBS) were investigated. In principle, rubber can be dispersed in the PP matrix as a separate microphase or as a shell embedding a PA6 core.

EPR-Modified PP–PA–PP-g-MA. In the first approach to improving the toughness of PP–PA6 (70/30), discrete EPR microphases were dispersed simultaneously with PA6. As a rule, when PP and EPR exhibited similar MFIs, it was possible to control EPR and PA domain sizes independently (*21, 22*). Although the PA6 domain size was affected exclusively by the volume fraction of PP-*g*-MA blend compatibilizer, the average EPR domain sizes increased primarily with increasing EPR volume fraction. The average EPR domain size varied between 0.1 and 10 μm. According to TEM studies, in contrast to the spherical PA6 microphases, EPR microphases exhibited irregularly shaped structures with small PP subinclusions. As is apparent in Figure 3, decreasing the size of PA6 domains by increasing the PP-*g*-MA volume fraction

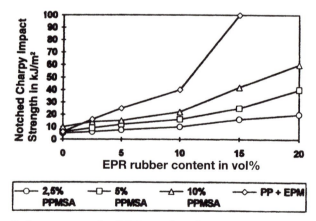

Figure 3. Impact strength of EPR-toughened PP–PA6 (70/30) blends containing 2.5, 5.0, and 10 vol% PP-g-MA compared with EPR-toughened PP, as a function of EPR volume fraction.

from 2.5 to 10 vol% gave improved notched Charpy impact resistance. However, impact performance of EPR-toughened PP–PA6 (70/30) was much poorer than that of EPR-toughened PP, especially at an EPR rubber volume fraction exceeding 10 vol%. This difference in impact performance could result from mutually overlapping stress fields of interconnected PA6 microparticles, depressing the efficiency of energy dissipation involving EPR microphases. Moreover, a small EPR rubber volume fraction is sufficient to reduce the stiffness of the PP–PA6 matrix and eliminate PA6-microparticle reinforcement.

In conclusion, the simultaneous dispersion of separate, discrete EPR and PA6 microparticles in the continuous PP accounted for antisynergistic blend properties combining poor stiffness with inadequate impact strength. Slightly improved impact strength was achieved at the expense of unacceptable losses of stiffness.

PP–PA6 Blends Containing Dispersed Core-Shell Microparticles. In the third type of PP–PA6 blend system, the PP-*g*-MA blend compatibilizer and rubber was completely substituted by maleic anhydride–grafted rubbers such as EPR-*g*-MA and SEBS-*g*-MA. As reported previously (*22, 23*) and schematically represented in Figure 4, imide-coupling at the PP–PA6 interface, and surface-tension gradient and immiscibility between PP, PA6, and rubber are responsible for the accumulation of the rubber at the PA6 microparticle surface, which results in microparticles with a PA6 core and a rubber shell. Like PP-*g*-MA blend compatibilizers, maleic anhydride-grafted rub-

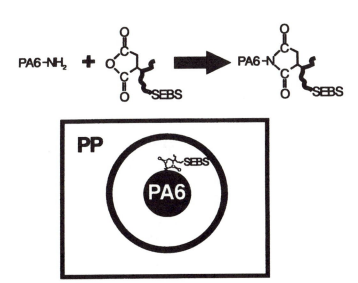

Figure 4. In situ formation of core-shell-type microparticles, in which a PA6 core is surrounded by a SEBS-g-MA shell.

bers are polymeric dispersing agents, improving PA6 dispersion in the PP matrix. This fact is reflected by small average PA6 domain sizes with increasing volume fractions of functionalized rubbers. Figure 5 shows TEM images of RuO$_4$-stained thin cuts, which indicate the absence of separate rubber microphases and the formation of microparticles with a PA6 core and a rubber shell. The dispersing-agent performance of SEBS-*g*-MA was much better than that of PP-*g*-MA. Less than 5 vol% SEBS-*g*-MA was sufficient to achieve PA6 domain sizes of 0.1 μm, whereas more than 15 vol% PP-*g*-MA would be required to achieve similar PA6 dispersion. Clearly, SEBS-*g*-MA yielded much smaller PA6 microparticles, which were encapsulated in an SEBS-*g*-MA shell.

In Figure 5 it is apparent that the much smaller PA6 core-shell microparticles, obtained in the presence of SEBS-*g*-MA, form large agglomerates. As reported previously (*23*), with increasing SEBS concentration a second cocontinuous phase consisting of large clusters of PA6 core-shell-type microparticles appears. Interestingly, even at small SEBS volume fractions, such cocontinuous phases of PP and SEBS–PA6 were detected in the PP matrix.

Most likely, this unusual morphological feature of PP–PA6 (70/30) blends compatibilized with SEBS-*g*-MA could account for the unexpected combination of high toughness, stiffness, and strength (see Figures 6 and 7). At an SEBS-*g*-MA volume fraction greater than 15 vol%, crack propagation during pendulum impact was stopped, and intense stress-whitening was observed near the crack tip. As reflected by high yield stresses, interfacial adhesion between the SEBS shell and PP matrix was excellent in spite of the immiscibility of both components. The SEBS-modified PP–PA6 gave higher stiffness than the EPR-modified PP–PA6. This difference could result from the much higher Young's modulus of the SEBS compared with that of the rather soft EPR.

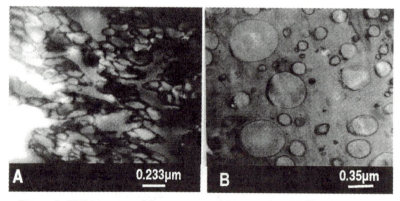

Figure 5. TEM images of thin sections of PP–PA6 (70/30) blends compatibilized with 5 vol% SEBS-g-MA (A) and 5 vol% PP-g-MA (B).

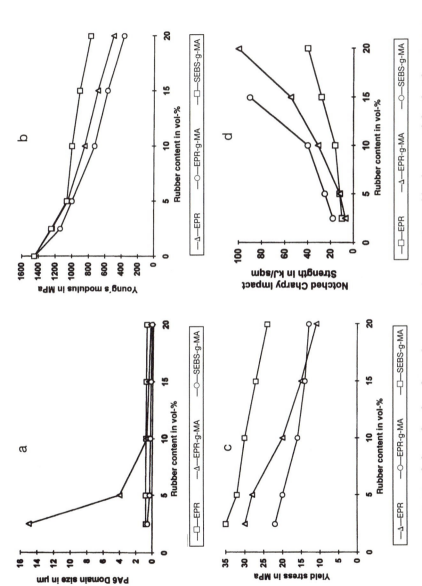

Figure 6. Morphological and mechanical properties of PP–PA6 (70/30) blends compatibilized with EPR, EPR-g-MA, and SEBS-g-MA as a function of the volume fraction of the volume fraction blend compatibilizer: PA6 domain size (a), Young's modulus (b), yield stress (c), and notched Charpy impact strength (d). In the case of nonfunctionalized EPR, 10 vol% isotactic PP-g-MA (M_n = 30,000 g/mol) was added.

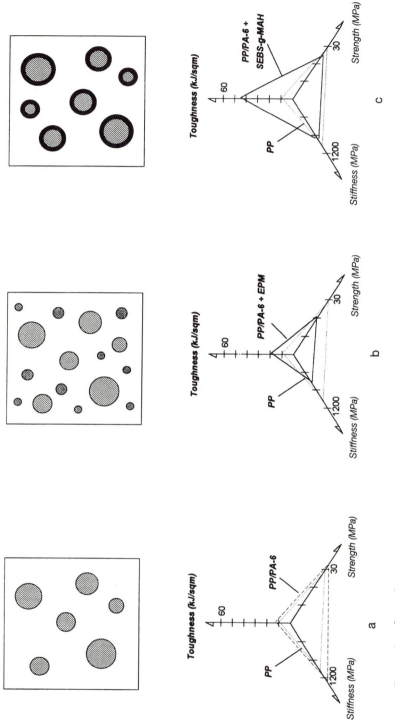

Figure 7. Relationship between morphology and mechanical properties of PP–PA6 (70/30) blends compatibilized with PP-g-MA (a), EPR (b), and SEBS-g-MA (c), as compared with the mechanical properties of PP.

Conclusions

The morphologies of PA6-reinforced PP–PA6 (70/30) can be tailored to enhance yield stress and impact strength without drastic losses of stiffness. In Figure 7, the basic correlations between PP–PA6 (70/30) blend morphologies and mechanical properties are summarized. Three different types of blend morphologies have been realized. The first type is characterized by discrete, spherical PA6 microparticles dispersed in the PP matrix. In situ formation of PP-*block*-PA6 at PP–PA6 interfaces accounts for both steric stabilization of the PA6 dispersion and interfacial adhesion via cocrystallization of PP with the isotactic PP segments of PP-*block*-PA6, provided the average PP segment length exceeds the entanglement molecular weight. In spite of significantly increased yield stresses with increasing PP-*g*-MA stereoregularities and molecular weights, improvement in impact strength is rather poor, and impact strength does not change with the PP-*g*-MA volume fraction. The second type of morphology seen in Figure 7 is characterized by simultaneously dispersed, discrete PA6 and EPR microphases, where the average domain sizes are controlled independently by EPR or PP-*g*-MA volume fractions. For example, at a constant rubber content, PA6 domain size can be reduced by increasing the PP-*g*-MA volume fraction. Obviously, dispersed, discrete EPR microphases fail to improve toughness, as expected from the behavior of PP–EPR blends. The best results in terms of the stiffness–toughness balance were obtained with the third type of PP–PA6 (70/30) blend morphology, which is characterized by colloidal, dispersed PA6 microparticles embedded in thin rubber shells. The key to morphologic control is the succinic anhydride rubber blend compatibilizers, especially SEBS-*g*-MA. When core-shell-type microparticles agglomerate to form a cocontinuous phase, unusual blend-property synergisms are achieved, especially substantially improved impact strength combined with high yield stress without drastic losses of stiffness, as expressed by Young's modulus. This result indicates that fine-tuning the molecular architectures of blend compatibilizers, and the controlled formation of structured microparticles and cocontinuous blend morphologies, plays a key role in developing novel PP materials exhibiting unusual combinations of properties.

Acknowledgments

We thank Bundesministerium für Forschung und Technologie and Badische Auilin und Soda Fabrik AG/Ludwigshafen for supporting the research on metallocene-catalyzed propene oligomerization and functionalization, and the Bundesminister für Wirtschaft and Arbeitsgemeinschaft Industrie-Forschung for supporting research on adhesion promoters utilizing the potential of functionalized olefin polymers.

References

1. Galli, P.; Haylock, J. C. *Makromol. Chem. Macromol. Symp.* **1992**, *63*, 19.
2. Neissl, W.; Ledwinka, H. *Kunststoffe* **1993**, *83(8)*, 577.
3. Schlumpf, H. P. In *Plastics Additives Handbook*, 3rd ed.; Gächter, R.; Müller, H., Eds.; Hanser; Munich, Germany, 1990; p 525.
4. Katz, H. S.; Milewski, J. V. *Handbook of Fillers for Plastics;* Van Nostrand Reinhold: New York, 1987.
5. Altendorfer, F.; Begemann, M.; Schwaiger, R.; Seitl, E. *Kunststoffe* **1992**, *82*, 758.
6. Wu, S. *Polymer Interface and Adhesion;* Marcel Dekker: New York, 1982.
7. Michler, H. G. *Kunststoff Mikromechanik;* Hanser: Munich, Germany, 1992.
8. Utracki, L. A. *Polymer Alloys and Blends;* Hanser: Munich, Germany, 1989.
9. *Polypropylene Structure, Blends and Composites. Vol. 2. Copolymers and Blends;* Karger-Kocsis, J., Ed.; Chapman & Hall: London, 1995.
10. Ide, F.; Hasegawa, A. *J. Appl. Polym. Sci.* **1974**, *18*, 963.
11. Holsti-Miettinen, R.; Seppälä, J.; Ikkala, O. T. *Polym. Eng. Sci.* **1992**, *32(13)*, 868.
12. Gheluwe, P. v.; Favis, B. D.; Chalifoux, J. P. *J. Mater. Sci.* **1988**, *23*, 3910.
13. Scholz, P.; Fröhlich, D.; Muller, R. *J. Rheol.* **1989**, *33(3)*, 481.
14. Lambla, M.; Killis, A.; Magnin, H. *Eur. Polym. J.* **1979**, *15*, 488.
15. Park, S. J.; Kyu Kim, B. Q.; Jeong, H. M. *Eur. Polym. J.* **1990**, *26*, 131.
16. Borggreve, R. J. M.; Gaymans, J.; Luttmer, A. M. Makromol. *Chem. Macromol. Symp.* **1988**, *16*, 195.
17. Mülhaupt, R.; Duschek, T.; Rösch, *J. Polym. Adv. Technol.* **1993**, *4*, 465.
18. Mülhaupt, R.; Fischer, D.; Setz, S. *Polym. Adv. Technol.* **1993**, *4*, 439.
19. Duschek, T.; Mülhaupt, R. *Am. Chem. Soc. Div. Polym. Chem. Prepr.* **1992**, *33(1)*, 170.
20. Mülhaupt, R.; Duschek, T.; Rieger, B. *Makromol. Chem. Macromol. Symp.* **1991**, *48/49*, 317.
21. Rösch, J.; Mülhaupt, R. *J. Appl. Polym. Sci.* **1995**, *56*, 1 599.
22. Rösch, J.; Mülhaupt, R. *Polym. Bull. (Berlin)* **1994**, *32*, 697.
23. Rösch, J.; Mülhaupt, R. Makromol. *Chem. Rapid Commun.* **1993**, *14*, 503.

20

Ductile-to-Brittle Transitions in Blends of Polyamide-6 and Rubber

R.J. Gaymans[1], K. Dijkstra[2], and M.H. ten Dam[3]

[1]University of Twente, P.O. Box 217, 7500 AE Enschede, Netherlands
[2]DSM Research, P.O. Box 18, 6160 MD Geleen, Netherlands
[3]National Starch, P.O. Box 13, 7200 AA Zutphen, Netherlands

Blends of polyamide-6 and rubber were prepared with various rubber concentrations and particle sizes. The blends were studied for (1) notched Izod impact behavior as a function of temperature (–50 to 80 °C), and (2) notched tensile impact behavior as a function of test speed (10^{-5} to 13 m/s) and temperature (–10 to 70 °C). The structure of the deformation zone was studied with electron microscopy. With the notched tensile test as a function of test speed, two ductile-to-brittle transitions are apparent: one at low test speeds and one at high test speeds. The rubber concentration affects both the low- and high-speed transitions, whereas the rubber-particle size only has an effect at high speeds. Plastic deformation of these materials is great next to the fracture surface. The plastic is largely dissipated as heat. In the high-speed regime, the deformation is virtually adiabatic, and the temperature seems to rise locally to a level higher than the melting temperature of the material. This melt formation at high test speeds is affirmed by scanning electron microscopic studies of the deformation zone. The presence of a melt layer ahead of a crack blunts the crack and allows more deformation to take place.

POLYMERS HAVE INTERESTING MECHANICAL PROPERTIES at low deformation rates, but under notched-impact conditions they fracture mainly in a brittle manner. At elevated temperatures most polymers become ductile. Polyamide-6 (PA-6) and PA-66 (dry) become ductile at 70 °C. An effective means of modifying the impact behavior of engineering plastics is by blending in rubber. In this way toughness increases manyfold, whereas the tensile strength and modulus decrease approximately in proportion to the rubber concentration (*1, 2*). Upon blending rubber into PA, a sharp ductile-to-brittle transition is apparent in the toughness–temperature curve, and at a lower temperature than in the neat PA (Figure 1).

On fractured samples at temperatures lower than the ductile-to-brittle transition, stress-whitening is apparent in the notch region but is nearly absent

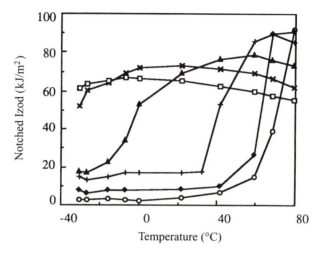

Figure 1. Notched Izod impact strength versus temperature of PA-6–polybutadiene blends with different rubber concentrations. Key: ○, *0 vol%;* ◆, *1 vol%;* +, *7.5 vol%;* ▲, *15 vol%;* x, *22.5 vol%;* □, *30 vol% (2).*

on the fracture surface. The stress-whitening is an indication of severe deformation (3). The energy absorption in this region is mainly from elastic and plastic deformation before the crack is initiated. At temperatures higher than the ductile-to-brittle transition, stress-whitening is apparent both in the notch region and in the fracture-propagation region. At the ductile-to-brittle transition, crack propagation changes from unstable to stable crack growth. The ductile-to-brittle transition is a critical parameter, moreso than the toughness value itself, and it can be used well for evaluation purposes.

The ductile-to-brittle transition in PA has been studied as a function of (1) materials parameters, including matrix molecular weight (4), type of polyamide (5, 6), and type of rubber (7), and interface (8), and (2) morphological parameters, including rubber concentration (9), particle size (9), ligament thickness (9, 10), use of very small particles (5, 11), and particle distribution (2). When one of these variables is studied, other variables are often changed too. Only in experiments with large series can some idea of the effect of a variable be obtained.

The materials and morphological parameters are usually studied as function of test temperature and test speed. The Izod method as function of temperature (9, 10) is standard. Also studied is the Charpy impact behavior, both as a function of test temperature and test speed (12). With a notched tensile impact test, both test speed and test temperature can easily be varied (1, 2, 13).

In the work described in this chapter, we studied the influence of test speed and test temperature in the notched tensile setup and compared the

data with those obtained using the notched Izod method. We also performed fractographic studies on the broken samples to gain insight into the fracture process.

Experimental Details

Materials.   The matrix polymer was PA-6 (Akulon K124, relative viscosity η_{rel} = 2.4, and Akulon M258, η_{rel} = 5.8, DSM) and a commercial PA-6–polybutadiene blend (Durethane BC303, Bayer). The blends were made by compounding with a twin-screw extruder (Berstorff, ZE 25, Hannover). The blends were injection-molded into Izod test bars (80 × 10 × 4 mm), and the notch was milled in (ISO 190/1A). Before testing, the samples were dried in a vacuum oven at 110 °C for 18 h. In the notched tensile impact test performed on single-edged notched Izod bars, a clamp distance of 60 mm was employed. The tensile tester used was a Schenck VHS hydraulic instrument, with a clamp speed ranging from 10^{-5} to 13 m/s.

The structural analysis of the blends and the deformation layer was performed with scanning electron microscopy (SEM) on cryomicrotomed samples. The weighted average particle size (d_w = $\Sigma n_i d_i^3/\Sigma n_i d_i$, where n is the number of particles and d is diameter) was determined from the micrographs with a particle-size analyzer (Zeiss TGZ 3). Transmission electron microscopy (TEM) was performed on microtomed samples that were stained with OsO_4 for 24 h at room temperature before being cut. The thin slices were then stained again for 48 h at room temperature and studied with a JEOL 200 CX instrument.

Results

Notched Tensile Test.   In the notched tensile impact test used, the clamp speed could be varied over a wide range (10^{-5}–13 m/s), and with an oven the temperature could be varied. The advantages of this notched tensile impact test are that the loading is simple and that, as the samples are clamped, signal noise is low. From each stress–displacement graph, the maximum stress was recorded and the energy supplied to the specimen during the test was calculated. The fracture energy was divided into an initiation part and a propagation part. The point of maximum stress was chosen as the boundary between crack initiation and crack propagation. In the case of brittle behavior, the stress falls almost instantaneously from the maximum stress to zero. Therefore, a brittle fracture is characterized by a low propagation energy, meaning that after the fracture has started, the stored elastic energy is sufficient to propagate the crack. By propagation energy, we mean the extra supplied energy necessary to fracture the sample. A clamp speed of 1 m/s in the notched tensile impact test is comparable to the speeds in the notched Izod test (2).

The stress–displacement curves for the PA–ethylene-propylene rubber (PA–EPR) blend (80/20) at different speeds (10^{-2}–10 m/s) are shown in Figure 2. The increase in maximum stress at the highest speed and the increase in fracture energy with increasing speed are noteworthy.

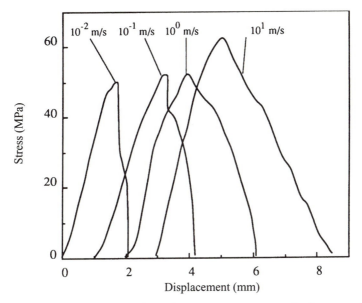

Figure 2. Results of notched tensile impact tests versus displacement for the PA-6–EPR blend (24 vol%, 0.3 μm), for different piston speeds (13).

Rubber Concentration

Notched Izod Test. The influence of rubber concentration on the notched Izod impact strength was studied in PA-6–polybutadiene blends with a small particle-size distribution (Figure 1) (1, 2). The Durethane was diluted with M258. The PA-6 has a low notched impact strength at room temperature and fractures in a brittle manner. A ductile-to-brittle transition is apparent near its glass transition. In polypropylene the material becomes ductile well above its glass-transition temperature, T_g (14), yet polycarbonate becomes ductile below its T_g. Thus the question is whether the ductile-to-brittle transition in PA is due to the glass transition or to a change in yield behavior that just happens near the glass transition of this material.

When blending rubber into PA (particle size $\approx$ 0.5 μm, the ductile-to-brittle transition strongly shifts with rubber concentration to lower temperatures (9). The impact energy in the low-temperature "brittle region" increases with increasing rubber concentrations (Figure 1). The impact energy in the "tough region," however, decreases with increasing rubber concentration. In earlier studies on PA–etheylene-propylene dimer monomer (PA–EPDM) and PA–EPR (Figure 3), performed using a broader particle-size distribution, the impact level in the tough region did not decrease with rubber concentration. The new data (Figure 1) are for a more uniform blend and thus better explain

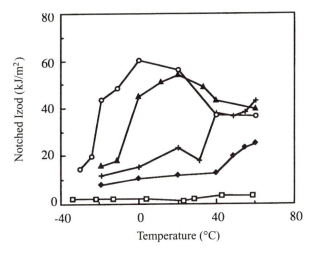

Figure 3. Results of notched Izod impact tests versus temperature for PA-6–EPR blends with different rubber contents at a constant rubber-particle size (0.3–0.4 μm). Key: □, 0 vol%; ◆, 6.3 vol%; +, 12.5 vol%; ▲, 18.4 vol%; and ○, 24.3 vol% (13).

what the rubber concentration does. The fall of impact energy in the tough region with rubber concentration is due to the decreasing PA content (Figure 1). At T_g of the PA in the blend, no transition in toughening behavior is observed. The position of the ductile-to-brittle transition in PA near T_g is thus just a coincidence and has little to do with T_g of the material. At low deformation speeds (notched three-point bending method), Lazzeri found that rubber concentration also had a clear effect on the ductile-to-brittle transition temperature, but at lower temperatures than at high speed (*12*).

Notched Tensile Impact Test.   The influence of test speed was studied with the notched tensile impact test on PA–EPR blends (*13*). The force–time signal was recorded, and from it were calculated the maximum stress, total energy absorption, and energy absorption after crack initiation (propagation energy). Because we are interested in the crack-propagation behavior of the blends, the propagation energies and maximum stress are given. At piston speeds of 10^{-5} to 10 m/s, PA-6 (K124) has a low crack-propagation energy (Figure 4). Thus PA underwent brittle fracture in the whole speed region.

The blends give a complex picture. The propagation energies in the blends all start high at low piston speeds, fall to near zero at 10^{-2} m/s, and show some increase again at higher rates. At low speeds with small amounts of rubber, the material is already tough at room temperature. The propagation energy at low speeds (in the ductile region) shows little dependence on rubber concentration. Surprisingly, the propagation energy of the 15 and 20% blends

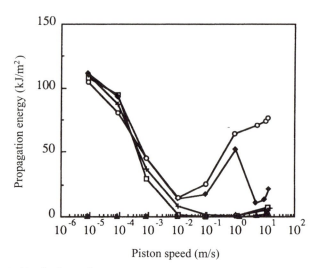

Figure 4. Notched tensile impact propagation energy versus piston speed for PA-6–EPR blends with different rubber contents. Key: ▲, 0 vol%; □, 6.3 vol%; +, 12.5 vol%; ◆, 18.4 vol%; and ○, 24.3 vol% (13).

increases again above 1 m/s, but the 15% blend is nearly brittle again at 5 m/s. The 15 and 20% blends clearly show discontinuous behavior, and this behavior was found several times. The 15% blend shows a low-speed ductile-to-brittle transition at 10^{-2} m/s and a high-speed ductile-to-brittle transition at 5 m/s. Another observed effect is the small increase in propagation energy at the highest test speed (10 m/s). This effect is puzzling.

The maximum stress in these samples is also complex (Figure 5). In PA-6, the stress first rises a bit with speed, a behavior that is expected for a sample that can still reach its yield point. At higher speeds, the maximum stress falls off and the material becomes more brittle. Surprisingly, at the highest speeds, 10 and 13 m/s, the maximum stress increases a bit again.

In the blends, the maximum stress at low rates decreases, as expected, with increasing rubber concentration. At high rates, the 15 and 20% blends show an unexpectedly strong upswing in the maximum stress. The maximum stress before fracture can increase if the stress-concentration factor ahead of the notch is lowered.

We also studied the notched tensile impact test behavior of PA–polybutadiene blends (Durethane diluted with M258). The high molecular weight PA (M258) undergoes brittle fracture at low speed and low temperatures and tough fracture at high temperatures (Figure 6) (2). The propagation energy of the blends at 10^{-3} m/s as a function of temperature shows a gradual increase in toughening with temperature. Above 50 °C, all the samples show strong plastic deformation. The gradual increase in propagation energy suggests that at low

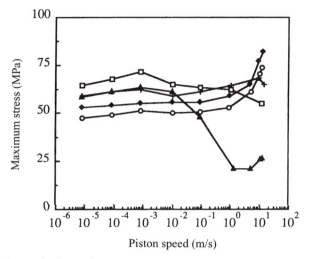

Figure 5. Notched tensile impact maximum stress versus piston speed for PA-6–EPR blends with different rubber contents at a constant rubber-particle size (0.3–0.4 µm). Key: ▲, 0 wt%; □, 6.3 wt%; +, 12.5 wt%; ◆, 18.4 wt%; and ○, 24.3 wt% (13).

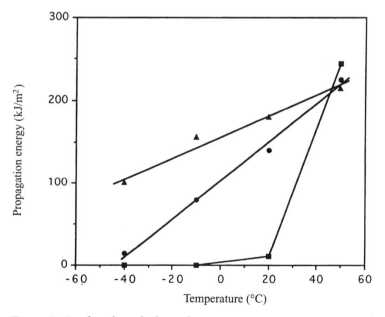

Figure 6. Results of notched tensile impact tests versus temperature for PA-polybutadiene at 10^{-3} m/s for different rubber concentrations. Key: ■, 0 vol%; ●, 15 vol%; and ▲, 30 vol% (2).

speeds there is not a sharp, discontinuous, ductile-to-brittle transition as there is at high speeds. The temperature at which the materials become brittle shifts with increasing rubber concentration to lower temperatures. At 50 °C, the neat PA seems to have a higher fracture-propagation energy than the blends. The ductile-to-brittle transition temperatures at 10^{-3} m/s as compared to 1 m/s are 30 °C lower for the PA and the 15% blends.

All these results indicate that the high-speed data cannot be obtained by extrapolation from the low-speed data. The extrapolated low-speed data should just give brittle fracture for all the samples. Thus the fracture process at high speeds must be different from the process at low speeds, which indicates that there is an extra deformation mechanism operative in the high-speed regime that enables crack propagation to become stabilized under these conditions. The high maximum stress in the high-speed tests suggests that this mechanism becomes operative before a crack is initiated. These findings suggest that there are two ductile-to-brittle transitions, one at low test speeds and one at high test speeds. Rubber concentration affects both ductile-to-brittle transitions.

Particle Size

Notched Izod Test.   The influence of particle size was studied in a PA–EPDM blend (Figure 7) (9). The ductile-to-brittle transition temperature decreases with decreasing particle size. The S-shape of the curve is so uniform

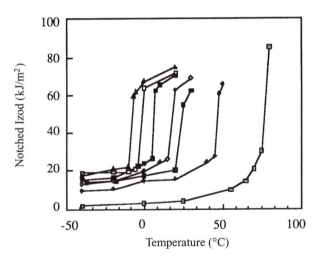

Figure 7. Notched Izod impact strength versus temperature for PA-6–EPDM blends with a constant rubber volume fraction (26.1 vol%) and different particle sizes. Key: □, nylon-6; ◆, 1.58 μm; ■, 1.20 μm; ◇, 1.14 μm; ■, 0.94 μm; □, 0.57 μm; and ▲, 0.48 μm (9).

that with a temperature shift a master curve can be constructed. A change in particle size seems to affect only the ductile-to-brittle transition temperature and have little influence on the impact energies. This effect is thus different from that of rubber concentration as seen in the preceding section.

Lazzeri also studied the influence of particle size on the ductile-to-brittle-transition at low rates in three-point bending tests on notched samples (*12*). Lazzeri found the ductile-to-brittle transition at low temperatures, near the T_g of the rubber. So a firm conclusion cannot be drawn from this study. Lazzeri and Bucknall (*15*) described the influence of particle size on the difference in the cavitation behavior of the rubber.

Notched Tensile Impact Test. The influence of particle size in a PA–EPDM blend (90/10) was also studied with the notched tensile impact test at 10^{-3} m/s and 1 m/s with different test temperatures (*16*). For comparison, the notched Izod impact strength of these blends was also studied (Figure 8). The different particle sizes were produced by changing the extruder-barrel temperatures during blending (290–260–240 °C). The ductile-to-brittle transition of this 10% blend as measured by the notched Izod test decreases with decreasing particle size. For this small change in particle size, a ductile-to-brittle temperature shift of 12 °C was observed. The propagation energies in the notched tensile impact test conducted at 1 m/s show a sharp increase with temperature [Figure 9 (top)]. The temperature shift in this ductile-to-brittle transition is now 10–12 °C too. This ductile-to-brittle transition is similar to that measured with the Izod method. However, if the samples are tested at a low speed (10^{-3} m/s), the propagation energies show a more gradual increase [Figure 9 (bottom)]. The data for the different particle sizes at low speeds fall

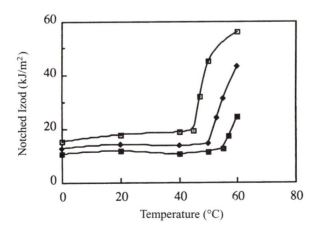

Figure 8. Notched Izod impact strength of PA–EPDM (13 vol%) for different particle sizes. Key: □, 0.8 μm; ◆, 0.9 μm; and ■, 1.0 μm (16).

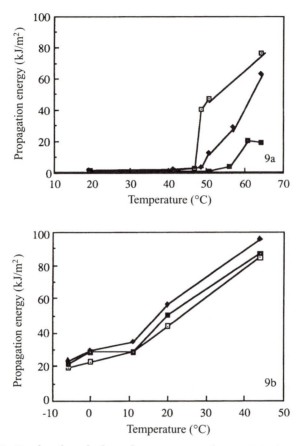

Figure 9. Results of notched tensile impact tests of PA–EPDM (13 vol%) for different particle sizes: □, *0.8 μm;* ◆, *0.9 μm; and* ■, *1.0 μm. (16). Top: 1 m/s. Bottom: 10^{-3} m/s.*

on top of each other. This result means that at low test speeds the particle size does not affect toughening. It is surprising that at low speeds the energy dissipation does not progress in a discontinuous manner and the particle-size effect is absent. These results also indicate that toughening at high speeds is different from that at low speeds.

Ligament Thickness. The ductile-to-brittle transition as measured in notched Izod, notched Charpy, and notched tensile impact tests is discontinuous and is dependent on both rubber concentration and particle size. These two parameters can be combined into a new morphological parameter that governs the ductile-to-brittle transition. The ligament thickness (interparticular distance), which is a function of rubber concentration and particle size, gives a good fit at high test speeds (9, 10). The impact energies in the ductile region decrease with increasing rubber concentration (Figure 1) and seem to depend little on the particle size (Figure 7). This result means that the impact energies at high test speeds cannot be correlated with ligament thickness.

At low test speeds, toughening shows a more gradual change with temperature [Figures 6 and 9 (left)]. With increasing rubber concentration, the ductile-to-brittle transition temperature is lowered, but, surprisingly, the influence of particle size on the ductile-to-brittle transition seems to be absent. This result means that at low test speeds there is no correlation between ligament thickness and ductile-to-brittle transition temperature or fracture-propagation energies.

The ligament thickness parameter in PA blends can therefore only apply to high-speed deformation. The meaning of ligament thickness as a parameter for toughening that is only applicable at high speeds is puzzling.

Fractographic Analysis

High-Speed Deformation.   The fractured blends have a large stress-whitened zone next to the fracture surface, and the structure of this zone has been studied (3, 9, 13, 17, 18). This stress-whitened zone is present both in the broken Izod and Charpy samples and in the samples of the notched tensile impact test. Ramsteiner and Heckman (3) observed a thick layer with cavitated rubber particles in PA–rubber blends, and next to the fracture surface a smaller layer with cavities and shear bands. Borggreve et al. (9) also observed a cavitation of the rubber particles in the stress-whitened zone, but no crazes. Oostenbrink et al. (17) observed on notched Izod samples, and Dijkstra et al. (13) and Janik et al. (18) on notched tensile impact samples tested at 1 m/s, a three-layer structure: far from the fracture surface (0.10 to 2 mm) a layer with reasonably round cavities, nearer to the fracture plane (5 to 100 μm from the fracture surface) a layer with strongly elongated cavities, and next to the fracture plant a 3- to 5-μm layer without cavities. The cavitation in this layer was in the rubber particles. The second layer, with strongly deformed cavities, suggests strong plastic deformation of the matrix material. The length-to-diameter ratio of the cavities in this layer is 3 to 5, and the angle to the fracture plane is 45° (18). The shape of the particles, but not so much the cavities, can be seen on a stained PA–polybutadiene sample studied with TEM (Figure 10). In the third layer, next to the fracture surface, no cavities can be seen on a ductile broken sample (13, 17). With TEM we see stained, round, rubber particles (Figure 10). The crystalline order in this layer is low, and orientation as observed with electron diffraction and polarized microscopy was absent (18). This result suggests that no cavities formed and no deformation took place in this layer or that the cavitated structure formed during the fracture process but subsequently disappeared. The cavitation can disappear, the orientation of the matrix can relax, and the crystallinity can be lowered if the material is heat-

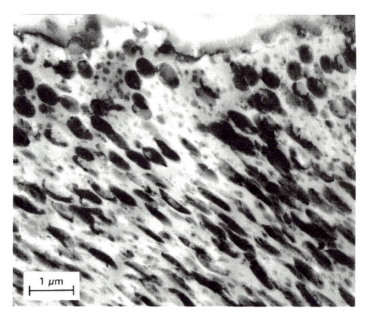

Figure 10. TEM image of stained PA–polybutadiene (15 vol%) after undergoing tough fracture in a notched tensile test at high speed (1 m/s). The section shown is next to the fracture surface (18).

ed to above the matrix melting temperature. Annealing a deformed blend at different temperatures did not change the cavitated structure up to 200 °C (Figure 11) (*19*). At 230 °C, above the melting temperature of PA-6, relaxation of the deformed structure had taken place.

A temperature increase of 140 °C was measured using a pyrometer (infrared detector) on an unnotched sample at high deformation rates (but which were much lower than the rates on the fracture surface) and with a large spot size (1 mm^2). All these results clearly indicate that the layer next to the fracture plane, without cavities and with no matrix orientation, must have been warm and that it subsequently relaxed. Relaxation of a deformed blend takes place in the melt. If a melt is present in the fracture plane ahead of a crack, the crack will blunt. So melt blunting takes place in ductile blend samples tested at high loading rates.

Low-Speed Deformation. Dijkstra et al. (*13, 19*) and Janik et al. (*18*) showed that in samples fractured in a slow-speed notched tensile impact, the stress-whitened zone has two layers. Far from the fracture plane a cavitated structure is present, with cavities in the rubber particles (Figure 12, top). In particular, the bigger particles seem to be cavitated and the particles less than 100 μm are not. In the layer next to the fracture plane, the cavities are strong-

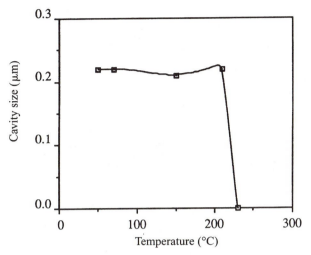

Figure 11. Cavity size versus temperature after heat treatment of a cavitated blend (19).

ly deformed, have a length-to-diameter (*L/D*) ratio of 5–10, and lie at an angle of less than 45° with the fracture plane (Figure 12, bottom). Also, polarized-light microscopic and diffraction studies in TEM show that the orientation in these samples increases as the fracture plane is approached (*18*). In a low-speed test, the cavities at the fracture plane are not relaxed and have a high *L/D/* ratio.

A schematic of the structure of the stress-whitened zone is given in Figure 13. The samples deformed at high speed and undergoing ductile breaking have a three-layer structure in which the layer next to the fracture plane is relaxed (molten) (Figure 13a). The samples deformed at low speed and undergoing ductile breaking have a two-layer structure with a strong deformation of the material next to the fracture plant (Figure 13b). Here, no melt layer is present.

Deformation Process.   The function of the rubber in PA–rubber blends is to create stable cavities upon loading. Because of cavitation, the von Mises effective stress in the matrix strongly increases and plastic deformation is possible (*1, 20*). The von Mises stress in a cavitated system is a function of cavity concentration. The cavity size does not play a role in this mechanism.

At the ductile-to-brittle transition, crack propagation changes from unstable to stable crack growth. Both at low and high speeds, a ductile-to-brittle transition can be observed. The high-speed transition cannot be obtained by extrapolating from the low-speed data. Also, the low-speed transition is not a discontinuous transition and depends on different structural parameters, like

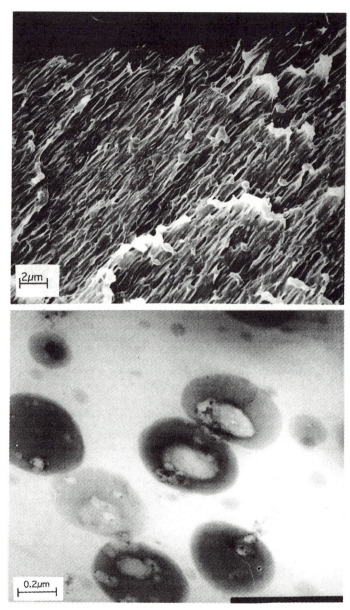

Figure 12. Structure of PA–polybutadiene (15 vol%) after undergoing tough fracture in a notched tensile test at low speed (10^{-4} m/s). Top: far from the fracture plane (TEM, stained). Bottom: next to the fracture plane (SEM, cryofractured).

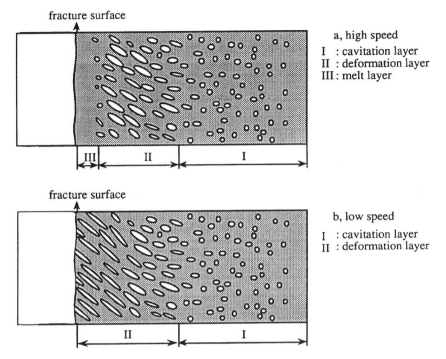

Figure 13. Schematic of the structure of the stress-whitened zone perpendicular to the fracture plane. a: Tough fracture at high speed. b: Tough fracture at low speed.

rubber concentration and particle size. So in these blends two ductile-to-brittle transitions must be present. The ductile-to-brittle transition temperature at low test speed depends on the rubber concentration but not on the particle size. The transition at high test speed depends on the rubber concentration and the particle size. The samples fractured at high speed have a melt layer next to the fracture plane. The difference between the low- and high-speed data is that at high speeds the process becomes adiabatic, and melt blunting can take place.

Summary

The toughening of PA depends on materials and on structural and test parameters. The toughening behavior at high speeds as a function of temperature gives an S-curve with a discontinuous transition. The materials and structural limit have a shift effect on this S-curve. Deformation occurs first by cavitation of the rubber, and then by plastic deformation of the cavitated matrix due to a change in the von Mises stress. Surprisingly, a relaxed layer is present in the fracture plane, and this layer is probably due to melting during plastic deformation. A melt layer in front of a crack induces crack blunting, as seen in the sudden increase of the maximum force of the samples tested at higher test speeds (Figure 5). The toughening with temperature at low speeds gives a gradual increase in fracture energy. Fracture strength is independent of particle size.

Rubber particle size or ligament thickness has no effect on the onset of plastic deformation but may affect the extent of plastic deformation (i.e., the maximum draw ratio). A small change in draw ratio is of little consequence under isothermal conditions, but under adiabatic conditions it may result in enough heat generation for local melting and, hence, melt blunting.

References

1. Dijkstra, K. Ph.D. Thesis. University of Twente, Netherlands, February 1993.
2. Dijkstra, K.; Wevers, H. H.; Gaymans, R. *J. Polymer* **1994**, *35*, 323.
3. Ramsteiner, F.; Heckmann, W. *Polym. Commun.* **1985**, *26*, 199.
4. Dijkstra, K.; Gaymans, R. J. Polymer **1994**, *35*, 333.
5. Oshinski, A. J.; Keskkula, H.; Paul, R. D. *Polymer* **1992**, *33*, 268.
6. Oshinski, A. J.; Keskkula, H.; Paul, R. D. *Polymer* **1992**, *33*, 284.
7. Borggreve, R. J. M.; Gaymans, R. J.; Schuijer, J. *Polymer* **1989**, *30*, 71.
8. Borggreve, R. J. M.; Gaymans, R. J. *Polymer* **1989**, *30*, 63.
9. Borggreve, R. J. M.; Gaymans, R. J.; Schuijer, J.; Ingen Housz, J. F. *Polymer* **1987**, *28*, 1489.
10. Wu, S. *Polymer* **1985**, *26*, 1855.
11. Oostenbrink, A. J.; Molenaar, L. J.; Gaymans, R. J. *Polymer Blends Conference Preprints, Cambridge, 1990*; Plastic and Rubber Institute (London); E3.
12. Lazzeri, A. Ph.D. Thesis, Cranfield Institute of Technology, UK, 1991.
13. Dijkstra, K.; Laak, J. van der; Gaymans, R. J. *Polymer* **1994**, *35*, 315.
14. Wal, A.; Gaymans, R. J. *Am. Chem. Soc. Div. Polym. Mater. Sci. Eng. Proc.* **1994**, *191*.
15. Lazzeri, A.; Bucknall, C. B. J. Mater. Sci. **1993**, *28*, 6799.
16. Gaymans, R. J.; ten Dam, M. H.; Dijkstra, K. to be published.
17. Oostenbrink, A. J.; Dijkstra, K.; Wal, A. van der; Gaymans, R. J. *Conference Proceedings on Deformation and Fracture of Polymers, Cambridge, 1991*; Plastic and Rubber Institute (London); p. 50.
18. Janik, H.; Gaymans, R. J.; Dijkstra, K. *Polymer*, **1995**, *36*, 4203.
19. Dijkstra, K.; Gaymans, R. J. *J. Mater. Sci.* **1994**, *29*, 3231.
20. Dijkstra, K.; Bolscher, G. H. ten *J. Mater. Sci.*, **1994**, *29*, 4286.

21

Failure Mechanisms in Blends of Linear Low-Density Polyethylene and Polystyrene

T. Li, V. A. Topolkaraev, A. Hiltner*, and E. Baer

Department of Macromolecular Science and Center for Applied Polymer Research, Case Western Reserve University, Cleveland, OH 44106

Mechanical models that explicitly consider microdeformation and failure mechanisms were developed for the high-strain tensile properties of uncompatibilized blends of linear low-density polyethylene (LLDPE) and polystyrene (PS). A decrease in ultimate tensile properties with increasing PS content resulted in a ductile-to-quasi-brittle transition. By scanning electron microscopy, it was determined that the irreversible microdeformation mechanisms that controlled the high-strain properties of these blends were interfacial debonding and void growth around particles of the dispersed PS phase. The void morphology was characterized by two parameters: the debonding angle (θ), and the local extension ratio of the void (λ). The yield-stress calculation was based on a modification of the effective-cross-section approach to incorporate the debonding angle; the true fracture stress was determined by considering the further reduction in effective cross section during void growth with, additionally, the strain-hardening characteristics of the LLDPE matrix. Because the true fracture stress decreased more rapidly than the yield stress with increasing PS fraction, a ductile-to-quasi-brittle transition was observed.

BLENDS OF IMMISCIBLE POLYMERS EXHIBIT a coarse and unstable phase morphology with poor interfacial adhesion. The ultimate properties of these blends are often poorer than those of either component. The poor mechanical properties can be improved with a small amount of an interfacial agent that lowers interfacial tension in the melt and enhances interfacial adhesion in the solid. High-strain properties, such as strength, tensile elongation, and impact strength, especially benefit from compatibilization (1, 2).

Mechanical models based on elastic assumptions are not generally applicable to high-strain deformation where specific failure mechanisms should be

---

*Corresponding author.

taken into account. The poor mechanical properties of incompatible blends on the one hand, and the synergistic properties achieved in compatibilized blends on the other, result from microdeformation mechanisms that are unique to the blend and not attributable to one component or the other. In uncompatibilized blends with poor interfacial adhesion, failure mechanisms might include debonding and void growth around inclusions of the dispersed phase. To describe the high-strain mechanical behavior of polymer blends, models are needed that explicitly consider microdeformation and failure mechanisms. When these models incorporate interfacial features along with tensile characteristics of the components, they can be the basis for understanding and subsequently optimizing high-strain properties.

Polyethylene and polystyrene (PS) represent an example of extreme immiscibility. Nevertheless, there is interest in their blends because of the potential synergistic effects of combining a thermoplastic with high modulus and strength with one that is ductile. In addition, these polymers comprise a large fraction of plastic scrap, and development of an effective compatibilizing strategy would open new opportunities for recycling the large volume of these polymers in the postconsumer waste stream. The beneficial effects of adding a graft or block copolymer of styrene and butadiene, or a similar hydrogenated copolymer, to blends of polyethylene and PS have been convincingly demonstrated. The application of structural mechanical models to the elastic properties of these blends has been demonstrated (3).

In preparation for formulating mechanical models that describe the improved high-strain properties obtained with compatibilization, the deformation and failure of the uncompatibilized blend were characterized and modeled. The mechanical models often used to predict ultimate tensile properties of blends with poor interfacial adhesion are based on an effective load-bearing cross section (4). These models are adequate if the amount of plastic deformation is small. They do not consider the growth of voids around inclusions of the dispersed phase when the continuous phase undergoes large plastic deformation accompanied by extensive strain-hardening. Furthermore, the mechanism of the ductile-to-quasi-brittle transition, which is typically observed as the amount of the dispersed phase increases, is generally overlooked.

Experimental Details

Linear low-density polyethylene (LLDPE) (GB502) was supplied by Quantum Chemical Corp. The number-average molecular weight was given as 2.7×10^4 and the polydispersity as 4.77. PS (Styron 623) was supplied by Dow Chemical Co. The PS, LLDPE, and 0.2% antioxidant (Irganox 1076 from Ciba-Geigy) were dry-mixed and then blended in a Banbury mixing head on a Brabender machine. The mixer was operated at 175 °C and 52 rpm. After melt-mixing for about 7 min, the blend was quenched in cold tap water and dried overnight at 40 °C in vacuum. Plaques were compression-molded by preheating the blend in the press at 170 °C for about 15 min, then applying a pressure of 3.85 MPa for 8 min. The press was water-cooled to room temperature with the plaque under pressure.

Rectangular bars were cut from the molded plaque with a jewelry saw, adhered to a Type I dog-bone template (ASTM D–638–84) and shaped with a high-speed cutting spindle. Tensile stress–strain curves were obtained by testing at least four specimens to fracture at a strain rate of 0.05 min^{-1} at room temperature. After straining to fracture, some of the tensile specimens were cryogenically fractured lengthwise at liquid-nitrogen temperature, coated with 60 Å of gold, and examined in the scanning electron microscope (SEM) (JEOL JSM–840A).

Results and Discussion

Stress–Strain Behavior.   The following description of the blend morphology of compression-molded, uncompatibilized LLDPE–PS blends is summarized from a previous study that focused on the low-strain mechanical response (*3*). When the PS concentration was 25% by volume or less, the dispersed PS domains were almost spherical and increased in size with increasing PS concentration from ~2 μm for the blend with 10% PS to ~7 μm for the blend with 25% PS. Because LLDPE had a lower viscosity than PS, the LLDPE component was the continuous phase in the blend with 50% PS except for a small core region where the phases were cocontinuous. The dispersed PS domains were largest in the 50% PS composition, averaging 28 μm, and they tended to be elongated in the plane of the molded sheet. The morphological evidence for a dispersed PS phase was corroborated by a bulk property, the measured tensile modulus, which in blends with 50% PS (or less) conformed closely to either Kerner's equation for monodispersed spherical inclusions or Hashin's lower bound.

The three types of tensile stress–strain behavior of LLDPE and uncompatibilized LLDPE–PS blends are shown in Figure 1. The stress–strain curve of LLDPE, identified as Type I, was characteristic of a ductile polymer. The initial linear region was followed by a region of gradually decreasing slope to a plateau stress. At higher strains, a work-hardening region was characterized by a gradually increasing slope until fracture at about 760% elongation. Deformation occurred by uniform extension of the entire gauge section. The stress–strain curves of blends with 5 (not included in Figure 1), 10, and 25% PS by volume were similar to that of LLDPE, although stress-whitening of the gauge section, first observed at about 1.5% strain, accompanied uniform drawing. The fracture strain decreased only slightly, although the fracture stress decreased significantly with increasing PS content.

The blend with 37.5% PS exhibited Type II stress–strain behavior with a stress drop at about 3% strain accompanied by formation of a localized neck. The neck was stable as a result of strain-hardening in the LLDPE matrix, and propagated along the gauge section. The necked region was uniformly stress-whitened; there was also some localized stress-whitening in the unnecked region of the gauge section. Fracture occurred during neck propagation at an

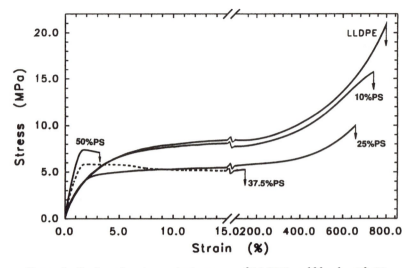

Figure 1. Engineering stress–strain curves of LLDPE and blends with PS.

engineering strain of about 150% after approximately a third of the gauge section had necked. As a result, the engineering fracture strain was much lower than in Type I behavior. The Type III behavior of the blend with 50% PS was characterized by the appearance of several regions of localized stress-whitening without global necking. Fracture occurred at one of these regions at a low strain of about 3.2%.

The yield stress (σ_y) and yield strain (ϵ_y) were defined from an idealized representation of the tensile stress–strain curve. The intersection of the line drawn from the initial linear portion of the stress–strain curve with the line drawn parallel to the strain axis through the plateau stress (Type I) or maximum stress (Types II and III) was taken as the yield point. Values of σ_y and ϵ_y are summarized in Table I. The yield stress of LLDPE decreased with increasing PS in the blend, from 8.3 MPa for LLDPE to 7.6 MPa and 5.9 MPa for

Table I. Tensile Properties of LLDPE–PS Blends

Composition (vol% PS)	σ_y (MPa)	ϵ_y (%)	σ_f (MPa)	ϵ_f (%)
0	8.3 ± 0.3	2.4 ± 0.2	171 ± 25	760 ± 30
5	8.2 ± 0.7	2.6 ± 0.2	130 ± 20	770 ± 90
10	7.6 ± 0.3	2.4 ± 0.2	100 ± 5	740 ± 10
25	5.7 ± 0.3	1.4 ± 0.3	35 ± 6	655 ± 40
37.5	5.9 ± 0.3	0.9 ± 0.1	8.5 ± 1.1	150 ± 55
50	7.5 ± 0.5	0.9 ± 0.1	7.0 ± 0.6	3.2 ± 1.0

LLDPE with 10% and 37.5% PS, respectively. The decrease in σ_y suggested that PS particles debonded from the LLDPE matrix and lost their load-bearing capacity before the yield point was reached. Debonding probably initiated near the onset of nonlinearity in the stress–strain curve.

Also included in Table I are the true fracture stress (σ_f) calculated from the cross section of the fractured specimen, and the fracture strain (ϵ_f). The fracture stress dropped significantly, from 171 MPa for LLDPE to 100 MPa and 35 MPa for 10% and 25% PS, respectively. The fracture stress of the 37.5% PS blend was even lower, 8.5 MPa, but was still higher than the σ_y value of 5.9 MPa, so the blend deformed in a ductile manner. The blend with 50% PS fractured in a quasi-brittle manner at a stress of 7.0 MPa, which was slightly lower than σ_y for this composition. The large decrease in σ_f with increasing PS concentration was consistent with debonded PS particles that were not load-bearing during plastic deformation.

Void Morphology of the Type I Blend. After fracture, tensile specimens were cryogenically fractured lengthwise and viewed in the SEM. The micrograph shown in Figure 2 shows elongated voids extending from the PS particles. The waviness is caused by retraction after fracture; the macroscopic strain recovery of the tensile specimen was about 10%. The voids were characterized by the local extension ratio (λ), defined as the total length of the void (ℓ) divided by the diameter of the PS particle. The parameter λ is not a material property of the matrix but describes the local elongation of the blend.

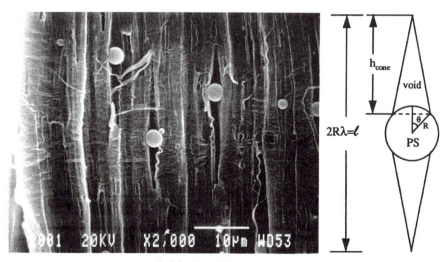

Figure 2. SEM image of a blend with 10% PS strained to fracture and then cryogenically fractured parallel to the direction of deformation. The parameters that describe the void geometry are defined in the schematic.

Therefore the value of λ at fracture is expected to depend on the composition. The magnitude of λ ranged from 7 to 12, with an average of 8.9 in the blend with 5% PS and 8.5 in the blend with 10% PS. The close proximity of PS particles in the blend with 25% PS made it difficult to measure λ; however, a value of about 7 was estimated. Measurements taken from the micrographs underestimated λ by the amount that the specimens recovered after fracturing, which was about 10%.

The PS particles were not completely detached from the LLDPE matrix. At the equator they remained joined to the matrix. A debonding angle (θ) was defined by the loading direction and the radius that intersected the interface at which debonding started. The debonding angle was estimated from the micrographs to be between 45° and 60°. In comparison, the debonding angle at spherical inclusions in polyethylene and polypropylene is about 60° (5).

From the debonding angle and the local draw ratio, it was possible to calculate a void content in the drawn blend. Subsequently, a density could be calculated for comparison with the measured density. Assuming that the void is conical in shape with a base area (A_{cone}) given by

$$A_{cone} = \pi (R \sin \theta)^2 \tag{1}$$

where R is the radius of the particle, and a height (h_{cone}) given by

$$h_{cone} = R(\lambda - \cos \theta) \tag{2}$$

then the volume of the void plus the PS particle ($V_{void+PS}$) is given by

$$V_{void+PS} = \frac{4}{3}\pi R^3 \left[\frac{2 \cos \theta + \lambda \sin^2 \theta}{2} \right] \tag{3}$$

It is then possible to define a parameter κ that is a function of both λ and θ:

$$\kappa = \frac{\text{volume (void + PS particle)}}{\text{volume (PS particle)}} = \frac{2 \cos \theta + \lambda \sin^2 \theta}{2} \tag{4}$$

The parameter κ is related to the change in density due to voiding according to

$$\frac{\rho_o}{\rho_f} = \frac{V_{PE} + \kappa V_{PS}}{V_{PE} + V_{PS}} \tag{5}$$

where ρ_o is the density of the undeformed blend, ρ_f is the density of the voided blend after fracture, and V_{PE} and V_{PS} are the volume fractions of LLDPE and PS, respectively. Values of κ calculated from λ for debonding angles of 45° and 60° are given in Table II together with κ values obtained from the bulk densities. The calculated values may be high because the geometry of the voids was not actually conical but somewhat concave; the measured values may also be low because some of the voids collapsed when the load was removed.

Table II. Local Extension Ratio (λ) and Void Fraction (κ)

Composition (vol% PS)	λ	κ^a ($\theta = 45°$)	κ^a ($\theta = 60°$)	ρ_o (g/cm^3)	ρ_f (g/cm^3)	κ_ρ^b
0	—	—	—	0.918 ± 0.001	0.918 ± 0.001	—
5	8.9 ± 1.4	2.9	3.8	0.924 ± 0.001	0.886 ± 0.001	1.8
10	8.5 ± 1.5	2.8	3.7	0.929 ± 0.001	0.844 ± 0.0101	2.0
25	7.0 ± 1.0	2.5	3.1	—	—	—

[a]From microscopy.
[b]From density.

Yield Stress. It was hypothesized that the processes of interfacial debonding, void formation, and void growth, which led to the highly elongated voids in drawn blends of uncompatibilized LLDPE and PS, proceeded during tensile deformation, as illustrated in Figure 3. In the elastic region (Stage I), the interface was intact and the PS particles were load-bearing (3). At some stress below the yield stress (σ_y), probably close to the onset of nonlinearity, the particles started to debond and voids initiated at the poles, where the tensile stress was highest. In Stage III, the voids lengthened in the loading direction until the ligaments between voids broke and coalescence of the voids led to final fracture. Debonding and void growth were responsible for the macroscopic stress-whitening. From this model, the yield stress, fracture stress, and

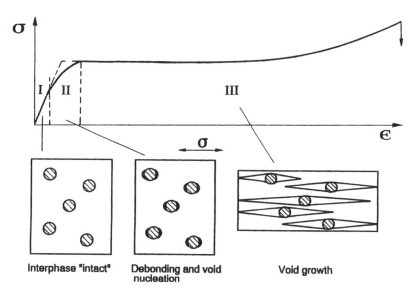

Figure 3. Schematic representation of particle debonding and void growth during deformation of Type I blends.

fracture strain were predicted, as well as the transition from Type I to Type II and Type III fracture.

The yield stress was estimated by assuming that the effective load-bearing cross section decreased as a result of interfacial debonding. For a cubic array of spherical PS particles that are partially detached from the LLDPE matrix (Figure 4), the effective load-bearing cross-sectional area (A_{eff}) is given by

$$A_{eff} = A_o \left(1 - \beta(V_{PS})^{2/3}\right) \tag{6}$$

where A_o is the initial cross section, V_{PS} is the volume fraction of the dispersed PS phase, and the geometric parameter β is given by

$$\beta = \pi \left(\frac{3}{4\pi}\right)^{2/3} \sin^2 \theta \tag{7}$$

where θ is the debonding angle. Assuming further that only the matrix is load-bearing at the yield point and that the matrix has the same σ_y as LLDPE, and neglecting any stress concentration at the PS particles, σ_y of the blend as obtained from Nielsen's approach is given by

$$\sigma_y = \sigma_y^o \frac{A_{eff}}{A_o} = \sigma_y^o \left(1 - \beta(V_{PS})^{2/3}\right) \tag{8}$$

where σ_y^o is the yield stress of LLDPE. This equation reduces to the well-known Nielsen's equation with $\beta = 1.21$ when $\theta = 90°$ and the dispersed spherical particles are completely debonded (4).

The effective cross section decreases as the debonding angle becomes larger. The effect of this phenomenon on σ_y of the blend is illustrated in Figure 5. When the debonding angle is small, σ_y is only slightly affected, even when the volume fraction of the dispersed phase is large. As the debonding an-

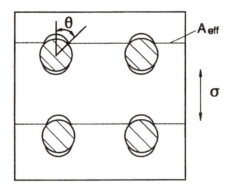

Figure 4. Schematic of the cubic array of partially debonded, spherical particles. The effective cross-sectional area (A_{eff}) is defined.

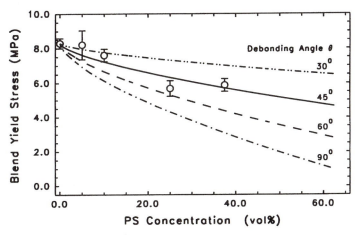

Figure 5. Effect of debonding angle (θ) on the calculated yield stress from equation 8. Data for LLDPE–PS blends are included as open circles.

gle increases toward 90°, the effect becomes pronounced. Measured values of σ_y of the uncompatibilized blends are scattered around the curve for $\theta = 45°$ and are always higher than predicted for a debonding angle of 60°. The corresponding values of β are 0.6 (45°) and 0.9 (60°). Similar values of β, in the range of 0.8 to 1.0, were reported for high-density polyethylene and polypropylene with spherical inclusions (5).

True Fracture Stress.   The elongated voids in the drawn blend contributed to the total cross-sectional area at the fracture but were not load-bearing. It was assumed that the load was carried by the drawn LLDPE ligaments between the voids, and that fracture occurred when these reached their true fracture stress. An estimate was made of the true fracture stress of the blend by considering a cubic array with a unit dimension L_o. The particle was represented by a linear element of length $2R$. The unit cell was assumed to deform uniformly to a final length L_d, with the extension ratio given by $\Lambda = L_d/L_o$. The extended linear element of length $\Lambda 2R$ was replaced with a cylindrical void of the same length and a cross-sectional area A_v (Figure 6a). If it is assumed that the void volume fraction is equal to κV_{PS} for low-to-medium PS content, then

$$A_v = \frac{V_v}{L_v} = \frac{\kappa V_{PS}(L_o)^3}{\Lambda 2R} \tag{9}$$

where L_v is the void length, and V_v is the void volume. The final cross-sectional area of the unit cell (A_d) is given by

$$A_d = \frac{V_d}{L_d} = \frac{(L_o)^2}{\Lambda} \tag{10}$$

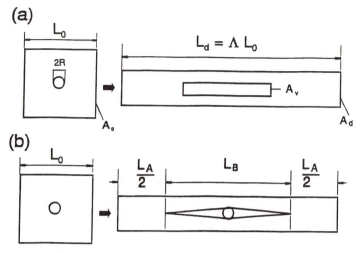

Figure 6. Schematic of a voided element of the rectangular array for the calculation of true fracture stress (a), and fracture strain (b).

The effective cross-sectional area of the drawn blend (A'_{eff}) is then given by

$$A'_{eff} = A_d - A_v = A_d(1 - \beta'(V_{PS})^{2/3}) \qquad (11)$$

where

$$\beta' = \kappa\left(\frac{\pi}{6}\right)^{1/3} \qquad (12)$$

and the true fracture stress of the blend (σ_f) is given by

$$\sigma_f = \sigma_f^o \frac{A'_{eff}}{A_d} = \sigma_f^o(1 - \beta'(V_{PS})^{2/3}) \qquad (13)$$

where σ_f^o is the true fracture stress of LLDPE.

The dependence of σ_f^o of LLDPE–PS blends on the parameter β' is illustrated in Figure 7. The parameter β' characterizes the amount of load-bearing LLDPE in the cross section relative to the area of voids. A higher extension at fracture, and therefore a smaller relative cross section of load-bearing LLDPE, corresponds to a larger β' and a lower σ_f^o. A β' value of 1.21, for the case when the spherical PS particles are completely debonded but there is no void growth, predicts σ_f^o values that are too high. A β' value of 1.85 gives the best fit with the measured σ_f^o values.

The parameter β' is a function of both θ and λ; in contrast, β depends only on θ, and it is therefore possible for β' to vary with blend composition. The observation that all the data points in Figure 7 fall close to the curve for

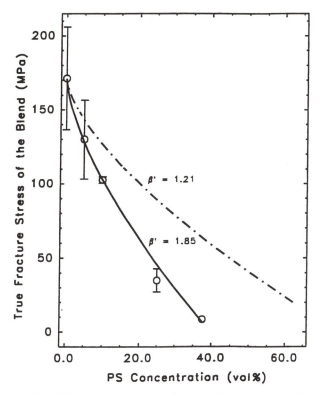

Figure 7. Effect of the parameter β' on the true fracture stress calculated from equation 13. Data for LLDPE–PS blends are included as the open circles.

β'= 1.85 reinforces the assumption that θ is independent of both composition and elongation and is consistent with the observation that λ is not strongly dependent on composition (Table I). From equation 12, this value of β' gives κ = 2.30; values of 2.5 to 2.9 are calculated from λ for a 45° debonding angle (Table II).

Fracture Strain

The engineering fracture strain of the LLDPE–PS blends demonstrated transitional behavior. When the PS content was low, less than 30%, high fracture strains were observed with only a gradual decrease in fracture strain as the PS content increased. The stress–strain behavior was characterized as Type I in this composition range. This was followed by a sharp decrease in the fracture strain in a narrow composition range as the stress–strain behavior changed from Type I to Type II and Type III.

It is often observed in filled ductile polymers that if adhesion is poor and deformation is accompanied by particle debonding followed by void formation and growth, fracture strains close to the fracture strain of the unfilled matrix can be achieved (6–8). This phenomenon is generally related to strain-hardening of the matrix. Thus, even when the load-bearing cross section is reduced by the debonded particles, the strain-hardening characteristics of the matrix ligaments make it possible for them to sustain the higher local stresses imposed during drawing. Transitional behavior from ductile drawing to quasi-brittle fracture at low strains occurs when the amount of matrix material in the cross section drops below a critical value.

The gradual and nearly linear decrease in fracture strain in the ductile region can be predicted using a rectangular array of voids. Following the previous approach, it is assumed that fracture occurs when the stress in the ligaments between voids reaches the true fracture stress of the matrix. The corresponding fracture strain can be estimated by considering the summation of elongations in regions A and B of the voided element (Figure 6b).

The stress–strain relationship in the strain-hardening region of the matrix is almost linear. Then, for blends that fracture in the strain-hardening region, the relationship between the true fracture stress (σ_f) and the fracture strain (ϵ_f) is given by

$$\sigma_f = \sigma_f^o - E_d(\epsilon_f^o - \epsilon_f) \tag{14}$$

where E_d is the slope in the strain-hardening region and is taken from the stress–strain curve of LLDPE, and σ_f^o and ϵ_f^o are the true fracture stress and fracture strain of LLDPE. Combining equations 13 and 14, the strain at fracture in region A (ϵ_{fA}) is given by

$$\epsilon_{fA} = \epsilon_f^o - \frac{\sigma_f^o \beta'(V_{PS})^{2/3}}{E_d} \tag{15}$$

while the fracture strain in region B (ϵ_{fB}) is equal to ϵ_f^o. The corresponding lengths of region A (L_{fA}) and region B (L_{fB}) at the fracture are given by

$$L_{fA} = (\epsilon_{fA} + 1)(L_o - 2R) \tag{16}$$

and

$$L_{fB} = \lambda 2R \tag{17}$$

where λ is the local extension ratio measured from the micrographs. The fracture strain of the blend (ϵ_f) is then given by

$$\epsilon_f = \frac{L_{fA} + L_{fB}}{L_o} - 1 = \left[\epsilon_f^o - \frac{\beta'\sigma_f^o(V_{PS})^{2/3}}{E_d} + 1\right]\left[1 - \left(\frac{6V_{PS}}{\pi}\right)^{1/3}\right] + \lambda\left(\frac{6V_{PS}}{\pi}\right)^{1/3} - 1 \tag{18}$$

Equation 18 applies to Type I behavior, when fracture occurs during strain-hardening, and it applies specifically to blends with 25% or less PS. The fracture strain (ϵ_f) calculated from equation 18 using $\beta' = 1.85$ and $E_d = 50$ MPa is compared with measured values in Figure 8. Calculations performed using λ values in Table II were lower than the observations. When λ was increased by 10% to account for recovery of the fractured specimens, the calculation satisfactorily predicted ϵ_f of blends with 5, 10, and 25% PS. When the PS content is less than 5%, the cubic-array model is no longer relevant and the meaning of equation 18 is lost.

Ductile-to-Quasi-Brittle Transition. The transition from Type I to Type II and Type III stress–strain behavior was accompanied by a sharp drop in ϵ_f. The large decrease in fracture strain with a relatively small change in composition has been called a ductile-to-quasi-brittle transition rather than a ductile-to-brittle transition, because even when the fracture strain is low,

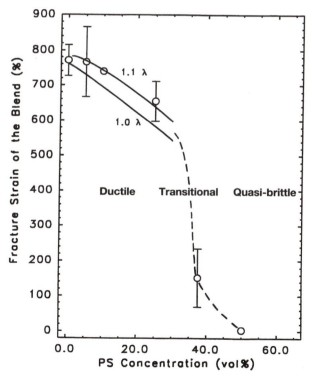

Figure 8. Fracture strain of LLDPE–PS blends. The calculated fracture strain in the Type I fracture region, where equation 18 applies, is given by the solid curves for 1.0λ and 1.1λ. Data for all the blend compositions are included as open circles.

some level of plastic deformation precedes fracture. The 50% PS composition with Type III stress–strain behavior was classified as quasi-brittle because it fractured at about 3.2% strain after localized stress-whitening. The composition with 37.5% PS and Type II stress–strain behavior was transitional. Although a stable neck formed in this composition, fracture during neck propagation indicated that the true stress in the propagating neck was close to the true fracture stress. In this case, fracture occurred when the propagating neck encountered a region of the gauge section in which the amount of load-bearing matrix material in the cross section was too low to sustain the neck-propagation stress.

The ductile-to-quasi-brittle transition occurs because the true fracture stress, as determined by β', decreases with increasing PS more rapidly than the yield stress, determined by β (Figure 9). The critical PS content (V_{PS}°) at the ductile-to-quasi-brittle transition can be determined from the condition that the true yield stress is equal to the true fracture stress. From equations 8 and 13,

$$V_{PS}^\circ = \left[\frac{\sigma_f^o - \sigma_y^o}{\beta'\sigma_f^o - \beta\sigma_y^o} \right]^{3/2} \tag{19}$$

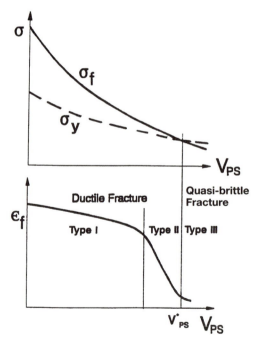

Figure 9. Schematic of the effect of PS content on yield stress (σ_y) and true fracture stress (ϵ_f) to give the ductile-to-quasi-brittle transition.

With $\beta = 0.60$ and $\beta' = 1.85$, the critical PS content is calculated to be 38%, which is close to the observed value of about 40%. The result from equation 19 shows that for this blend system, the critical filler content can be predicted from a model that considers only the difference between the true fracture stress and the yield stress of the LLDPE matrix. The only requirement of the analysis is that the PS phase be dispersed as debonded, approximately spherical particles. Surprisingly, in this instance, the fracture behavior was modeled satisfactorily without specific reference to particle size. However, particle size may become a factor when matrix yielding is constrained by a more severe triaxial stress state. Equation 19 predicts that ductility is maintained with a higher filler content in matrix materials that show a greater amount of strain-hardening. The other factors that affect the critical filler content are the debonding angle (θ) and the void content (κ) through the parameters β and β'.

Summary

The irreversible microdeformation mechanisms that control the high strain properties of uncompatibilized blends of LLDPE with dispersed PS were interfacial debonding and void growth around the PS particles. Characterization of the void morphology by two parameters, the debonding angle (θ) and the local extension ratio of the void (λ), was sufficient to derive models for true stress and strain predictions. The yield stress calculation was based on a modification of the effective-cross-section approach to incorporate the debonding angle; the true fracture stress was determined by considering the further reduction in effective cross section during void growth with, additionally, the strain-hardening characteristics of the LLDPE matrix phase. The model predicted that the true fracture stress should decrease more rapidly than the yield stress with increasing fraction of dispersed PS phase. This accounted for the observed ductile-to-quasi-brittle transition. In this poor interfacial adhesion model, the maximum concentration of dispersed phase that can be tolerated before ductility is lost is determined primarily by the strain-hardening characteristics of the continuous phase. Further modification of the model is needed to consider blends with good adhesion. However, a compliant interface that would provide stress transfer and also permit yielding and plastic deformation of the matrix might be attractive.

Acknowledgments

This research was generously supported by the National Science Foundation (Grant EEC93-20054) and the Edison Polymer Innovation Corp. (EPIC) of Brecksville, OH.

References

1. Heikens, D.,; Hoen, N.; Barensen, W.; Piet, P.; Ladan, H. J. *Polym. Sci. Symp*. 1978, 62, 309.
2. Fayt, R.; Jérôme, R.; Teyssié, Ph. *J. Polym. Sci.: Part B: Polym. Phys*. **1989**, 27, 775.
3. Li, T.; Topolkaraev, V. A.; Hiltner, A.; Baer, E; Ji, X. Z.; Quirk, R. P. *J. Polym. Sci. Part B: Polym. Phys*. **1995**, 33, 667.
4. Nielsen, L. E. *J. Appl. Polym. Sci*. **1966**, 10, 97.
5. Zhuk, A. V.; Knunyants, N. N.; Oshmyan, V. G.; Topolkaraev, V. A.; Berlin, A. A. *J. Mater. Sci*. **1993**, 28, 4595.
6. Topolkaraev, V. A.; Gorbunova, N. V.; Dubnikova, I. L.; Paramzina, T. V.; D'Yachkovskii, F. S. *Polym. Sci. U.S.S.R*. **1990**, 32, 2124.
7. Bazhenov, S.; Li, J. X.; Hiltner, A.; Baer, E. *J. Appl. Polym. Sci*. **1994**, 52, 243.
8. Li, J. X.; Silverstein, M.; Hiltner, A.; Baer, E. *J. Appl. Polym. Sci*. **1994**, 52, 255.

Failure Mechanisms in Compatibilized Blends of Linear Low-Density Polyethylene and Polystyrene

Tao Li[1], Cosimo Carfagna Jr.[2], Vasily A. Topolkaraev[1], Anne Hiltner[1]*, Eric Baer[1], X.-Z. Ji[3], and Roderic P. Quirk[3]

[1]Department of Macromolecular Science and Center for Applied Polymer Research, Case Western Reserve University, Cleveland, OH 44106
[2]Sniaresearch Società consortile per Azioni, Via Pomarico, 75010 Pisticci (MT), Italy.
[3]Institute of Polymer Science, University of Akron, Akron, OH 44325

The tensile failure of blends of linear low-density polyethylene (LLDPE) and polystyrene (PS) compatibilized with block copolymers of styrene (S) and butadiene (B) or hydrogenated butadiene (EB) has been studied. Compatibilizers were compared at the 5 wt% level in blends with equal amounts of LLDPE and PS by volume. Because of the lower viscosity of LLDPE, the morphology consisted of spherical PS particles dispersed in an LLDPE matrix. The stress–strain curve of the compatibilized blends was composed of an initial linear region, followed by a region of decreasing slope to a second linear region with a small positive slope. The yield point was defined by the intersection of the two linear regions. A modified yield-strain approach was used to predict the yield stress of the compatibilized blends. As a result of good adhesion, yielding of the matrix was constrained by the rigid PS particles to a region determined by the yielding angle, φ. Most of the blends exhibited a yield strain in the range of 1.2 to 1.6%. The calculated yielding angle of 70° or less was in accord with predictions for a rigid sphere in a plastic matrix. The higher yield stress of blends with crystalline styrene–hydrogenated butadiene and styrene–hydrogenated butadiene–styrene compatabilizers, as compared to blends with noncrystalline compatibilizers, resulted from their higher modulus. The exception was the blend compatibilized with Kraton G, which had a yield strain of 1.9% and a yielding angle of 79°.

H IGH-STRAIN PROPERTIES OF POLYMER BLENDS, such as strength, tensile elongation, and impact strength, benefit from compatibilization. These

---

*Corresponding author.

0-8412-3151-6   © 1996 American Chemical Society

advantages are specifically attributed to finer phase dispersion and improved interfacial adhesion. Elastic properties of blends with isotropic, dispersed morphologies are amenable to analysis with mechanical models in which good adhesion of the phases is assumed. Mechanical models that assume good adhesion without considering the structure of the interface are less satisfactory for describing high-strain deformation. Previously, we modeled high-strain properties of uncompatibilized blends using an approach based on observed microdeformation mechanisms, specifically, particle debonding and void growth (see Chapter 21). Approaches based on structural or morphological models that explicitly incorporate deformation and failure mechanisms of the interface or interphase are needed for compatibilized blends.

The beneficial effects of adding a graft or block copolymer of styrene (S) and butadiene (B), or a similar hydrogenated copolymer, to blends of polyethylene and polystyrene (PS) have been demonstrated (1–8). A previous study performed at this laboratory focused on the modulus of the compatibilized blend, which lay within Hashin's upper modulus bound and Kerner's equation for spherical voids (1). Two interfacial models resulted: a core–shell model similar to that described by others (9, 10), and an interconnected-interface model. The core–shell model with a compatibilizer coating the dispersed PS particle was used to calculate the decrease in modulus observed with rubbery compatibilizers. Using this model, it was possible to estimate the amount of compatibilizer at the interface, which varied from 5 to 50%, with the remaining compatibilizer presumed to be dispersed in the continuous polyethylene phase. The modulus of blends compatibilized with crystalline copolymers was calculated from an interconnected-interface model in which the blocks selectively penetrated the polyethylene and PS phases to provide good adhesion without forming a rubbery coating on the PS particle. The interconnection provided improved stress and strain transfer between the phases, and as a result the modulus of the blend approached Hashin's upper bound. These structural–mechanical models are now extended to incorporate interfacial-failure mechanisms for application to high-strain deformation and failure of compatibilized blends.

Experimental Details

Linear low-density polyethylene (LLDPE) and PS resins were the same as described previously (Chapter 21). The various block copolymers that were used as compatibilizers have also been described (1). A series of crystalline copolymers (Q series) was prepared by hydrogenation of diblock and triblock copolymers of styrene and butadiene [styrene–hydrogenated butadiene (SEB) and styrene–hydrogenated butadiene–styrene (SEBS)] (1). Triblock copolymers of styrene and butadiene [styrene–butadiene–styrene (SBS)] and a noncrystalline hydrogenated block copolymer (SEBS) (Kraton) were supplied by Shell Chemical Co. Diblock copolymers of styrene and butadiene [styrene–butadiene (SB) (Vector)] were obtained from Dexco Polymers. The characteristics of the resins are given in Table I.

Table I. Properties of Resins Used

Polymer	Number-Average Molecular Weight, $M_n (\times 10^{-3})$	Styrene-to-Rubber Ratio (wt/wt)	Yield Stress, $\sigma_y (MPa)$	Engineering Fracture Stress, $\sigma_{f,engrg} (MPa)$	Fracture Stress, $\sigma_f (MPa)$	Fracture Strain, $\epsilon_f (\%)$
LLDPE (GB502)	27		8.3 ± 0.3	18.8 ± 1.0	171 ± 25	760 ± 30
PS (Styron 623)	130			37.3 ± 3.9		1.3 ± 0.2
SEB copolymers (crystalline)						
Q29	30	67/33				
Q26	21	53/47				
Q292	53	19/81	12.5 ± 0.9	15.8 ± 0.5	96 ± 3	346 ± 19
Q302	75	20/80	11.1 ± 0.3	24.3 ± 0.8	133 ± 12	451 ± 15
Q304	117	19/81	9.8 ± 0.2	29.9 ± 1.7	174 ± 23	536 ± 29
SEBS copolymers (crystalline)						
Q293	60	33/67				
Q303	90	33/67				
Q307	139	32/68	12.1 ± 0.3	19.3 ± 1.6	88 ± 24	252 ± 40
SEBS copolymer (noncrystalline)						
Kraton G-1652M	115	29/71		22.3 ± 2.6	130 ± 3	575 ± 30
SB copolymers						
Vector 6000–D	150	30/70		0.9 ± 0.1	1.7 ± 0.3	98 ± 56
Vector 2320–D	70	25/75		0.2 ± 0.1	0.3 ± 0.1	110 ± 36
Vector 6010–D	130	12/88		0.3 ± 0.1	0.5 ± 0.1	65 ± 11
SBS copolymers						
Kraton D-4122P (plasticized)	137	48/52		7.1 ± 2.2	77 ± 33	937 ± 238
Kraton D–1101	154	31/69				
Kraton D–1102	88	28/72		14.3 ± 1.8	139 ± 8	1109 ± 13

Equal amounts of LLDPE and PS by volume were dry-mixed with 5 wt% compatibilizer and 0.2 wt% antioxidant (Irganox 1076), and blended in a Banbury mixer for about 7 min at 175 °C. The blend was quenched in cold tap water and dried overnight at 40 °C in vacuum. Plaques were compression-molded by pre-heating the blend in a press at 170 °C for about 15 min, then applying a pressure of 3.85 MPa for 8 min. The press was water-cooled to room temperature with the plaque under pressure.

Type I dog-bone tensile specimens (ASTM D–638–84) were cut from the compression-molded sheet. The stress–strain curve was obtained with a strain rate of 0.05 min^{-1}. A gauge length of 90 mm was used to calculate the strain. After straining to fracture, some of the tensile specimens were cryogenically fractured lengthwise at liquid-nitrogen temperature, coated with 60 Å of gold, and examined using a scanning electron microscope (JEOL JSM–840A).

Results and Discussion

Stress–Strain Behavior.   The effect of various compatibilizers at the 5% level was examined in blends with equal amounts of LLDPE and PS on a volume basis. The following description of the blend morphology of the com-pression-molded LLDPE–PS blends is summarized from our previous study, where the focus was on the low-strain mechanical response (1). Because of the lower viscosity of LLDPE, the morphology consisted of spherical PS particles dispersed in an LLDPE matrix. The large, somewhat elongated PS domains of the uncompatibilized blend became smaller and more spherical with the addi-tion of compatibilizer. The particle size was reduced from ~28 μm to ~4 μm with 5% Kraton G. The crystalline SEB and SEBS compatibilizers (Q series) were also effective in reducing the particle size. Unsaturated SB and SBS copolymers were not as effective as the hydrogenated compatibilizers; they de-creased the particle size only slightly, to ~20 μm.

Tensile deformation of the uncompatibilized blend with 50% PS was characterized by the appearance of several regions of localized stress-whiten-ing in the gauge section without global necking. Fracture occurred at one of these regions at a relatively low strain, about 3.2%. This behavior is character-ized as quasi-brittle rather than brittle, because some level of plastic deforma-tion precedes fracture even though the fracture strain is low (Chapter 21).

The stress–strain curves of most of the compatibilized blends consisted of an initial linear region, followed by a region of decreasing slope to a second lin-ear region with a small positive slope (Figure 1). Stress-whitening was notice-able at about 2% strain, and deformation proceeded with uniform extension of the entire gauge section. The slope of the initial linear region of the stress–strain curve reflected the tensile modulus, and it could either increase or decrease with respect to the uncompatibilized blend. Noncrystalline com-patibilizers lowered the tensile modulus of the blend by as much as 50%. Cal-culations based on a core–shell model with the rubbery compatibilizer coating the PS particle satisfactorily described the decrease in modulus and made it

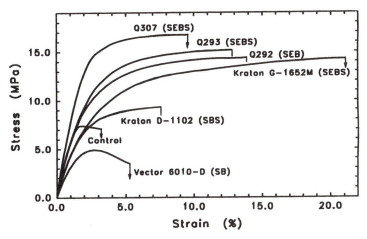

Figure 1. Effect of 5 wt% compatibilizer on the stress–strain curve of a blend with equal volumes of LLDPE and PS.

possible to estimate the fraction of compatibilizer in the interfacial coating. The increased modulus of blends compatibilized with crystalline, nonrubbery SEB and SEBS copolymers approached Hashin's upper bound. An interconnected interface model was proposed in which the blocks selectively penetrate the LLDPE and PS phases to provide good adhesion and improved stress and strain transfer between the phases (*1*).

The general shape of the stress–strain curve is the same for all the blends with Q-series and Kraton compatibilizers. This shape is described by two tangent lines drawn from the initial elastic region and the plastic region, respectively. The intersection of the lines is defined as the yield point and is described by a yield stress (σ_y) and an apparent yield strain (ϵ_y). The stress (σ) and strain (ϵ) in the plastic region are related by

$$\sigma = \sigma_y + E_p(\epsilon - \epsilon_y) \tag{1}$$

where E_p is the slope of the stress–strain curve in the plastic region.

All the SEB and SEBS copolymers were effective in increasing the yield stress and improving the fracture stress and strain of the blend (Table II). Compatibilization with the SBS copolymers resulted in some increase in the yield stress and fracture stress, but the increases were not as large as those achieved with the SEB and SEBS copolymers. There was enough variation in the Q-series copolymers to suggest some relationships between the styrene fraction in the copolymer and the properties of the compatibilized blend. The SEB copolymers with the highest styrene fraction (Q29 and Q26) formed blends with slightly higher yield stress and fracture stress, and lower fracture

Table II. Properties of the Blends Studied

Compatibilizer (5 wt%)	Average PS Domain Size[a] (μm)	σ_y (MPa)	ϵ_y (%)	σ_f (MPa)	ϵ_f (%)	Toughness[b] (J/m^3 × 10^{-6})	$\sigma_f-\sigma_y$ (MPa)	E_p × 10^{-2} (MPa)
None		7.5 ± 0.5	0.9 ± 0.1	7.0 ± 0.6	3.2 ± 1.0			
SEB copolymers (crystalline)								
Q29	5	16.4 ± 0.4	1.4 ± 0.1	17.3 ± 0.8	4.2 ± 0.3	0.56	0.9	0.32
Q26	1	16.0 ± 0.6	1.4 ± 0.1	17.3 ± 0.3	4.5 ± 0.4	0.61	1.3	0.42
Q292	3	12.9 ± 1.0	1.6 ± 0.2	15.6 ± 0.2	13.7 ± 2.4	1.67	2.7	0.22
Q302	15	13.5 ± 0.2	1.6 ± 0.1	13.7 ± 0.3	8.2 ± 1.4	0.92	0.2	0.03
Q304	15	13.7 ± 0.8	1.6 ± 0.1	14.4 ± 0.6	8.9 ± 2.0	1.05	0.7	0.10
SEBS copolymers (crystalline)								
Q293	6	14.9 ± 0.4	1.9 ± 0.1	16.1 ± 0.4	12.8 ± 1.4	1.62	1.2	0.11
Q303	8	16.2 ± 0.6	1.7 ± 0.1	16.4 ± 0.6	9.4 ± 0.7	1.25	0.2	0.03
Q307	9	15.6 ± 0.9	1.6 ± 0.1	16.3 ± 0.7	9.6 ± 1.3	1.35	0.7	0.09
SEBS copolymer (noncrystalline)								
Kraton G-1652M	4	10.1 ± 0.4	2.6 ± 0.6	12.2 ± 1.4	20.4 ± 1.7	2.44	2.1	0.12
SB copolymers								
Vector 6000D	15	6.3 ± 0.1	2.0 ± 0.1	5.7 ± 0.3	3.3 ± 0.8			
Vector 2320D	18	4.7 ± 0.1	2.6 ± 0.2	3.9 ± 0.4	9.4 ± 6.2			
Vector 6010D	20	4.9 ± 0.1	2.7 ± 0.1	3.4 ± 0.7	5.2 ± 1.2			
SBS copolymers								
Kraton D–4122P	20	8.6 ± 0.1	1.2 ± 0.1	10.2 ± 0.8	16.2 ± 1.1	1.29	1.6	0.11
Kraton D–1101	20	8.9 ± 0.5	1.4 ± 0.2	9.6 ± 0.5	10.1 ± 1.6	1.00	0.7	0.08
Kraton D–1102	20	8.1 ± 0.6	1.5 ± 0.1	8.8 ± 0.6	7.6 ± 0.6	0.60	0.7	0.11

[a]From reference 1.
[b]Area under the stress–strain curve.

strain, than the copolymers with a lower styrene fraction (Q304, Q302, and Q292). Increasing the molecular weight of SEB copolymers with 20/80 (S/EB) composition caused the copolymer yield stress to decrease and the fracture stress and strain to increase significantly. These trends did not carry over to the blends compatibilized with these copolymers. The only difference in the properties of the blends that might be significant was the slightly higher fracture strain of the blend compatibilized with Q292, the copolymer with the lowest molecular weight. Comparing SEB to SEBS copolymers as compatibilizers did not reveal any striking differences between the diblock and triblock architecture. Adding a second styrene block to an SEB copolymer did not affect the blend properties except for a slight increase in the yield stress and possibly also the fracture stress, effects that were also consistent with increasing styrene fraction. Blends with the SB copolymers had the lowest yield stress and fracture stress, even lower than those of the uncompatibilized control. In this instance, the low strength of the SB copolymers carried over to the poor properties of the blends.

The parameter E_p was greater than zero, and it was greatest for blends with the two low-molecular-weight SEB copolymers with high PS content. A range of values, from 0.03 to 0.22, was obtained for blends with the other crystalline SEB and SEBS copolymers; no clear trends with molecular weight or differences between diblock and triblock copolymers were seen. Blends with the noncrystalline SEBS and SBS all had E_p values near the middle of the range cited.

The higher fracture stress and strain in the compatibilized blend, together with the positive slope in the plastic region, indicated that improved adhesion between PS particles and the LLDPE matrix suppressed the processes of debonding and subsequent void growth that were observed in the uncompatibilized blends (Chapter 21). With good adhesion of the PS particles to the matrix, the local strains at the interface would have been high because of the constraint imposed by the undeformed PS particles. The fractured tensile specimens were cryogenically fractured lengthwise to reveal the internal morphology of the deformed blend. Two examples, in Figures 2 and 3, of blends compatibilized with 5% Kraton G and 5% Q292, respectively, show that the PS particles were not debonded from the matrix. Instead, they remained connected to the matrix by microfibrils that formed when the interfacial region was drawn out during tensile deformation. Subtle differences in the texture of the particle surface are discerned at higher magnification. The PS particles on the fracture surface of the blend with Kraton G appear to have a coating of the compatibilizer, which is suggestive of adhesive fracture between a rubbery shell and the matrix. In contrast, the PS particles in the blend with the crystalline SEB copolymer have a rougher texture, as if a fibrous interface has fractured.

The processes of interfacial stretching and microfibrillation that accompanied tensile deformation are shown schematically with the stress–strain

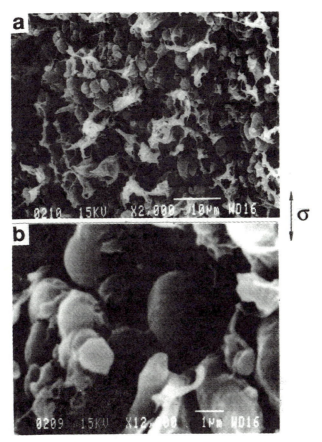

σ

Figure 2. Scanning electron micrographs of a blend with 5% Kraton G–1652M that was drawn to fracture and cryogenically fractured lengthwise. The direction of stress is shown by the arrow.

curve in Figure 4. In the elastic region, Stage I, the interface is intact. As the yield point is approached, in Stage II, the interface (or interphase) begins to undergo large strains that culminate in microfibrillation. In Stage III, the microfibrils stretch out until fracture of the interface leads to catastrophic failure of the blend. This concept is thought to be applicable to both the core–shell model and the interconnected-interface model of compatibilization. Microfibrils in the blend compatibilized with Kraton G would have formed by drawing of the rubbery shell of the core–shell particle. When the blend fractured, a rubbery coating remained on the PS particles. In the blend compatibilized with Q–292, with the interconnected interface, it was more likely that microfibrils formed by drawing of the LLDPE matrix connected to the PS parti-

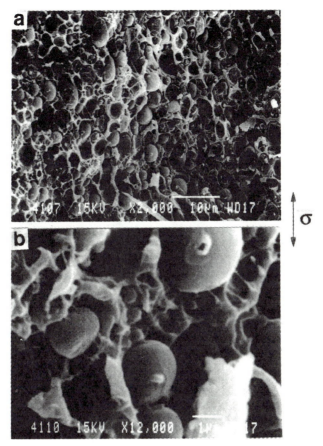

Figure 3. Scanning electron micrographs of a blend with 5% Q292 that was drawn to fracture and cryogenically fractured lengthwise. The direction of stress is shown by the arrow.

cles. The need for significantly higher loads to deform blends compatibilized with Q-series copolymers supports this hypothesis.

Prediction of Yield Stress. A modified yield-strain approach was used to predict the yield stress of the compatibilized blends. Instead of assuming interfacial debonding as in the previous analysis of uncompatibilized blends (Chapter 21), we assumed that the PS particles were well-connected to the LLDPE matrix. As a result, yielding of the matrix was constrained by the rigid PS particles to a region determined by the yielding angle (ϕ), as defined in Figure 5. Nielsen's approach was used to obtain the effective deformation length of the matrix (L_{eff}) with the constraint from the rigid PS particles (*11*):

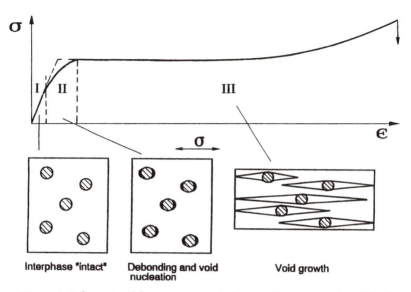

Figure 4. Schematic of the stress–strain behavior of a compatibilized blend with good adhesion.

$$L_{\text{eff}} = L_0[1 - \alpha(V_{\text{PS}})^{1/3}] \tag{2}$$

where L_0 is the unit cell dimension of the cubic array of particles, V_{PS} is the volume fraction of PS, and α is a geometric factor that depends on the yielding angle. The factor α is given by

$$\alpha = 2\left(\frac{3}{4\pi}\right)^{1/3} \cos\phi \tag{3}$$

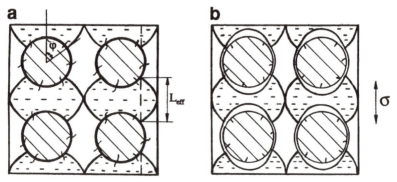

Figure 5. Schematic of constrained yielding: (a) interpenetrating-interface model; and (b) core–shell interface model with a rubbery shell. L_{eff} is the effective deformation length of the matrix.

Assuming further that the matrix has the same effective yield strain as LLDPE ($\epsilon_y^o = 2.4\%$), the yield strain of the compatibilized blends is given by

$$\epsilon_y = \epsilon_y^o \frac{L_{\text{eff}}}{L_0} = \epsilon_y^o [1 - \alpha(V_{\text{PS}})^{1/3}] \qquad (4)$$

The yield strain calculated from equation 4 is plotted in Figure 6a for several yielding angles. A yielding angle of 90° corresponds to the unconstrained condition. Depending on the yielding angle, the decrease in yield strain can be

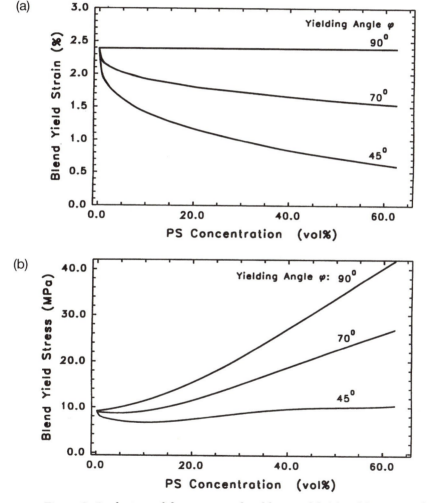

Figure 6. Predictions of the constrained-yielding model: (a) yield strain; and (b) yield stress.

dramatic. Analysis of the stress and strain field around a rigid sphere in a plastic matrix suggests that the yielding angle should be about 70° (12).

The yield stress in the idealized representation of the stress–strain curve is then given by

$$\sigma_y = E\epsilon_y \qquad (5)$$

where E is the modulus of the compatibilized blend and ϵ_y is the yield strain from equation 4. The yield stress calculated from equation 5 for several yielding angles is plotted in Figure 6b using Hashin's upper modulus bound for spherical PS particles dispersed in LLDPE. In equation 5, the PS particles have two opposing effects on the yield stress: they constrain yielding, which decreases the yield strain, and they also have a reinforcing effect that increases the modulus. At a low PS content, the effect of the constraint is more important and the yield stress decreases; however, as the PS content increases, the reinforcing effect dominates, and with a reasonable value for the yielding angle (70°), significant increases in the yield stress are predicted for blends of LLDPE and PS.

The yield stress taken from the stress–strain curves is compared with predictions in Figure 7. The upper bound is the prediction from equation 5 for a yielding angle of 70° and Hashin's upper modulus bound. The lower bound in Figure 7 is obtained from Nielsen's approach for completely debonded parti-

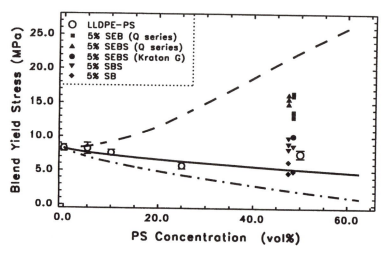

Figure 7. Yield stress of the compatibilized blends compared with calculations. Key: —, a yielding angle of 70° and Hashin's upper modulus bound; ---, Nielsen's approach for debonded particles; and _ _, Nielsen's approach with a debonding angle of 45°. Data for the uncompatibilized blends are from Chapter 21.

cles by considering the decrease in load-bearing cross section that results from interfacial debonding. A modification of Nielsen's approach, developed to consider partial debonding of the spherical particles, describes the yield stress of uncompatibilized blends of LLDPE and PS (Chapter 21). This prediction, which incorporates a debonding angle of 45°, is included as the solid curve. The uncompatibilized blend with V_{PS} of 0.5 falls slightly above the solid curve because the morphology of this composition did not conform with the assumptions of the model as well as compositions with a lower PS fraction; specifically, the PS particles tended to be elongated rather than spherical. The yield stress of blends with SB copolymers falls on the same curve as the uncompatibilized blends. This result is attributed to interfacial failure before yielding in blends with the low-strength SB copolymers. Blends with the Q-series compatibilizers are the closest to the upper bound; blends with SBS compatibilizers are intermediate between the uncompatibilized blend and blends with Q-series compatibilizers. Differences in the yield stress among these compatibilized blends primarily reflect the effect of compatibilization on the blend modulus.

The yield strain obtained from the point of intersection of the two linear regions of the stress–strain curve is compared in Table III with the yield strain calculated from the yield stress and the modulus. There are slight differences between measured and calculated values; the calculated values are considered

Table III. Yield Parameters of Compatibilized Blends

Compatibilizer (5 wt%)	Yield Stress, σ_y (MPa)	Modulus[a] (GPa)	Measured Yield Strain, ϵ_y (%)	Calculated Yield Strain, ϵ^o_y (%)	Yielding Angle, ϕ (deg)
SEB copolymer (crystalline)					
Q29	16.4 ± 0.4	1.25 ± 0.07	1.4 ± 0.1	1.3	62
Q26	16.0 ± 0.6	1.06 ± 0.06	1.4 ± 0.1	1.5	68
Q292	12.9 ± 1.0	0.90 ± 0.04	1.6 ± 0.2	1.4	65
Q302	13.5 ± 0.2	0.92 ± 0.03	1.6 ± 0.1	1.5	66
Q304	13.7 ± 0.8	1.03 ± 0.06	1.6 ± 0.1	1.3	63
SEBS copolymer (crystalline)					
Q293	14.9 ± 0.4	0.94 ± 0.02	1.9 ± 0.1	1.6	70
Q303	16.2 ± 0.6	1.09 ± 0.04	1.7 ± 0.1	1.5	67
Q307	15.6 ± 0.9	1.03 ± 0.02	1.6 ± 0.1	1.5	68
SEBS copolymer (noncrystalline)					
Kraton G–1652M	10.1 ± 0.4	0.52 ± 0.07	1.9 ± 0.2	1.9	79
SBS copolymer					
Kraton D–4122P	8.6 ± 0.1	0.71 ± 0.04	1.2 ± 0.1	1.2	59
Kraton D–1101	8.9 ± 0.5	0.67 ± 0.04	1.4 ± 0.2	1.3	62
Kraton D–1102	8.1 ± 0.6	0.52 ± 0.03	1.5 ± 0.1	1.6	69

[a]From Reference 1.

more accurate because they are based on careful determinations of modulus. The yielding angle obtained from equation 4 with a V_{PS} of 0.5 is also included in Table III. Most of the blends exhibit a yield strain in the range of 1.2 to 1.6% and a yielding angle of 70° or less. The exception is the blend compatibilized with Kraton G, where a yield strain of 1.9% gives a yielding angle of 79°.

The Q-series compatibilizers are crystalline SEB and SEBS block copolymers. From analysis of the modulus of LLDPE–PS blends compatibilized with Q-series copolymers, and examination of electron micrographs, these compatibilizers are thought to impart good adhesion to the interface between LLDPE and PS by interpenetrating both phases without creating a rubbery interphase (1). Because they provide good adhesion, the stress and strain fields around the particle are expected to resemble those obtained analytically (12). This expectation was confirmed when the calculated yielding angle in blends compatibilized with Q-series copolymers was found to be close to the 70° predicted angle. Therefore, the yield stress of blends with Q-series compatibilizers is determined by the blend modulus.

In contrast to the Q-series compatibilizers, the noncrystalline Kraton compatibilizers are thought to provide good adhesion by forming a rubbery shell around the PS particles (1). One manifestation of the rubbery shell is the lower modulus of blends compatibilized with the Kratons. The Kratons differ in the amount of compatibilizer in the shell; in a blend with 5% compatibilizer, it is estimated that about 50% of Kraton G is in the shell, whereas only 5–15% of a Kraton D coats the particles, with the rest assumed to be dispersed in the LLDPE matrix. With only a thin shell of Kraton D coating the PS particle, the stress and strain fields should not be affected as much as with a thicker shell of Kraton G. This is borne out in the yielding characteristics; the yield strain and corresponding yielding angle calculated for blends compatibilized with Kraton D's are similar to those with Q-series compatibilizers. On the other hand, the higher yield strain and yielding angle of the blend compatibilized with Kraton G are consistent with a rubbery, deformable shell on the PS particle.

Fracture Stress and Strain. Yielding and plastic deformation in the schematic representation of tensile deformation were associated with microfibrillation at the interface and stretching of the microfibrils. Because this representation was assumed to apply to both the core–shell and interconnected-interface models of compatibilization, the constrained-yielding approach was used without specific reference to the microstructure of the interface. In extending the discussion to fracture, however, it is useful to consider the interfacial-deformation mechanisms. Tensile deformation culminated in catastrophic fracture when the microfibrillated interface failed. This was inferred from the quasi-brittle fracture behavior of the uncompatibilized blend with V_{PS} of 0.5, which indicated that the reduced load-bearing cross section after interfacial debonding could not support plastic deformation. Accordingly, the ultimate properties of the compatibilized blend depended on interfacial characteristics such as adhesive strength, ductility of the fibrillating material, and strength of the microfibrils. These were controlled by the microstructure of the interface.

Microfibrils in the blend compatibilized with Kraton G probably formed by drawing of the rubbery shell of the core–shell particle. Important factors would have been the amount of rubber in the shell, the strength of the rubber, and the strength of adhesion to LLDPE. All these factors may have contributed in some degree to the high fracture stress and strain of the blend with Kraton G. The amount of compatibilizer in the shell differed for the various Kratons; the thicker coating was certainly one of the reasons Kraton G gave better properties to the compatibilized blend than the Kraton D compatibilizers.

Microfibrils formed at the interconnected interface of the Q-series compatibilizers by drawing out of the attached LLDPE matrix. Interconnection, possibly involving cocrystallization of the EB block (*13*), was thought to create a modified LLDPE region around the PS particles. This modified LLDPE, which probably deformed more easily than LLDPE, would have drawn out into microfibrils. Even the lowest molecular weight EB block, nominally 10,000 in Q26 and Q29, was sufficient to give good interconnections between the phases. Blends with these compatibilizers exhibited the largest values of E_p, the slope of the stress–strain curve in the plastic region, and the fracture stresses were at least as high as in blends with other Q-series compatibilizers. The blends with Q26 and Q29 did have fairly low fracture strains, possibly because the modified LLDPE layer was relatively thin. The molecular weights of the crystallizable EB blocks in the other Q-series compatibilizers were nominally 40,000, 60,000, and 95,000. The blends with these compatibilizers showed modest effects of the molecular weight of the EB block. The SEB and SEBS with EB blocks that had a molecular weight of 40,000 imparted slightly higher fracture strains to the blends, but there were no differences between diblock and triblock architectures. Possibly, the molecular weight of the EB blocks was too high to show the effect of architecture.

Summary

Structural-mechanical models of compatibilization of LLDPE blends with PS have been extended to take interfacial-failure mechanisms into account. The core–shell model with a coating of the rubbery compatibilizer on the PS particle is exemplified by blends with Kraton G. The interconnected-interface model with the blocks selectively penetrating the LLDPE and PS phases to provide good adhesion without a rubbery coating on the PS particle apply to the Q-series compatibilizers. Yielding of both interfacial models is described by processes of interfacial stretching and microfibrillation. Analytically, a modified yield-strain approach was successful. Yielding of the matrix was assumed to be constrained by the rigid PS particles to a region determined by a yielding angle, φ. Yielding of blends with an interconnected interface conformed well with predictions for rigid spheres in a plastic matrix. The presence of a rubbery coating relaxed the constraint somewhat to allow for a larger yielding angle and hence a higher yield strain. Ultimate properties of the compatibilized blends were determined by fracture of the microfibrillated interface. This in turn was controlled by interfacial characteristics such as adhesive strength, ductility of the fibrillating material, and strength of the microfibrils. Although these factors were expected to depend on compatibilizer structure, including molecular weight of the blocks and diblock versus triblock architecture, the expected variations in ultimate properties were not observed.

Acknowledgments

This research was generously supported by the National Science Foundation (Grant EEC93–20054) and the Edison Polymer Innovation Corp. (EPIC) of Brecksville, OH.

References

1. Li, T.; Topolkaraev, V. A.; Hiltner, A.; Baer, E.; Ji, X.-Z.; Quirk, R. P. *J. Polym. Sci.: Part B: Polym. Phys.* **1995**, *33*, 667.
2. Fayt, R.; Jérôme, R.; Teyssié, Ph. *J. Polym. Sci.: Polym. Phys. Ed.* **1982**, *20*, 2209.
3. Fayt, R.; R. Jérôme, R.; Teyssié, Ph. *J. Polym. Sci.: Part B: Polym. Phys.* **1989**, *27*, 775.
4. Fayt, R.; Jérôme, R.; Teyssié, Ph. *Makromol. Chem.* **1986**, *187*, 837.
5. Lindsey, C. R.; Paul, D. R.; Barlow, J. W. *J. Appl. Polym. Sci.* **1981**, *26*, 1.
6. Heikens, D.; Hoen, N.; Barentsen, W.; Piet, P.; Ladan, H. *J. Polym. Sci. Symp.* **1978**, *62*, 309.
7. Barentsen, W. M.; Heikens, D. *Polymer* **1973**, *14*, 579.
8. Heikens, D.; Barentsen, W. *Polymer* **1977**, *18*, 69.
9. Theocaris, P. S. *The Mesophase Concept in Composites;* Springer-Verlag, Berlin, Germany, 1987; Chapter III.
10. Maurer, F. H. J. In *Polymer Composites;* Sedlacek, B., Ed.; W. De Gruyter: Berlin, Germany, 1986; p. 399.
11. Nielsen, L. E. *J. Appl. Polym. Sci.* **1966**, *10*, 97.
12. Zuck, A. V.; Gorenberg, A. Ya.; Topolkaraev, V. A.; Oshmyan, V. G. *Mech. Compos. Mater.* **1988**, *5*, 533.
13. Duvall, J.; Sellitti, C.; Myers, C.; Hiltner, A.; Baer, E. *J. Appl. Polym. Sci.* **1994**, *52*, 195.

Toughened Plastics of Isotactic Polystyrene and Isotactic Polypropylene Blends

Guangxue Xu and Shangan Lin

Institute of Polymer Science, Zhongshan University, Guangzhou 510275, China

Toughened engineered polyolefin plastics with improved heat re-sistance, increased impact strength, and no loss of stiffness were prepared by blending (1) isotactic polystyrene (iPS), (2) isotac-tic polypropylene (iPP), and (3) styrene–propylene diblock copoly-mer (iPS–b–iPP, styrene/propylene-40/60) with crystalline isotactic structure in each block. iPS–b–iPP was synthesized by sequential copolymerization of styrene and propylene using a NdCl₃-modified Ziegler–Natta catalyst. The stiffness/toughness and heat resistance of the iPS–b–iPP–iPS–iPP polyblends were higher than those of typ-ical high-impact polystyrene or iPP when iPS–b–iPP copolymer in the blends was 25 wt%. iPS–b–iPP copolymer was proven to be an effective compatibilizer of iPS–iPP blends by differential scanning calorimetry, dynamic mechanical thermal analysis, and morpholog-ical study (scanning electron microscopy). iPS–b–iPP enhanced in-terfacial interaction between phases of blends and reduced particle dimensions of the dispersed phase, thus improving their mechanical and thermal properties.

POLYOLEFINS HAVE HAD AN IMPORTANT POSITION among synthetic polymers because of their low cost, versatile properties, and growing applications as polymeric films, containers, and pipes. However, polyolefins have limited ap-plication in several technologically important fields because of their extremely poor stiffness and toughness and low heat resistance. Accordingly, there has been strong interest in recent years in the modification of polyolefins as engi-neered polyolefin plastics (*1–3*). Among the polyolefins, engineered isotactic polypropylene (iPP) is attracting more and more interest (*4*). The balance be-tween impact and stiffness, as well as the heat-distortion temperature of iPP, is far inferior to that of most engineering plastics. Blending iPP with high-perfor-mance engineering resins represents an attractive route to making tailor-made materials with improved heat resistance, rigidity, and toughness, and a good performance/cost ratio (5). Several PP-based blends with engineering resins, such as polycarbonate, poly(ethylene terephthalate), poly(phenylene oxide), and nylon-6, have been reported, but the results indicate that they have poor mechanical behavior, severe delamination, and weak weld-line strength in their extruded or injection-molded parts. Admittedly, poor adhesion at the interface, coarse phase dispersion, and coalescence were considered to be mainly responsible for the inferior performance (6–9).

Relating the final morphology of a blend to its end-use properties is an interesting and challenging aspect of this research area. Among the parameters governing phase morphology and its stability against coalescence of polyblends are interfacial tension, which generates a small phase size (10, 11); the nature of the interactions, or interfacial adhesion, that transmit applied force effectively between the phases (12); the viscoelastic properties of each component of the blend; the thermal history; the blending procedure; and the characteristics (molecular weight and composition) of the polymers (13, 14).

One possible way of reducing interfacial tension and improving phase adhesion between PP-based blend phases is to use a selected copolymeric additive that has similar components to the blend, as a compatibilizer in the blend system. Well-chosen diblock copolymers, widely used as compatibilizing agents in PP-based blends, usually enhance interfacial interaction between phases of blends (15, 16), reduce the particle dimensions of the dispersed phase (16, 17), and stabilize phase dispersion against coalescence (16–18) through an "emulsification effect," thus improving the mechanical properties (15–19).

To our knowledge, nothing has been reported recently on the blend of iPP and high-performance isotactic polystyrene (iPS) as a toughened plastic. iPS is an attractive semicrystalline polymer characterized by low cost and versatile performance. The high melting point (~230 °C), above-ambient glass-transition temperature (100 °C), exceedingly high stiffness or rigidity, and excellent heat resistance (heat-distortion temperature $\geq$ 170 °C) add to the favorable profile of properties. So there is reason to believe that the improved balance between impact and stiffness and the improved heat resistance of semicrystalline iPP could be achieved by blending with iPS. It is necessary, however, to find means of (1) avoiding macroscopic phase separation of the iPP and iPS, and (2) grafting or otherwise improving the strength of the interfaces between the iPP and iPS phase. As was just discussed, these two requirements are met by the addition of block copolymer emulsifiers to the homopolymer blends; little is known about the corresponding approach for the semicrystalline iPS–iPP blend system.

The recent successful synthesis in our laboratory of styrene–propylene diblock copolymer (iPS-b-iPP), with crystalline isotactic structure in each block (20, 21), makes it possible to attempt toughening of the iPS–iPP blend by the method of block copolymer emulsification of the diblock copolymer. This chapter presents our recent work on the toughened plastics of the iPS-*b*-iPP–iPS–iPP polyblend, and the compatibilizing effect of iPS-*b*-iPP diblock copolymer on the mechanical properties, thermal behavior, and morphologies of iPS–iPP blends.

Experimental Details

Synthesis of Homopolymers and Styrene–Propylene Copolymer.   The isotactic polypropylene (iPP, molecular weight M_w = 300,000) and isotactic polystyrene (iPS, M_w = 280,000; isotactic index $\geq$ 99%) used in this study were prepared in a laboratory-scale reactor using a modified Ziegler–Natta catalyst ($MgCl_2/TiCl_4/NdCl_x(OR)_y/Al(iBu)_3$, where iBu is isobutyl), which was developed in our laboratory (*20*). Styrene–propylene block copolymer was synthesized by sequential copolymerization of styrene and propylene using the modified Ziegler–Natta catalyst; the synthesis is similar to that of the corresponding homopolymer and gives a high yield. The copolymer was separated from unwanted homopolymer species in the reaction products using a successive fractionation procedure by means of the difference of solubility and crystallizability in different solvents (*21*). In contrast experiments on two homopolymers (corresponding homopolymer blends and copolymerization products), atactic polystyrene (aPS), atactic polypropylene (aPP), and iPS were separated from the pure copolymer by successive extraction fractionation using boiling methyl ethyl ketone, heptane, and chloroform, respectively, and iPP homopolymer in copolymerization products was also isolated from pure copolymer by crystallization at 110 °C in α-chloronaphthalene. Thus the remaining soluble fraction in α-chloronaphthalene at 110 °C was confirmed to be pure copolymer species, with 40.5% by weight of the original reaction products. The copolymer contained 40% PS as determined by IR spectroscopy and elemental analysis. Gel permeation chromatographic (GPC) experiments gave the copolymer molecular weight as 340,000 g/mol.

Preparation of iPS-*b*-iPP–iPS–iPP Polyblends.   The blends of iPS-*b*-iPP, iPS, and iPP were prepared by completely dissolving all polymers and an antioxidant into *o*-dichlorobenzene at 165 °C, and using a 1:1 mixture of acetone and methanol as a precipitant. After being thoroughly washed and dried under vacuum at 60 °C for 20 h, the precipitated powders were compression-molded at 300 °C into sheets or plates suitable for cutting specimens for mechanical testing and morphologic study.

Measurements.   The structure of the block copolymer was determined by NMR using a JEOL FX-90Q instrument. The glass-transition temperature (T_g) and melting temperature (T_m) of the copolymer and the blends were measured by differential scanning calorimetry (DSC) (Perkin Elmer, DSC-2) at a heating rate of 20 °C/min. The dynamic mechanical properties were determined by dynamic mechanical thermal analysis using a Rheovibron DDV-II EA instrument. Measurements were made at an operating frequency of 110 Hz and a heating rate of 3 °C/min under a liquid-nitrogen atmosphere. The morphologies of the polymer blends or the copolymer were studied with a scanning electron microscope (SEM, Hitachi S-570). Fracture surfaces were prepared at either room temperature or liquid-nitrogen temperature.

Tensile properties, Young's modulus, Izod impact strength, and Rockwell hardness of the compression-molded specimens were determined by the standard procedures described in ASTM D638, D790, D256, and D785, respectively. The heat-distortion temperature (HDT) was measured under 980 and 9.8×10^4 Pa at heating rates of 2 and 20 °C/min, respectively.

Results and Discussion

Characterization of the Styrene–Propylene Copolymer. The synthesis of styrene–propylene block copolymer with crystalline isotactic structure in a polystyrenic block has been an interesting research topic from both academic and practical points of view. However, little is found about styrene–propylene block copolymer synthesis in the literature (22), and the structure and properties of the block copolymer have not yet been clarified. Thus, prior to studying its effect as a compatibilizer on iPS/iPP blends, extensive structural characterization was carried out on the purified styrene–propylene copolymer.

Figure 1 presents the DSC scan of the copolymer. The two glass transitions at $T_g = -2$ and 98 °C indicate segregated PP and PS microdomains. Also, the DSC analysis of the copolymer exhibits the two strong first-order transitions at 162 and 220 °C, suggesting that considerable amounts of crystallinity exist in respective iPP and iPS blocks or segments. The melting temperature (T_m) of each segment in the copolymer was found to be lower than those of corresponding isotactic homopolymers determined under similar conditions. On the basis of the DSC analysis and wide-angle X-ray scattering, we estimate the crystallinity of PP and PS blocks in the copolymer to be 35.1 and 14.6%, respectively.

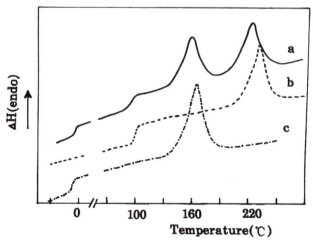

Figure 1. DSC curves of (a) pure styrene–propylene copolymer, (b) iPS, and (c) iPP.

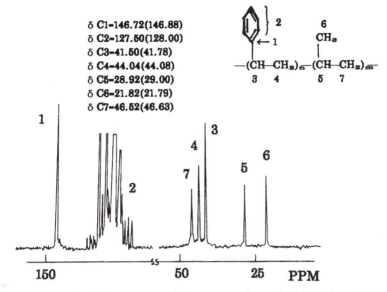

Figure 2. *13C-NMR spectrum of the pure copolymer (iPS-b-iPP) at 130 °C in o-dichlorobenzene. Numbers in parentheses denote corresponding carbon peaks in the homopolymer.*

As shown in Figure 2, ^{13}C-NMR analysis of the copolymer gave unambiguous signals of a block copolymer with two long sequences of essentially pure iPS and iPP (iPS-*b*-iPP). Isotactic triads of each block were about 98%. Because of little measurable shift of peak locations, which might be interpreted in terms of the repeat unit interaction of a styrene–propylene sequence as compared with respective homopolymers, the copolymer characterized in this study is considered to be a simple mixture of the two homopolymers. But the results of successive fractionation, as discussed in the section "Experimental Details," could eliminate the possibility of homopolymer mixtures.

In the transmission electron microscopic (TEM) studies, we found that it was exceedingly difficult to obtain ultramicrotomed sections of iPS–iPP blends, whereas the iPS-*b*-iPP diblock copolymer could be cut with relative ease. This result exhibits one major difference between the diblock copolymer and the corresponding homopolymer blend. Unfortunately, owing to the difficulty of finding a selective staining technique, the sample of diblock copolymer did not display visible contrast or obvious structural features in the TEM studies. However, the results of SEM studies do reveal a clear difference between the blend and the diblock copolymer; the macrophase separation is revealed on the etched surface of the blend and is not present in the copolymer (Figure 3). The diblock copolymer exhibits only a finely dispersed and continuous submicron structure throughout the field of view, as expected.

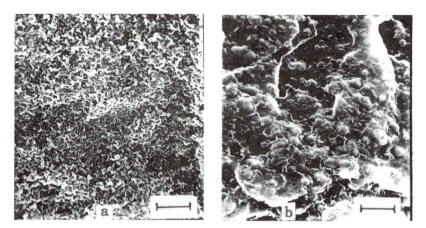

Figure 3. SEM images of cast films of (a) iPS-b-iPP copolymer (40/60), and (b) iPS–iPP blend (40/60). The surface of the casting films was etched with allylamine vapor at room temperature for 1 h. The scale is 20 μm.

As compared with the homopolymers and the corresponding homopolymer blend, the copolymer also shows significant differences in dynamic mechanical behavior (Figure 4). The blend shows two distinct transitions at 0 and 98 °C, which result from the glass transitions of the respective homopolymer components. The blend exhibits a rapid drop in modulus with increasing temperature, whereas the copolymer exhibits a more gradual drop in modulus, suggesting that the phases of two blocks or segments are welded together to a greater extent in the copolymer.

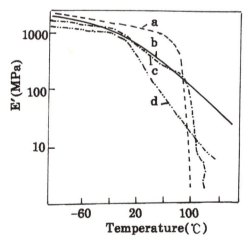

Figure 4. Dynamic mechanical spectrum of (a) iPS, (b) iPS-b-iPP copolymer (40/60), (c) iPS–iPP (40/60) blend, and (d) iPP.

Significant differences in mechanical properties are also seen in the copolymer by comparing corresponding homopolymers or homopolymer blends (Table I). As seen in the table, the diblock copolymer has superior tensile strength and hardness to iPP. The Izod impact strength of the copolymer is higher than that of homopolymers and homopolymer blends. The results are perhaps interrelated with the stereoregular structure, crystallinity, and microphase morphologies of the copolymer. Evidence that the phases are welded together and "interpenetrated" or cross-linked to a greater extent in the copolymer may be inferred from the increased modulus, Izod impact strength, and ultimate elongation.

Taking all the fact presented in this section into account, together with the synthesis method and fractionation results, we conclude that the purified copolymer separated from reaction products is an iPS-*b*-iPP diblock copolymer consisting of iPS and iPP blocks; it is definitely not a simple blend of homopolymers. On the other hand, the distinctive characteristics of the copolymer's crystallization kinetics also indicate that, compared with homopolymers and the iPS–iPP blend, the purified copolymer is a true iPS-*b*-iPP diblock copolymer (*23*).

Toughened Plastics of iPS-*b*-iPP–iPS–iPP Polyblends.   The molecular weight and block composition of the diblock copolymer are certainly key thermodynamic criteria in designing the most efficient compatibilizer for polymer blends as toughened plastics (*24*). Accordingly, we deal here with the iPS-*b*-iPP diblock copolymer with a composition (styrene/propylene, S/P) of 40/60 and a molecular weight of 340,000 g/mol. To establish a connection between mechanical or thermal properties and observed morphological changes, ternary blends were made by following the isopleth of the triangular phase diagram (Figure 5). Along the isopleth, all blend materials contain equal weight percentages of propylene and styrene repeat units, while the copolymer content varies from 0 to 100%.

Table I. Mechanical Properties of the Styrene–Propylene Diblock Copolymer

Sample	Young's Modulus (MPa)	Unnotched Izod Impact Strength (KJ/M²)	Tensile Strength at Yield (MPa)	Hardness (Rockwell R Scale)	Ultimate Elongation (%)	Heat-Distortion Temperature[a] (°C)
iPS	2250	2.5 – 4.0	42.9	102	3–5	210
iPS-*b*-iPP	1890	38.5	35.8	80	302	168
iPS–iPP blend (40/60)	1595	4.1	—	84	2–3	150
iPP	1050	27.5	23.4	70	898	140

NOTE: The molecular weight of the copolymer is 340,000 g/mol (by GPC).
[a]Measured under 980 Pa at a heating rate of 2 °C/min.

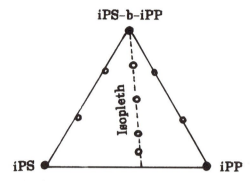

Figure 5. Triangular phase diagram showing the composition of the iPS-b-iPP–iPS–iPP blends examined.

Mechanical Properties. Table II shows the mechanical properties (i.e., Young's modulus, Izod impact strength, tensile strength, Rockwell hardness, elongation at break, and HDT) of several iPS-*b*-iPP–iPS–iPP polyblends. The incorporation of the diblock copolymer in a 5% amount does make the impact strength of the iPS–iPP blend rise enormously. In addition to noting the clear trend of increased impact resistance with increased copolymer content, we note that the impact strength of the iPS–iPP blend containing 25% copolymer already exceeds that of high-impact polystyrene (HIPS), and at high copolymer contents the impact strength is approximately double that of a typical HIPS. These results indicate that the dispersed phase may not only adhere to

Table II. Mechanical Properties of iPS-*b*-iPP–iPS–iPP Polyblends

Weight ratio of iPS-b-iPP– iPS–iPP	Young's Modulus (MPa)	Unnotched Izod Impact Strength (KJ/M²)	Tensile Strength at Yield (MPa)	Hardness (Rockwell R Scale)	Ultimate Elongation (%)	Heat- Distortion Temperature[a] (°C)
0/40/60	1590	4.1	6.3	84	2–3	150
5/38/57	1601	8.4	12.1	84	7.6	150
10/36/54	1586	16.0	16.2	84	16.5	150
25/30/45	1580	24.3	23.3	82	27.5	153
50/20/30	1595	29.1	29.0	82	38.2	155
75/10/15	1608	33.5	30.0	82	48.2	160
100/0/0	1880	38.5	35.8	80	302.2	168
0/100/0	2120	3.5	42.9	102	3–5	210
0/0/100	1030	27.5	23.4	70	808	140
HIPS[b]	1160	18.5	19.2	65	25.5	80

NOTE: The weight ratio of two blocks in iPS-*b*-iPP copolymer is 40/60 (iPS/iPP).
[a]Measured under 980 Pa at a heating rate of 2 °C/min.
[b]From Jinling Chemical Co., Ltd., China.

the surrounding glassy polymer but also be cross-linked (25). The improved adhesion in blends with a small amount of diblock copolymer retards mechanically induced slippage between the iPS and iPP phases, thus enhancing the impact strength of the polyblends.

Enhancement of mechanical properties is of interest only if it is not accompanied by a loss of other important properties of the blend. Of particular concern for such polymer blends is stiffness, because most means of increasing impact strength also reduce stiffness (14–19). But this is not the case for the iPS-*b*-iPP–iPS–iPP blends studied here as seen in Table II. It is clear that the enhancement in toughness just described is not accompanied by a loss of stiffness, but it is essentially unaffected by the compatibilizer. And the stiffness of iPS-*b*-iPP–iPS–iPP is higher than that of iPP and HIPS. The impact-modulus behavior seems to be due to the "tough" (or rigid) characteristics, morphologies of phases, and semicrystalline isotactic structure of each block in the iPS-*b*-iPP diblock copolymer.

The addition of the diblock copolymer to the iPS–iPP blends also significantly enhances both tensile strength and ultimate elongation (Table II). As both iPS and iPP have higher tensile strengths than the 40/60 iPS/iPP blends, it is feasible to suppose that the phase boundary is the weakest spot in these blends. The fact that the addition of the copolymer contributes to enhancing tensile strength and ultimate elongation also suggests improved interfacial adhesion of the blend through the "emulsification effect" of the diblock copolymer.

Of considerable interest is the fact that added iPS in PP resin can improve the heat resistance of the materials and thus upgrade their endurance at elevated temperature. Also, as seen in Table II, the HDT of iPS–iPP blends is, as expected, higher than that of common HIPS and iPP. The results seem to be interrelated with the isotactic structure and crystallinity of the iPS component. The iPS-*b*-iPP that is added contributes to a further improvement in the HDT of the iPS–iPP binary blend. Figure 6 shows that, at lower measured pressure, the HDTs increase slowly with increasing iPS-*b*-iPP copolymer. The HDTs of the blends, however, increase rapidly at higher measured pressure as iPS-*b*-iPP copolymer is increased.

Thermal Properties. Modifications of the thermal behavior of polymer systems, particularly in the temperature and breadth of various transitions of state, are often used to show changes in their morphology and miscibility (26). Differential scanning calorimetric (DSC) thermograms of iPS-*b*-iPP–iPS–iPP ternary blends are shown in Figure 7. Binary blends (iPS–iPP) are clearly immiscible, as evidenced by the presence of two distinct glass-transition temperatures (T_gs) of the respective homopolymer. In Figure 7 it is seen that, owing to the addition of iPS-*b*-iPP diblock copolymer in the blends, T_{g1} of iPS decreases, whereas T_{g2} of iPP increases slightly, and then T_{g1} and T_{g2} tend to become closer as the amount of the diblock copolymer is increased. These results

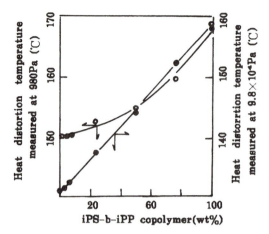

Figure 6. Heat-distortion temperature of iPS-b-iPP–iPS–iPP polyblends.

suggest the occurrence of mutual "dissolution" or cross-linking of iPS and iPP segments of the homopolymers with the diblock copolymer in the blends.

It is well known that the third components, although small in quantity, might also bring about other significant changes such as changes in crystalline behavior, morphology, and dynamic mechanical properties, which influence the blend properties for end use. Table III lists the data obtained by DSC for iPS-*b*-iPP–iPS–iPP ternary blends. The results indicate that PP crystallization

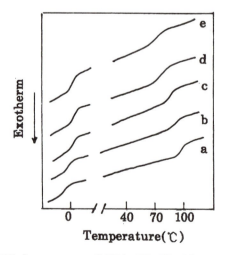

Figure 7. DSC thermograms of iPS-b-iPP–iPS–iPP ternary blends: (a) 0/40/60; (b) 5/38/57; (c) 10/36/54; (d) 25/30/45; and (e) 50/20/30.

**Table III. DSC Melting and Crystallization Results for Homopolymers and
iPS-*b*-iPP–iPS–iPP Blends**

Sample and Composition	Melting Point, T_m (°C)	Crystallization Peak, T_c (°C)	Degree of Undercooling $(T_m - T_c)$
iPP	165.0	107.0	58.0
iPP–iPS (90/10)	164.2	107.0	57.2
iPP–iPS (75/25)	164.7	107.1	57.6
iPP–iPS (60/40)	163.5	106.8	56.7
iPP–iPS–iPS-*b*-iPP (80/15/5)	164.0	116.8	47.2
iPP–iPS–iPS-*b*-iPP (57/38/5)	164.2	120.2	44.0
iPP–iPS–iPS-*b*-iPP (45/30/25)	164.1	122.4	41.7
iPP–iPS–iPS-*b*-iPP (30/20/50)	164.3	124.0	40.3

is not affected by the presence of an iPS phase in the binary blend, but that PP crystallization is affected by the presence of an iPS phase in ternary blends. That is, the melting temperature (T_m) as well as the crystallization temperature (T_c) of the PP phase in a binary blend are unchanged with increasing iPS content, which indicates that there is no miscibility or entanglement between the two phases. However, the addition of iPS-*b*-iPP copolymer to the blends can affect the PP crystallization: T_c of the PP phase increases significantly, while T_m is almost unchanged, resulting in a lower degree of supercooling and higher crystallization. It is believed that iPS-*b*-iPP copolymer can function as a fair nucleation agent to cocrystallize with the PP phase and then increase the PP crystallization temperature. Of course, the diblock copolymer could also possibly act as a nucleating agent in the role of an impurity to increase T_c of the PP and reduce the degree of undercooling of PP crystallization. As seen from Table III, at a given amount of iPS-*b*-iPP copolymer added, T_c of the PP phase also tends to increase with increasing iPS content. These results provide insight into the level of interaction between the blend components and may also suggest the mutual interaction of the iPS and iPP segments of the homopolymer with the iPS-b-iPP diblock copolymer.

Dynamic mechanical thermal analysis (DMTA) is often more sensitive than DSC in identifying the glass-transition temperature (T_g) of blends (27). The DMTA data can be assumed to closely reflect the phase behavior of blends. Figure 8 shows the results of DMTA for the 40/60 iPS–iPP blend. The complex modulus in DMTA (E') of the unmodified blend is slightly enhanced with increasing iPS-*b*-iPP copolymer content. At 25% copolymer, the modulus decreases slightly but remains at the modulus of the unmodified blend. As shown in Figure 8, two distinct tan δ peaks are discernible. The lower one is due to T_g of iPP, while the peak at 98 °C is due to T_g of iPS. The diblock copolymers slightly modify the two transition temperatures. As the iPS-*b*-iPP copolymer content increases, the tan δ peaks of two individual components

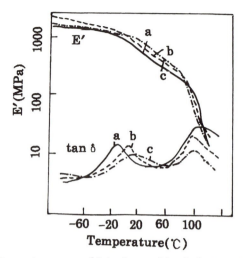

Figure 8. Dynamic spectra of (a) a binary blend of iPS–iPP (40/60), (b) a poly-blend of iPS-b-iPP–iPS–iPP (5/38/57), and (c) an iPS-b-iPP–iPS–iPP blend (25/30/45).

shift progressively toward each other, and the peaks become broader. This effect may be attributed to emulsification of the diblock copolymer, thus promoting partial miscibility of iPS–iPP blends.

Morphology. The change in the thermal and mechanical properties of the blends containing the diblock copolymer may be closely related to their morphology. Figure 9 shows SEM images of fracture surfaces of (1) the iPS–iPP binary blend, and (2) the iPS-*b*-iPP–iPS–iPP (5/38/57) ternary blend. Both samples were prepared by impact fracture at room temperature and then etched with allylamine vapor for 1 h to remove the PS phase. As seen in Figure 9a, the PS domains have a defined spherical shape with diameters ranging from 1 to 40 μm. In addition, through selective etching, many domains are pulled away from their previous positions and the surfaces of holes left by the PS phases removed appear to be clear. The lack of adhesion between iPS and iPP is obvious in Figure 9, where dispersed particles do not adhere to the matrix and leave cavities with a smooth surface. The morphology is strikingly modified by adding as little as 5 wt% iPS-*b*-iPP copolymer (Figure 9b). The component iPP no longer forms dispersed particles but a continuous thin network (entanglement) or fine lamellar bundle firmly anchored into the iPS matrix by emulsification of iPS-*b*-iPP copolymer. This result indicates that the diblock copolymer exhibits a pronounced interfacial activity that reduces the particle size of individual phases and enhances interfacial adhesion between iPS and iPP.

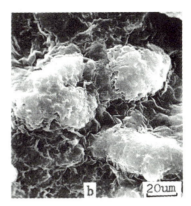

Figure 9. SEM images of the fracture surface of (a) a binary blend of iPS–iPP (40/60), and (b) a polyblend of iPS-b-iPP–iPS–iPP (5/38/57). Both specimens were etched with allylamine vapor for 1 h. The scale is 20 μm.

The SEM results further support these findings (Figure 10). The observed surface morphologies, obtained by fracturing specimens in liquid nitrogen, reveal that smooth surfaces of iPS–iPP blend are seen on the larger protrusions and craters formed during fracturing of the surface (Figure 10a). After the addition of iPS-b-iPP copolymer, the surface morphologies of blends reveal a different behavior; a pattern of ever-smaller scale or roughness is observed as copolymer content increases (Figure 10b–e). At higher concentrations of the diblock copolymer (25 and 50%), the surface of these materials exhibits a level of roughness (finer scale) that is not seen at lower copolymer contents. This roughness or smaller particle size results from the stretched and broken fibrils of material, which form during the fracturing process. Stretched and broken "fibrils" of materials appear to span the interfaces between regions of iPS and iPP. The dispersed particles have been broken in the plane of fracture, but no slippages have occurred. Apparently, the sequences are firmly anchored into the domains they penetrate. When the percentage of diblock copolymer increases, it is impossible to identify individual phases in Figures 9 and 10. Similar observations have been made by Heikens and co-workers (28) for a PS–PS-g(b)-PE–LDPE blend and by Del Giudice et al. (29) for an iPS–iPS-b-iPP–iPP blend system. These results demonstrate the compatibilizing effect of iPS-b-iPP on iPS–iPP blend.

Morphological study, together with DMTA and DSC results, confirms the expectation of miscibility of the diblock copolymer with each component of the blend. This miscibility occurs at the interphases between the components of blends, allowing enhanced interphase interactions and better stress transfer in the blend system. This is probably due to the anchoring of each sequence of the block with its corresponding component of the blend, which is in good

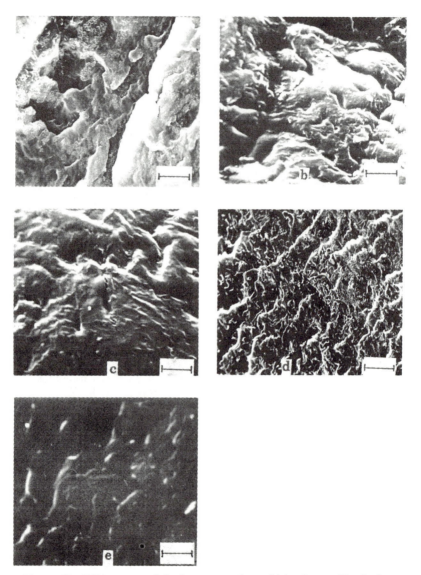

Figure 10. SEM images of the fracture surface of (a) a binary blend of iPS–iPP (40/60), (b) a polyblend of iPS-b-iPP–iPS–iPP (5/38/57), (c) an iPS-b-iPP–iPS–iPP blend (10/36/54), (d) an iPS-b-iPP–iPS–iPP blend (25/30/45), and (e) an iPS-b-iPP–iPS–iPP blend (50/20/30). The scale is 20 μm.

agreement with what is inferred from the mechanical properties of the iPS-*b*-iPP–iPS–iPP polyblends.

Summary and Conclusions

By sequential copolymerization of styrene and propylene using a modified Ziegler–Natta catalyst, $MgCl_2/TiCl_4/NdCl_x(OR)_y/Al(iBu)_3$, which was developed in our laboratory, a styrene-propylene block copolymer is obtained. After fractionation by successive solvent extraction with suitable solvents, the copolymer was subjected to extensive molecular and morphological characterization using ^{13}C-NMR, DSC, DMTA, and TEM. The results indicate that the copolymer is a crystalline diblock copolymer of iPS and iPP (iPS-*b*-iPP). The diblock copolymer contains 40% iPS as determined by Fourier transform infrared spectroscopy and elemental analysis.

A toughened polyolefin-engineered plastic with improved heat resistance and increased impact strength, and with no loss of stiffness, was prepared by blending (1) iPS, (2) iPP, and (3) iPS-*b*-iPP diblock copolymer (styrene/propylene ratio is 40/60). The mechanical properties and stiffness of the iPS-*b*-iPP–iPS–iPP polyblends were higher than those of iPP and HIPS when iPS-*b*-iPP copolymer in the blends was 25 wt%. Because of the outstanding tensile modulus and heat resistance of iPS, added iPS in the iPP resin containing a given amount of iPS-*b*-iPP diblock copolymer can enhance the heat resistance. The thermal properties are also superior to those of HIPS and iPP.

Enhanced interphase interactions, deduced from thermal and dynamic mechanical properties and morphology observed by SEM, demonstrate the efficient compatibilizing effect of iPS-*b*-iPP copolymer on iPS–iPP blends. Each sequence of the iPS-*b*-iPP diblock copolymer can probably penetrate or easily anchor its homopolymer phase and provide important entanglements, improving the miscibility and interaction between the iPS and iPP phases. This is in good agreement with what is inferred from the mechanical properties of the iPS-*b*-iPP–iPS–iPP polyblends.

Acknowledgments

This work was supported by the National Natural Science Foundation of China and the Foundation of Specialties Opened to Doctoral Studies of the State Education Committee, China.

References

1. Xu, G.; Lin, S. *J. Macromol. Sci., Rev. Macromol. Chem. Phys.* **1994,** *C34(4),* 555.
2. Dyachkovskii, F. S. *Trip* **1993,** *1(9),* 274, (Review).
3. Sain, M. M.; Kokta, B. V. *J. Appl. Polym. Sci.* **1993,** *48,* 2181.
4. Liang, Z.; Williams, H. L. *J. Appl. Polym. Sci.* **1991,** *43,* 379.

5. Bataille, P.; Boisse, S.; Schreiber, H. P. In *Advances in Polymer Blends and Alloys Technology;* Kokudic, M. A., Ed.; Technomic: Lancaster, PA, 1989; Vol. 1, p. 165.
6. Delgiudice, L. U.S. Patent 4,713,416, 1987.
7. Shinichi, Y. Jpn. Patent 63,105,022, 1988.
8. Takeaki, M. Jpn. Patent 63,215,714, 1988.
9. Favis, B. D.; Chalifoux, J. P. *Polymer* **1988,** *29,* 1761.
10. Barlow, J. W.; Paul, D. R. *Polym. Eng. Sci.* **1984,** *24,* 525.
11. Anastasiadis, S. H.; Cancarz, I.; Koberstein, J. T. *Macromolecules* **1989,** *22,* 1449.
12. Fayt, R.; Jerome, R.; Teyssie, P. *J. Polym. Sci., Polym. Lett. Ed.* **1986,** *24,* 25.
13. Brahimi, B.; Ait-Kadi, A.; Ajji, A.; Fayt, R. *J. Polym. Sci., Polym. Phys.* **1991,** *29,* 945.
14. Datta, S.; Lohse, D. J. *Macromolecules* **1993,** *26,* 2064.
15. Heikens, D.; Hoen, N.; Barentsen, W.; Piet, P.; Ladan, H. *J. Polym. Sci. Polym. Symp.* **1978,** *62,* 309.
16. Fayt, R.; Jerome, R.; Teyssie, P. *Makromol. Chem.* **1986,** *187,* 837.
17. Barentsen, W. M.; Heikens, D.; Piet, P. *Polymer* **1974,** *15,* 119.
18. Chen, C. C.; White, J. L. *Polym. Eng. Sci.* **1993,** *33,* 923.
19. Barentsen, W. M.; Heikens, D. *Polymer* **1973,** *14,* 579.
20. (a) Zhou, X.; Zhao, Z.; Lu, Y.; Lin, S. *Chin. J. Polym. Sci.* **1987,** *4,* 309. (b) Xu, G.; Lin, S. *Makromol. Chem., Rapid Commun.* **1994,** *15,* 873.
21. (a) Xu, G.; Lin, S. *First China–Korea Bilateral Symposium on Polymer Science and Materials;* Preprints, Beijing, China, October 15, 1993; p. 21. (b) Xu, G.; Lin, S. *Chin. J. Polym. Sci.* **1996,** in press.
22. Nocci, R.; Attalla, G.; Del Giudice, L.; Cohen, R. E.; Bertinotti, F. Eur. Patent Application EP 99,271, 1984.
23. (a) Xu, G.; Lin, S. *Chem. J. Chin. Univ.* **1994,** *16,* 139. (b) Xu, G.; Lin, S. *Chem. J. Chin. Univ.* **1995,** *16,* 1972.
24. Fayt, R.; Jerome, R.; Teyssie, P. *J. Polym. Sci., Polym. Lett. Ed.* **1981,** *19,* 79.
25. Xu, G.; Lin, S. *Polymer* **1996,** *37,* 421.
26. Kim, W. N.; Burns, C. M. *J. Appl. Polym. Sci.* **1990,** *41,* 1575.
27. Stoelting, J.; Karasz, F. E., Macknight, W. *J. Polym. Eng. Sci.* **1970,** *10,* 133.
28. Coumans, W. J.; Heikens, D.; Sjoeradsma, S. D. *Polymer* **1980,** *21,* 103.
29. Del Giudice, L.; Cohen, R. E.; Attalla, G.; Bertinotti, F. *J. Appl. Polym. Sci.* **1985,** *30,* 4305.

INDEXES

Author Index

Affiliation Index

Subject Index

Abbreviations Used in This Index

ABS, acrylonitrile–butadiene–styrene polymer
ATBN, amino-terminated butadiene–acrylonitrile
BCB–MI, 1,2-dihydrobenzocyclobutene and maleimide
CE, cyanate ester
CET, cross-linkable epoxy thermoplastic
CSR, core–shell rubber
CTBN, carboxy-terminated poly(butadiene–acrylonitrile)
DCBA, 4,4'-dicyanato-2,2-diphenylpropane
DDM, 4,4'-diaminodiphenylmethane
DDS, 4,4'-diaminodiphenyl sulfone
DGEBA, diglycidyl ether of bisphenol A
DPEDC, 4,4'-dicyanato-1,1-diphenylethane
DSC, differential scanning calorimetric
ECO, epoxidized crambe oil
EPDM, ethylene–propylene dimer monomer
EPR, ethylene–propylene–rubber
ESR, epoxidized soybean rubber
ETBN, epoxy-terminated polybutadiene
HIPS, high-impact polystyrene

IPN, interpenetrating polymer network
iPP, isotactic polypropylene
iPS, isotactic polystyrene
LLDPE, linear low-density polyethylene
MA, methacrylic acid
MBS, methacrylate–butadiene–styrene
PA, polyamide
PC, polycarbonate
PES, poly(ether sulfone)
PMMA, poly(methyl methacrylate)
PP, polypropylene
PS, polystyrene
RTPMMA, rubber-toughened poly(methyl methacrylate)
SAXS, small-angle X-ray scattering
SEM, scanning electron microscopy
TEM, transmission electron microscopy
TGDDM, tetraglycidyldiaminodiphenyl-methane
TP, thermoplastic
VER, vinyl ester resins
VR, vernonia rubber